ANIMALS AND ARCHAEOLOGY IN NORTHERN MEDIEVAL RUSSIA

Zooarchaeological studies in Novgorod and its region

Mark Maltby and Mark Brisbane

with contributions from

Ellen Hambleton, Sheila Hamilton-Dyer,
Olga A. Krylovich, Mikhail V. Sablin,
Arkady B. Savinetsky and Andrei V. Zinoviev

OXBOW | books
Oxford & Philadelphia

Published in the United Kingdom in 2020 by
OXBOW BOOKS
The Old Music Hall, 106–108 Cowley Road, Oxford, OX4 1JE

and in the United States by
OXBOW BOOKS
1950 Lawrence Road, Havertown, PA 19083

Hardcover Edition: ISBN 978-1-84217-277-3
Digital Edition: ISBN 978-1-78925-418-1 (epub)

A CIP record for this book is available from the British Library

Library of Congress Control Number: 2019955880

Printed in Malta by Gutenberg Press

Typeset in India for Casemate Publishing Services. www.casematepublishingservices.com

For a complete list of Oxbow titles, please contact:

UNITED KINGDOM
Oxbow Books
Telephone (01865) 241249
Email: oxbow@oxbowbooks.com
www.oxbowbooks.com

UNITED STATES OF AMERICA
Oxbow Books
Telephone (800) 791-9354, Fax (610) 853-9146
Email: queries@casemateacademic.com
www.casemateacademic.com/oxbow

Oxbow Books is part of the Casemate Group

THE ARCHAEOLOGY OF MEDIEVAL NOVGOROD

SERIES PREFACE

Novgorod is one of the most intensively and continuously studied urban sites in northern Europe. Systematic excavations began in 1932 and have continued every year since then except during World War II. The excellent preservation of organic and inorganic material in its anaerobic soils, including the structural remains of streets, properties and buildings, has made it possible to study entire quarters of the town as well as the activities of its inhabitants. With deposits up to 8 m deep in places and with well-dated sequences from the early/mid-10th century, its importance to the study of both medieval Russia and the development of Europe cannot be over emphasised. In addition, excavations have recovered many examples of the organic remains normally lost to archaeologists, including a stunning collection of birch bark letters, unique written documents of the medieval period, which now number over a thousand separate inscriptions. Because of this the site has received attention from scholars with a wide range of specialisms and from differing fields including medieval archaeology, history, architecture, botany, zoology and linguistics.

This publication series presents some of the results obtained from international, multidisciplinary projects supported by various European universities and institutions into the origins and development of the medieval town of Novgorod and its hinterland. With the support of EU funding via INTAS (the International Association for the Promotion of Scientific Collaboration between the EU and former Soviet Union countries), a number of projects were initiated that used the Novgorod area as a test bed for wider issues concerning urban origins, town-hinterland relationships, environmental analyses, trade connections, accurate chronologies, innovative artefact studies, the development of accounting systems and the spread of written language.

This series of publications are the outcome of collaborative projects that have their origins in the early 1990s when funding was obtained from INTAS to set up an international collaboration into aspects of medieval towns and their hinterlands in north-west Russia. Much of the field work took place from 1993 to 2004 within Novgorod itself, but also included discoveries from other key sites in the area such as Ryurik Gorodishche, Staraya Russa, Pskov and sites, such as Minino, in the Byeloozero region on the northern margin of the territory of Novgorod (a territory that comprised the city's own medieval state, known as Novgorod Land, which at its height covered an area larger than modern day France).

The volumes in this series cover some of the topics investigated by the Novgorod Archaeological Research Centre with the support of INTAS-funded projects and focus on the following aspects of medieval Novgorod and its region:

- The Pottery from Medieval Novgorod and its Region (published 2006)
- Wood Use in Medieval Novgorod (published 2007)

- Animals and Archaeology in Northern Medieval Russia: zooarchaeological studies in Novgorod and its region (this volume)
- The Archaeology of Medieval Novgorod in Context: studies in centre/periphery relations (published 2012)

The first two volumes contain papers on key materials, namely pottery and wooden artefacts. Whilst elsewhere throughout Europe pottery tends to take the lion's share of attention and wood less so, partially due to its lack of survival, in Novgorod this position is reversed. Wood survives in abundance and what is more it was used prolifically for artefacts, fuel, buildings, fences, and even streets, making it the key means of dating site levels through the extensive use of dendrochronology. As pottery has never been relied on for dating purposes, its typological and scientific study has lagged behind ceramic studies in Western Europe. For this reason the pottery volume in this series attempted to set out some preliminary findings as well as discussing differences in methodology, sampling and analysis.

The synthetic volume entitled *The Archaeology of Medieval Novgorod in Context* examines the environmental background for the settlement pattern that developed from the 9th to 15th centuries in north-west Russia and then investigates the role that various natural resources had in contributing to that pattern. It includes insights into the natural environment based on recent palynological studies commissioned as part of the project and presents data from three study areas (the Byeloozero area to the north-east of Novgorod; the immediate hinterland of Novgorod and within Novgorod itself). It considers what, where and how certain natural resources were exploited during the medieval period in these areas. Where possible, it attempts to explain the processes by which these resources were transformed into commodities (via craft production, centralised workshops, household production, specialised settlements, etc) and places the evidence from the three other volumes on ceramics, wood use and zooarchaeology into a wider context, concentrating on the exploitation, manufacture and consumption of these and other materials including plants, leather, textiles and metals.

Turning to this volume on zooarchaeological evidence from sites in Novgorod and Novgorod Land, it follows on from the volumes on pottery and wood in that it too raises fundamental issues to do with recording, sample selection and methodology. It then proceeds to discuss the evidence for domesticated and wild species, as well as the integration of animal studies into social and economic contexts (for example, the fur trade and butchery practices) and investigates the differences and similarities in the material from the town, its immediate hinterland and its wider territory. This work is important not only in a Russian setting, but in a pan-European one, as the quantity and quality of material is exceptional, as is the extent of its preservation. The implications of this study to urban-rural relationships across northern medieval Europe are significant and should be seen in this wider context.

Mark Brisbane
Bournemouth, August 2019

CONTENTS

LIST OF ILLUSTRATIONS

LIST OF ILLUSTRATIONS

LIST OF ILLUSTRATIONS

LIST OF ILLUSTRATIONS

LIST OF TABLES

LIST OF TABLES

PREFACE

This book is the last of the four to be published in the Archaeology in Novgorod Series. The programme of zooarchaeological work envisaged in the original INTAS-funded project (1993–1996) was merely intended to evaluate the potential of studying animal bones from the excavations in Novgorod and make recommendations about their future study. As described in Chapter 2, this initial assessment gradually transformed into a much more extensive and ambitious project, which involved not only studying large numbers of bones from several sites in Novgorod itself but also analysing assemblages from several sites excavated by Evgenii Nosov in Novgorod's hinterland. Eventually, we also had an opportunity to incorporate evidence from Nikolai Makarov's excavations in Minino, near the borders of Novgorod's vast northern territory. Therefore a brief assessment report has been expanded into a monograph.

When the project began, Novgorod's importance as an international trading centre was quite well known to medieval historians and archaeologists beyond Russia, through the general works of Thompson (1965) and Martin (1986) amongst others. The significance of the recent investigations into the superbly preserved deposits in Novgorod and sites in its hinterland was also beginning to become better known as communications between the West and the Soviet Union improved during the glasnost era in the late 1980s (Brisbane 1992). However, little had been published about the animal bones since the 1950s (Tsalkin 1956). The INTAS projects have significantly expanded our understanding of the archaeology of this important centre and facilitated a much broader dissemination of the evidence. This volume contributes to that expansion by providing a detailed discussion of the exploitation of animals in Novgorod and its lands.

This book has a different approach to most traditional studies of animal bones from medieval sites in Russia by placing less emphasis upon zoological aspects of the animals themselves, such as their morphology and stature. The focus here has been to examine the relationships between animals and the human inhabitants of Novgorod and its territory. This is social zooarchaeology not archaeozoology. Animal bone data, however, does form the core of this research and most zooarchaeologists will be familiar with the methodologies used in the analyses carried out in Chapters 3–10.

The recording of bones from Novgorod, Gorodishche and other hinterland sites took place over several visits to Novgorod between 1993 and 2003. Led by Mark Maltby, Sheila Hamilton-Dyer and, in later years, Ellen Hambleton, the data have been recorded by 20 people. Indeed, assemblages from Novgorod provided exceptional opportunities for training students in identification and recording methods.

There were many logistical challenges in obtaining and storing the assemblages and we are grateful to all the excavation directors for collecting the bones for us to study. We have been able to examine bones collected from a complete sequence of deposits from one Novgorod site (Troitsky XI) and obtained substantial assemblages from two adjacent sites (Troitsky IX and X). In addition, we have been fortunate to be able to include another large assemblage from Novgorod, from the Desyatinny site analysed by Andrei Zinoviev. He employed the same methods of analysis and quantification. This has allowed us to make broad chronological comparisons within and between sites.

Studies of urban archaeology must be seen in conjunction with studies carried out beyond the town's boundaries in order to seek answers to a wide range of questions. For example, did the foundation of Novgorod have profound effects on animal husbandry and hunting practices in its region? Did the diets of the urban inhabitants differ from those in the farms and villages in the region? Did changes in agricultural practices and forest clearances stimulate changes in animal husbandry? From where did the Novgorodians get their huge supplies of pelts that they acquired for international trading?

We were fortunate therefore to be able to examine assemblages from several settlements in the near hinterland of Novgorod. These provided substantial amounts of 9th- and 10th-century material with which to compare the assemblages from Novgorod itself. The high status settlement at Gorodishche, in particular, has produced a large, well-preserved assemblage, which included substantial amounts of fish bones that had been retrieved from an extensive sieving programme. In addition to the material we looked at ourselves, we have been able to include assemblages from more recent excavations at Gorodishche, which have been examined by Mikhail Sablin.

Novgorod at its height controlled a huge territory and it was important that we also included material from sites deeper in the forest zone, from where most of the animals hunted for their fur would have been acquired. We were able to obtain comparative data from the medieval settlement complex at Minino on Lake Kubenskoye, some 500 km to the north-east of Novgorod. Olga Krylovich and Arkady Savinetsky have supplied a report on the mammal bones for this volume. This assemblage contrasts markedly with those from Novgorod and the hinterland, particularly with respect to the abundance of bones of wild species. Sheila Hamilton-Dyer's analysis on the fish bones from Minino also revealed some variations from the ones she studied from Gorodishche and Novgorod. Our research has clearly demonstrated the value of comparing assemblages from different types of settlements and areas to get a better understanding of human-animal relationships.

Zooarchaeology is not the only discipline that can supply information on the use of animals in Novgorod and its territory. The exceptional preservation conditions in Novgorod have supplied us with a wealth of environmental materials and artefacts that can shed light on various aspects of how animals were exploited in life and after death. We make reference to these throughout the book but particularly in the final chapter. The opportunity has also been taken in places to compare the zooarchaeological evidence with information that can be gleaned from documentary evidence. Some of the birch-bark documents found in Novgorod and some of the entries in the Chronicles of Novgorod have added further insights into the role of animals in Novgorodian lives.

The book ends with an evaluation of the progress we have made in developing our understanding of animal exploitation in medieval Novgorod. It recognises that there are limitations in the dataset we have acquired, particularly in relation to sampling strategies and we need to obtain good faunal assemblages from other areas of the town and from other sites in the countryside. There are, however, many opportunities to carry out targeted multidisciplinary research to deepen the knowledge we have already gathered on humans and animals in northern medieval Russia.

ACKNOWLEDGEMENTS

The studies published in this book were part of three projects co-ordinated by Bournemouth University and funded by INTAS from 1993–1996 (INTAS 93-463), 1997–2000 (INTAS 96-099) and 2001–2004 (INTAS 2000-154). The fieldwork was also assisted by grants from the Societies of Antiquaries of London, UCL Institute of Archaeology Grants Sub-Committee, the School of Applied Sciences of Bournemouth University, and from the Russian side by several grants from the Russian Humanitarian Scientific Fund. The Russian texts by Mikhail Sablin, Olga Krylovich and Arkady Savinetsky in this volume were translated by Katharine Judelson, to whom the editors owe a huge debt of gratitude for her commitment to these projects over many years.

Over the past 30 years of collaborations, we have hugely benefitted from the support and friendship of our Russian colleagues. With respect to this volume in particular, special thanks are due to the excavators of the sites who supplied us with most of the assemblages, Evgenii Nosov, Peter Gaidukov, Alexander Sorokin and Nikolai Makarov. Lyuba Holden was an enormous help in the early years of the project, acting as our translator alongside her normal duties on site. All the above and many others, including Elena Rybina, Valentin Yanin, Andrey Zaliznyak, Mikhail Petrov, Olga Tarabardina, Sergei Troianovsky and Alexander Khoroshev provided us with many insights into the archaeology of Novgorod. This volume would not have been possible without the contributions and moral support of Sheila Hamilton-Dyer and Ellie Hambleton. To everyone who helped record the bones (you are all mentioned in Chapter 2), many thanks and we hope you have happy memories of Novgorod. Mikhail Sablin would like to acknowledge the support of ZIN RAS (Grant No. AAAA-A17-117022810195-3) and Mark Brisbane would like to thank the Leverhulme Trust for the award of an Emeritus Fellowship.

Finally, we would like to acknowledge and thank Julie Gardiner and Jessica Scott of Oxbow Books and Mark Dover formerly of Bournemouth University for their assistance in the production of this volume.

Mark Maltby would also like to thank his wife, Julia, and daughter, Hannah, for their forbearance during the many occasions he went to Russia without them and for their tolerance in putting up with his endless pacing during periods of writer's block and noisy typing in more productive moments. At least they have acquired a taste for Russian champagne, which was brought back after each visit!

CONTRIBUTORS

Mark Brisbane
Dept of Archaeology and Anthropology,
Faculty of Science and Technology,
Bournemouth University, Fern Barrow,
Poole, Dorset BH12 5BB, UK.
mbrisbane@bmth.ac.uk

Ellen Hambleton
Dept of Archaeology and Anthropology,
Faculty of Science and Technology,
Bournemouth University, Fern Barrow,
Poole, Dorset BH12 5BB, UK.
ehambleton@bmth.ac.uk

Sheila Hamilton-Dyer
Dept of Archaeology and Anthropology,
Faculty of Science and Technology,
Bournemouth University, Fern Barrow,
Poole, Dorset BH12 5BB, UK.
shd@shdbones.plus.com

Olga A. Krylovich
Severtsov Institute of Ecology and Evolution,
Russian Academy of Sciences, Moscow,
Russia.

Mark Maltby
Dept of Archaeology and Anthropology,
Faculty of Science and Technology,
Bournemouth University, Fern Barrow,
Poole, Dorset BH12 5BB, UK.
mmaltby@bmth.ac.uk

Mikhail V. Sablin
Zoology Institute of the Russian Academy of
Sciences, St Petersburg, 199034 Russia.
msablin@yandex.ru

Arkady B. Savinetsky
Severtsov Institute of Ecology and Evolution,
Russian Academy of Sciences, Moscow,
Russia.
arkadybs@rambler.ru

Andrei V. Zinoviev
Dept of Biology, Tver State University, Tver,
Russia.
Zinovev.AV@tversu.ru

DEDICATION

This volume is dedicated to the memory of Evgenii Nikolaevich Nosov, Director of the Institute for the History of Material Culture, St Petersburg (from 1998–2015), Professor at St Petersburg State University, Corresponding Member of the Russian Academy of Sciences, Visiting Research Fellow at Bournemouth University, renowned archaeologist, director of excavations at Ryurik Gorodishche and other important early medieval sites in the vicinity of Novgorod, and a true friend and supporter of our collaborations from 1989 up to the time of his death in February 2019.

Without his encouragement, support and friendship it would not have been possible for us to develop the close academic links with our Novgorod colleagues that Evgenii so ardently helped to foster and develop. One of us (MB) had the great pleasure of meeting him for the first time in the summer of 1989 as part of a collaboration initiated by Vadim Masson between first the University of Southampton, and subsequently Bournemouth University, and the Institute for the History of Material Culture in St Petersburg. From the initial tentative steps of what could be fairly characterised during those first few years as a 'trail-blazing cultural exchange', the following ten years saw this grow into a full programme of academic, international research activities involving over 60 researchers and numerous students from at least six countries (UK, Ireland, Denmark, Sweden, Germany and Russia). Throughout all this time, and indeed right up to 2018, Evgenii was central to our work, helping where he could, offering advice and information that made us stop and think, and always with a smile and a funny story to encourage us to keep going.

He wrote over 460 publications, published in Russia and abroad, and will be well remembered for his original concepts on the formation of the earliest urban centres of Medieval Russia, as well as his study of how eastern silver penetrated into the Baltic region.

He will be deeply missed as an international medieval scholar and superb archaeological project organiser, but above all as a true and abiding friend.

The editors would also like to pay tribute to Dr Eileen Reilly, who died tragically young and while in the midst of a vibrant, inspirational career in environmental archaeology.

Eileen studied Archaeology and Geography at University College Dublin (UCD), graduating with Honours in 1992. For her Postgraduate Masters degree she studied environmental archaeology at Sheffield University where she specialised in the study of beetles under Paul Buckland. In 1998 she returned to UCD to complete a Diploma in

Environmental Impact Assessment, working as a palaeo-environmental archaeologist on the Lisheen Mines Project, one of the first of numerous archaeological projects in the commercial sector to which she contributed.

Throughout her career she was involved in major multi-disciplinary projects in both Ireland and other European countries building up a significant international reputation. She joined the Novgorod team in the early 2000s where she made an enormous contribution working on INTAS projects. She was a significant contributor to the work described here and in a previous Novgorod Archaeology volume.

Eileen was a pioneer in environmental archaeology. She single-handedly created the study of archaeo-entomology in the Republic of Ireland. She was awarded her PhD from Trinity in 2008 which examined woodland history and how natural and human-driven change through time affected insect biodiversity. This work influenced the course of Eileen's research focusing on how beetle analysis could shed light on how people lived in the past. In 2011 Eileen secured funding to assess, catalogue and re-pack over 1000 samples from the iconic Viking site of Fishamble Street, Dublin. The legacy of this is a hugely important archive of environmental material which will be available for generations of scientists to come.

Eileen worked extensively on a range of sites and made herself an international name through her publications on the study of insect assemblages from medieval urban and Viking age sites in Dublin, Waterford, Antwerp and Novgorod. She secured prestigious Government of Ireland funding in 2013 to pursue her post-doctoral research at UCD examining medieval urban living conditions in Ireland and Europe. She was heavily involved in the establishment of the Experimental Archaeology Centre in UCD, where she was appointed Adjunct Research Fellow in the School of Archaeology. Her book, *Living Conditions in Early Medieval Europe: A Case Study from Viking Age Fishamble Street, Dublin,* is due for publication along with various other papers from her research. Collectively Eileen's work has helped shed a light on our past, and her research has helped to reconstruct past environments and living conditions from Skellig Michael in the West to Novgorod in the East. With her thoughtful, analytical mind she was well on the way to becoming one of the great scholars of palaeoecology and environmental archaeology in Europe.

This volume is dedicated to the memory of

Evgenii Nikolaevich Nosov (21 August 1949–25 February 2019)
Eileen Reilly (15 May 1970–27 July 2018)

Evgenii Nosov at Ryurik Gorodishche, 2000.

Eileen Reilly.

RUSSIAN SUMMARY: РУССКОЕ РЕЗЮМЕ

Наша книга анализирует зооархеологические данные, полученные с помощью раскопок в Новгороде, на Рюриковом Городище и других памятниках в окрестностях Новгорода, а также в Минино, рядом с Кубенским озером. В ней рассматриваются основные виды домашних и диких животных – с точки зрения их численности, возраста (когда они были убиты), способов разделки туш и размера. Привлекаются и другие виды источников (берестяные грамоты, летописи, артефакты и иные свидетельства связей человека с окружающей средой), которые проливают свет на отношения между людьми и животными в Древней Руси. Используя целостный подход, при котором результаты зооархеологического анализа дополняются другими видами данных, мы также рассматриваем отношения между жителями Новгорода и его окрестностей и местными животными по отдельным видам, один за другим. Наша работа посвящена не археозоологии как таковой, а социальной зооархеологии. Тем не менее, данные о костных остатках составляют основное ядро наших исследований, и большинству зооархеологов известны методологии, применявшиеся для анализа, результаты которого описаны в Главах 3-10.

В Главе 1 описывается работа в проектах ИНТАС и общие направления работы археологов. В Главе 2 мы рассмотрели роль результатов изучения окружающей среды в исследованиях, предпринятых участниками наших проектов, её развитие и расширение в течение почти 20-и лет. Следующие две главы посвящены подробному анализу костных остатков домашних млекопитающих, которые мы получили как из памятников вокруг Новгорода (Глава 3), так и из самого Новгорода и Городища – главным образом из раскопов Троицкий IX, X и XI, а также из Десятинного раскопа (Глава 4). В Главе 5 рассматриваются свидетельства, связанные с обработкой туш домашних животных в Новгороде и на Городище. Глава 6 посвящена метрическому анализу домашних животных. В Главе 7 мы анализируем зооархеологические данные, связанные с обработкой туш диких млекопитающих в Новгороде и его окрестностях. Глава 8 посвящена зооархеологическим свидетельствам эксплуатации млекопитающих из Минино, и обсуждение данной темы проливает свет на интересные контрасты с городскими комплексами этого же времени в Новгороде. В Главах 9 и 10 рассматриваются данные, связанные с использованием птиц и рыб на всех анализируемых памятниках. В последней главе мы рассматриваем данные в более широком контексте средневековых памятников Северной Европы, зооархеологические данные которых уже опубликованы, а также вовлекаем другие источники в нашу дискуссию.

Во всей книге мы старались проводить широкие хронологические сравнения внутри памятников и между ними, учитывая при этом широкий диапазон вопросов. Например, оказало ли основание Новгорода сильное воздействие на практику животноводства и охоты в данном районе? Отличалась ли диета горожан от диеты жителей хуторов и деревень в данном районе? Привели ли изменения в сельскохозяйственной практике и вырубка лесов к изменениям в животноводстве? Откуда новгородцы раздобыли огромное количество шкур, которые им нужны были для международной торговли мехом? Наши исследования чётко показали, что сравнения зооархеологических комплексов из поселений и районов разных типов нам помогают лучше понять отношения человека с животными.

Основными выводами нашего исследования стали следующие: (1) Крупный рогатый скот был самым важным видом, использованным в Новгороде: от 60% до 70% костных остатков, определённых нами, были кости крупного рогатого скота, хотя при методах сбора костных остатков, использованных там, такой результат был ожидаем. Первоначальная разделка многих туш крупного рогатого скота происходила в южной части Городища. В Новгороде обработка крупного рогатого скота для мяса, кожи и рогов обычно происходила в пределах боярских усадеб. Методы, употребляемые для их забоя, показывали, что получение мяса - и в меньшей мере молочных продуктов – было главной целью эксплуатации крупного рогатого скота. Скот был маленького размера и в позднем средневековье средний размер данных животных уменьшался. (2) Свиньи были на втором месте среди животных, представленных в Новгороде. Их было меньше чем в слоях Городища, относящихся к 9-ому и 10-ому векам: может быть, это было из-за вырубки лесов в данной местности. Свиньи также плохо представлены в Минино. Возможно, что в период позднего средневековья в усадьбы Новгорода привозили больше свинины.(3) Мелкий рогатый скот мало представлен в Новгороде и его окрестностях, но он был более распространённым в Минино. Там было больше коз, чем в большинстве средневековых городов Европы. Методы, употребляемые для их забоя, показывают, что главной целью их эксплуатации было производство мяса. (4) Лошади ценились главным образом в качестве транспорта как в гражданской жизни, так и в военных целях. Иногда, однако, их туши также разделывались в Новгороде для получения мяса, шкур и костей. Были найдены следы ритуального использования лошадиных голов в Городище. (5) Дикие млекопитающие мало представлены в Новгороде и его окрестностях. Среди обнаруженных видов были лось, бобёр, медведь, куница, белка, заяц и кабан. Значение Новгорода в международной торговле мехом отражается скорее в документальных источниках, чем в костных остатках. В этом отношении мы видим сильный контраст с поселением Минино, где дикие млекопитающие – в особенности белка и бобёр – встречались часто. Есть свидетельства и из Новгорода, и из Минино, что популяции бобров сильно сократились в позднем средневековье – скорее всего из-за чрезмерой добычи меха. (6) Из птиц в Новгороде и его окрестностях больше всего охотились на диких уток. Курицы играли более важную роль в позднем средневековье как источник яиц и мяса. Иногда держали гусей в качестве домашних птиц. Что касается пернатой лесной дичи, охотились за глухарями и куропатками. Костные остатки тетеревятников (самок) и ястребов-перепелятников (самок) указывают на использование соколиной охоты. За орлами, вероятнее всего, охотились ради их оперения. (7) Поскольку при раскопках не использовали промывку или просеивание слоя, наши знания об использовании рыб в самом Новгороде достаточно ограничены. Однако костные остатки как из Новгорода, так и с Городища показали, что пескарь, окунь, щука и судак местного улова были самыми распространёнными видами. В Минино преобладали судак и пескарь. Кости дорогих рыб, таких как осётр или сом, обнаружены только в маленьком количестве.

Книга заканчивается оценкой успехов, которых мы достигли в своём понимании об эксплуатации животных в средневековом Новгороде. Очевидно, что имеются недостатки в собранных нами данных, в методах сбора образцов. Эти методы требуют немедленного улучшения. Также необходим сбор хороших комплексов костных остатков животных из других районов Новгорода и его окрестностей. Имеется, однако, много возможностей для проведения комплексных целенаправленных исследований, которые приведут к углублению уже полученных знаний о человеке и о животных в Северной Руси средневекового периода.

-1-

THE INTAS PROJECTS AND THE ARCHAEOLOGICAL BACKGROUND TO THE ZOOARCHAEOLOGICAL STUDIES

Mark Brisbane

THE INTAS PROJECTS

This chapter begins with an account of the internationally-supported collaborations into aspects of Novgorod archaeology that involved numerous individuals and institutions over a 20-year period (1989–2009).

Following the author's initial visits to Novgorod in 1989 and 1990, it was apparent how much important archaeological and historical research was being undertaken by Russian archaeologists, who devoted their summers (and more) to the systematic investigation of Novgorod and nearby sites. Almost none of this important work was reaching western archaeologists except in a very limited and often out-of-date manner. Also at that time, some of this work was not even reaching fellow Russian archaeologists due to problems with the production of publications. Thanks to the generosity of the UK's Society of Medieval Archaeology, a monograph on the archaeology of Novgorod was produced (Brisbane 1992) and subsequently an internationally supported programme of archaeological collaboration began in 1993. This collaboration, supported by INTAS (The International Association for the Promotion of Cooperation with Scientists from the New Independent States of the Former Soviet Union), was initially for a three-year programme of investigation into the palaeo-environment of medieval Novgorod and its hinterland, focusing on the animal and plant remains from four key sites, two in the hinterland (Georgii and Ryurik Gorodishche) and two from within the town itself (Troitsky on the west side of the River Volkhov and Fedorovsky on the east side). It is with this project that this volume has its origins.

The first INTAS project (93-463)

The project set out to establish a recovery and sampling strategy for the collection of environmental data from excavations in Novgorod and its hinterland. It brought together specialists from Western Europe with Russian archaeologists to determine what type of questions could be addressed through the collection of this material. Its objectives included the provision of training where necessary to enable the study of faunal remains by Russian archaeologists to continue after the completion of the project, and the establishment of a reference collection of animal bones in Novgorod for comparative purposes.

Its primary aim was the investigation of the archaeological evidence for the palaeo-environment of the medieval town of Novgorod and its hinterland from the late 8th to 15th centuries AD. The sites were selected in order to set up an integrated study of the environmental evidence, predominantly the evidence from animal bones (Mark Maltby and his team) and macroscopic plant remains (Mick Monk from University College Cork (UCC) for the town and, as part of a wider early Slavic agriculture study, Almuth Alsleben from the Institut für Ur- und Frühgeschichte of the Christian-Albrechts University in Kiel for the hinterland sites). The collaborating participants were the former Department of Conservation Sciences (later School of Applied Sciences, now Department of Archaeology and Anthropology) at Bournemouth University, the Department of Archaeology at UCC, the Institute for the History of Material Culture in St Petersburg, the Department of Archaeology of Moscow State (Lomonosov) University, Novgorod State Museum and the Centre for the Organisation of Archaeology in Novgorod.

The work initially concentrated on identifying and implementing sampling strategies and methodologies for both faunal remains and plant material and then moved on to begin to analyse and interpret the evidence provided by the material itself. This work produced clear variations between sites, some of which were due to marked differences in collection methods (*e.g.* sieving versus hand recovery). In some places these recovery methods were changed in an appropriate manner in response to these findings. A programme of sieving archaeological deposits specifically for environmental data was instigated on sites in Novgorod and its hinterland, with notable improvements in recovery at the excavations at Ryurik Gorodishche (henceforth referred to as Gorodishche) and Troitsky site XI.

The project's general results may be summarised as follows:

- It established the presence/absence of animal species (both wild and domesticated) at a number of sites, adding to the work undertaken by Russian zoologists in the 1950s and 1960s (*e.g.* Tsalkin 1956; Sychevskaya 1965).
- It enabled a preliminary comparison of meat procurement and consumption between the town and earlier/contemporary rural settlements in its hinterland.
- It provided an initial outline of the history of animal and plant exploitation in Novgorod through several centuries of the town's history.
- Sieving experiments demonstrated that fish caught in the area were a common component of the diet: these included primarily Cyprinidae, pike and zander.
- It showed that there was great potential in studying variability in the composition of the animal bone assemblages within a site, for example between properties or between indoor and outdoor areas. Such studies would provide insights into the organisation of the activities within and between properties, variations in diet, and whether there were specialists involved with carcass processing. It was

recognised, however, that problems with site recording methods and logistical issues were challenging (see Chapter 2).
- The link between artefact production using bone, antler and ivory and the procurement of the raw material for this artisan activity was also studied. This examined whether, for instance, antlers were collected separately from deer carcasses or were a by-product of the butchery process (Smirnova 2005, 10–14).

The second INTAS project (96-099)

In 1997 a second project was approved by INTAS that was designed to study the chronological framework for early medieval towns in north-west Russia based on the evidence from ceramics and dendrochronology. The test bed for this project was Novgorod, its hinterland sites, and other historic towns in the region, notably Pskov and Stara Russa. It brought together the original partners from the first project with colleagues from the Institute of Archaeology at University College London (UCL), the Department of Medieval and Modern Europe from the British Museum, the Laboratory for Ceramic Research of the University of Lund, and the dendrochronology laboratory at the Natural Research Unit of the National Museum of Denmark.

The main objectives of the project were the following:

- To assist in standardising the chronological framework for early medieval towns in north-west Russia based on the evidence from ceramics and dendrochronology.
- To integrate the results in an innovative manner building upon successful examples, such as Lubeck and London.
- To develop and implement strategies and methodologies for the examination and study of the artefactual evidence for ceramics and wood, placing these two material types into their wider economic and social context.
- To develop and support the dendrochronology laboratory in Novgorod, providing appropriate training where required.
- To enhance existing research capabilities in the study of medieval ceramics in north-west Russia, providing appropriate training where required.

Specific work carried out on this project included the following:

Ceramics

This part of the project concentrated on the sorting and analysis of ceramic assemblages from a number of urban sites. Material from the Troitsky XI excavations in Novgorod was selected to provide a type and fabric series for the project. These included handmade and wheel-thrown vessels from 10th- to 15th-century contexts. A variety of sherds were selected for petrological analysis to test the emerging fabric series and to answer a number of technological issues. Work also identified European imports within the Museums' reserve collection together with material held by the Centre for the Organisation of Archaeology in Novgorod. As well as articles in

Russian journals, the main results of these studies were published as Volume One in this series (Orton 2006).

Wood

Identification of the wood species was carried out for various artefacts from Troitsky and other excavations in Novgorod and the surrounding area, including Gorodishche, and from Novgorod State Museum's reserve collection, These identifications, combined with studies of the technology of woodworking and turning through the examination of finished objects, waste material and surviving woodworking tools, provided insights into the relationship between artisans and the exploitation of the natural resources of the forest. The results of this and other work on the use of wood in medieval Novgorod were subsequently published as Volume Two in this series (Brisbane and Hather 2007) and it is also the subject of on-going work on forest exploitation (Brisbane *et al.* in prep.).

The Project was also successful in bringing this work to a wider audience of western archaeologists and specialists through seminars, staff exchanges and conferences. Of particular note was a symposium devoted to the latest developments in Novgorod archaeology, which was held as part of the European Association of Archaeologists (EAA) conference at Bournemouth in September 1999, the papers of which were subsequently published (Brisbane and Gaimster 2001). There were 27 contributors to the symposium, 13 of whom were Russian.

The third INTAS project (00-154)

In June 2001 a third project was approved by INTAS. This was entitled 'Craft Production, Environment and Landscape: an archaeological study of centre/periphery relationships based on the evidence of the exploitation and processing of natural resources in medieval Novgorod and its region'. Its main objective was an examination of the relationship between the expanding mercantile centre of Novgorod and its region. To help achieve this objective, the project set out to examine and analyse an expanded range of source materials (as well as wood, plants, clay and animal products, it now also included metals) that were obtained from a variety of landscapes in the local and regional territories of medieval Novgorod from the 10th to 15th centuries. It attempted to place each of these materials into a landscape and environmental context, linking particular areas within the territory of Novgorod, known as Novgorod Land, with the specialised procurement of natural materials and resources. In addition to the personnel and organisations already mentioned in the earlier projects, Eileen Reilly from University College Dublin (UCD) investigated the insect remains from town sites. The metal analyses were led by Thilo Rehren of UCL and the pollen analyses were conducted by E.A. Spiridonova and her team from the Moscow Institute of Archaeology (MIA).

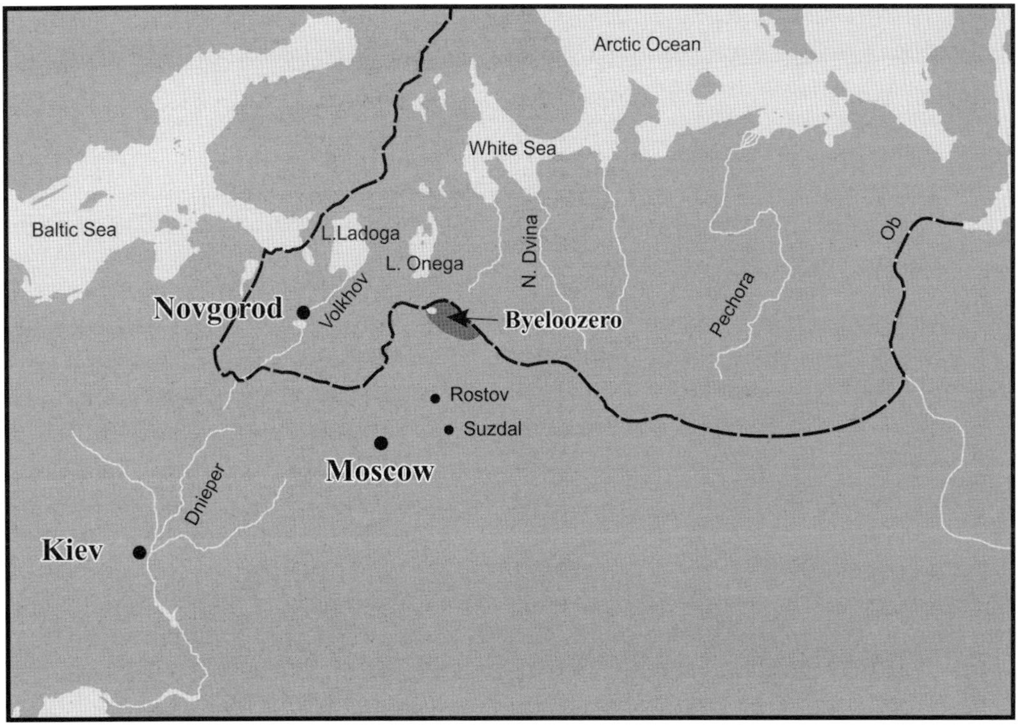

Figure 1.1: Novgorod and Novgorod Land around AD 1400 showing the Byeloozero region within which the Minino sites are located. Drawn by Mark Dover

In order to examine the remoter edges of Novgorod Land, the Project also included material from sites around the rural settlement complex of Minino on Lake Kubenskoye, which was being investigated by Nikolai Makarov and his team from MIA (see below, this chapter). These sites are located approximately 500 km to the north-east of Novgorod (Figure 1.1) and provide a useful comparator to the sites located in the heartland of Novgorod territory (Makarov 2007, 365).

The main objectives of the project were as follows:

- To determine how certain materials and resources were selected, exploited and transported to Novgorod, then turned into finished products.
- To investigate the location and quantify the scale of production, measuring the extent of craft specialisation in a variety of domestic and workshop settings.
- To develop a new model for measuring and evaluating the extent of standardisation within particular products, especially those produced for a mass market such as pottery, in order to isolate and better understand both conservatism and innovation.

- To examine the level of production and its relationship to the social and economic structure of Novgorod, more specifically to determine the impact that boyars and merchants had on the organisation and control of craft production.
- Using previously identified workshops from excavations within the town (*e.g.* the metalsmith at Property A, Troitsky), to analyse the material in order to determine the nature of craft activity and the level of skill of the artisan.
- To integrate the results of this work into a European context, comparing Novgorod Land with other areas within Europe, where towns acted as centres for the exploitation of a region (*e.g.* London, York and Lubeck).

The results of this project were published in Volume Four of this series (Brisbane *et al.* 2012) to which 28 people contributed, 16 of whom were Russian, across 24 chapters.

THE ARCHAEOLOGICAL BACKGROUND

Medieval archaeology offers the opportunity to bring together scientists from differing backgrounds to study technological aspects of past society and relate them to both social structure and the environment. However, rarely does an archaeological site produce enough organic remains to allow a study of perishable materials such as leather and wood to be studied in detail and integrated with the study of inorganic finds such as those of pottery, metal, slag, glass, etc. The superb preservation of medieval Novgorod combined with the evidence now being produced from sites within its region offers an excellent opportunity to undertake such studies and to put them into the context of how an urban centre affects the exploitation of natural resources in its region and the impact of that exploitation on the natural environment itself.

Of crucial importance to the understanding of the medieval period is the study of the extent that people dominated their environment by exploiting numerous natural resources, turning them into products. These products were sometimes for domestic or extended household consumption only. At other times they were highly valued products that were traded or exchanged between social elites, sometimes over large distances. Alternatively, they may have been mass-produced and intended for the marketplace to support individual artisans and their families. While medieval historians have commented extensively on the generalities of production and the social relationships that it implies, archaeology has been consistently building up a detailed picture of the physical reality of artisans, their products and their workshops.

There can be few sites in medieval Europe that offer so much potential to carry through a detailed investigation into the process of exploitation and production as Novgorod, its hinterland and its wider region. It is becoming possible to integrate these discoveries in ways that examine centre/periphery relations and the role of exploitation/production (Brisbane *et al.* 2012). The fact that Novgorod occupied the position of a Kontor within the Hanseatic League means that results here have historical, social and economic implications around the Baltic and across northern Europe in the high to late medieval periods. Equally important are the pre-Novgorod and early Novgorod contacts with the Scandinavian world, which brought with it trading, cultural and political implications from the Viking world to the Black Sea and the Caspian.

The Origins of Novgorod

The beginnings of Novgorod are closely linked to its natural setting. Some 5 km to the south lies the north shore of a large lake, Lake Ilmen, approximately 40 km (max) by 32 km (max), whose size varies depending on the extent of spring floodwaters. Many streams and rivers flow into the lake including four major rivers (the Msta, Pola, Lovat and Shelon) but only one, the Volkhov, flows out (Figure 1.2). Ilmen is a relatively shallow lake and it floods in the springtime, covering the surrounding area with sandy silt. The shallow lake also has the effect of creating a microclimate slightly warmer than the surrounding regions to the east, west and north (Sokolov 1926). As the floodwaters recede in May/June, this leaves land highly suitable for the growing of crops and hay in a short growing season that lasts only until September or early October. Due to local climatic conditions, the forest in this area is a mixture of boreal (largely pine, spruce and birch) and broad-leaf deciduous woodland including oak, elm and ash: a contrast to the mainly coniferous trees that dominate the forests on the thin sandy soils away from the Novgorod area (Atlas of the Novgorod Region 1982, 17).

It was into this environment that Eastern Slavs and local Finnish tribes established settlements in the 8th, 9th and early 10th centuries, particularly in a 4 to 5 km wide band along the lakeside. In one area, known as Poozerie (literally, 'the land beyond the lake'), lying along the north-west lakeshore, at least 20 settlements have been found that date to this period (Yeremeyev 2012, 145). Most are located on low hills that would have been virtual islands during the spring floods. In addition to these natural defences some, such as Georgii, had a substantial bank and ditch to protect them further (Nosov 1992, 16–19).

Taking the Poozerie and the upper reaches of the Volkhov together, the foremost of the settlements in this area is known today as Gorodishche (Figure 1.2). The defended site is located on the highest hill overlooking the very point where the River Volkhov flows out of the lake and starts its journey 220 km northwards towards Lake Ladoga, passing the trading site of Staraya Ladoga on the west bank of the Volkhov a few kilometres before it enters the lake. Any traveller along this route needed only to join the River Neva that flows out of Lake Ladoga to reach the Gulf of Finland and the Baltic, a total journey of approximately 400 km by water from Gorodishche.

In such a commanding location dominating this rich agricultural area, Gorodishche became a fortified trading centre that soon attracted Viking traders as early as the 9th century (Nosov 1992, 46–55). Its accessibility via the river routes meant that it could be reached from the Baltic, but it also sat at the junction of two further river routes:

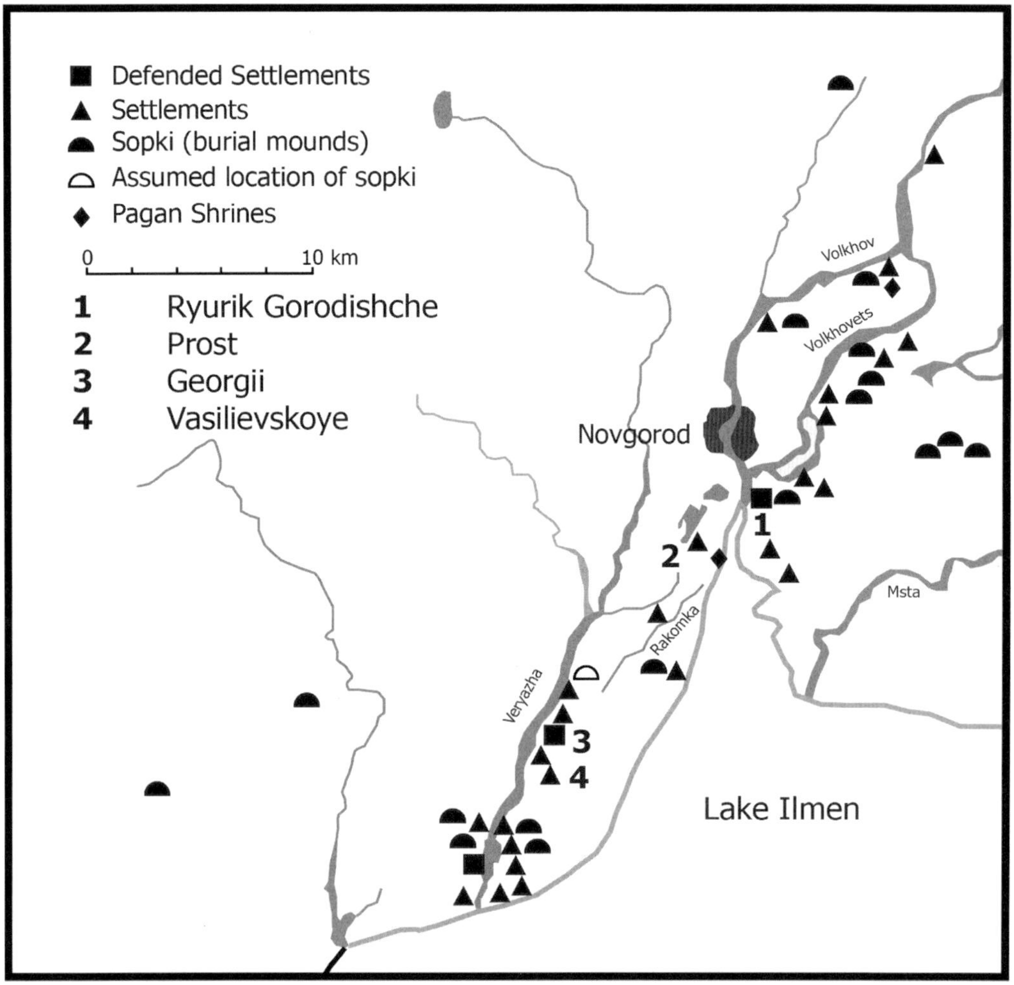

Figure 1.2: *Novgorod and its hinterland showing sites mentioned in the volume. Drawn by Mark Dover*

one southwards via the Dnieper to Kiev, the Black Sea and thence to Byzantium and the other route to the south-east via the Volga to the Caspian Sea and the Arab world. Situated at such an important node in this long distance trading system, Gorodishche was the direct predecessor to Novgorod and its success was part of the reason for the expansion of this area politically, economically and demographically in the 9th and 10th centuries (Nosov 2001, 8–9).

Novgorod

Novgorod (literally, 'the new town', most likely as successor to Gorodishche although other explanations have been offered) was founded, according to one version of the

The sites are numbered in the order they were excavated and are as follows: 1 - on Slavensky Hill. 2 - on Borkovaya Street. 3 + 5 - in Yaroslav's Court. 4, 25, 28 + 35 - in the Kremlin. 6 - on the eastern rampart. 7 - on Chudintsevaya Street. 8 - on Kholopya Street. 9 - Nerevsky. 10 - Ilyinsky. 11 - Buyany. 12 - Slavensky. 13 - Gotsky. 14 - Tikhvinsky. 15 - Mikhailovsky. 16 - Torgovy. 17 - Rogatitsky. 18 - Kirovsky. 19 - Lyudogoshchinsky. 20 - Troitsky. 21 - Kosmodemyansky. 22 - Dmitrievsky. 23 - Duboshin. 24 - Nutny. 26 - on Bolshevikov Street. 27 - Molotkovsky. 29 - Mikhailo-Arkhangelsky. 30 - Fedorovsky. 31 - Ipatevsky. 32 - Lukinsky. 33 - Konyukhov. 34 - Andreevsky. 36 - Polsolsky. 37 - Dobrynin.

Figure 1.3: *Plan of Novgorod showing the five Ends (Districts), the ramparts (cross-hatched), the street layout (known medieval streets in black and modern grid pattern dotted) and the location of excavations. Based on plans supplied by the Novgorod State Museum*

Primary Russian Chronicle, in AD 859, on a site only 2 km downstream (north) of Gorodishche, on low hills overlooking the River Volkhov on both sides of its river-banks (Figure 1.3). In contrast to the date in the Chronicle, archaeological evidence has yet to reveal any settlement deposits prior to the early 10th century (920s/930s based on dendrochronology) despite many years of archaeological investigation. What they do show is a town that developed extremely quickly from the mid-10th century to become a thriving and densely populated place, on both sides of the River Volkhov, with established street and property layouts (Khoroshev *et al.* 2001, 23–25). Within a year or so of the coming of Christianity to Russia in AD 988, the town built one of the very first churches in Kievian Rus (second only to Kiev itself) and established a defended area (first referred to in an entry in The Chronicle of Novgorod in AD 1044 as the town's Detinets, or Kremlin) around the cathedral and archbishop's residence (Nosov 1992, 61; Troianovsky 1998). Add to this, a marketplace, a thriving aristocracy (the boyars), intense artisan activity, town defences and a prince's residence (originally at Gorodishche but later within Novgorod's east side), and the town was well positioned to dominate not just its immediate area, but most of north-west Russia throughout the medieval period. This dominance, which included being a major kontor of the Hanseatic League, lasted until challenged by the rise of Moscow in the 14th and 15th centuries and was ultimately eclipsed by Moscow when Ivan III (the Great) took direct control of Novgorod in 1478. Subsequently, Ivan IV (the Terrible) sacked Novgorod and completely destroyed its aristocratic and mercantile power base in the 16th century.

Over the last hundred years or more, the question of who the first settlers of Novgorod were has been much debated. Put crudely, were they Slavs or were they Vikings? This became known as the Normanist debate, and during Soviet times arguments that advocated Viking settlers into Mother Russia were seen as dangerously colonialist. For reasons of self-censorship as much as political convictions, academics found it difficult to consider impartially the archaeological evidence that was coming out of Novgorod from the first modern excavations of 1932 onwards. This, of course, has changed over recent years, but for reasons of continued 'national conservatism' it is still the case that evidence for the establishment of Novgorod by Viking settlers in the 10th and 11th centuries remains contentious. However, recent work re-examining the collections has clearly shown that there is a significant distribution of Scandinavian finds throughout sites dating to the 10th and 11th centuries indicating widespread Viking influence (Musin 2018).

Town layout and development

Under Catherine the Great (Empress of Russia 1762–1796), many towns in Russia were re-planned including Novgorod, which under the Great Plan of 1778 was given a more regularised street pattern replacing the former medieval one (Khoroshev and Sorokin 1992, 107). Archaeological excavations and observations over many years have confirmed that much of the previous road system had its origins in the early years of the settlement in the 10th and 11th centuries. For instance, at the Troitsky excavation, the wooden paving of streets made with split logs appeared at almost the same time as the laying out of properties with the earliest wooden surfaces dating to the 950s or 960s (Khoroshev and Sorokin 1992, 133). Eventually the growing town's street pattern took the form of an organic, topographically related layout providing access to the riverside (Figure 1.3). An early sequence of deposits at the Troitsky site (Trenches X and XI directed by Peter Gaidukov) revealed a remarkable sequence

(Aleksandrovskaya *et al.* 2001). Beneath 6 m of deposits dating from the 10th to 20th centuries was clear evidence of plough marks (or perhaps more accurately, ard marks). A combination of archaeological evidence, soil analyses, palaeo-botanical and geochemical investigations suggested that a meadow and a field existed on the site immediately prior to the initial laying out of the town in the 930s when the first phase of wooden structures appear on the site. These have been dated by dendrochronology with a sequence of some 50 dates ranging from AD 926 to 939 (Aleksandrovskaya *et al.* 2001, 18–20), some of the earliest dates to be confirmed in the archaeological record of Novgorod.

Properties, boyars and craft production

It would appear that once property boundaries (wooden fence lines, usually substantial) were established within the town, these stayed more or less fixed over the following centuries and certainly for the period from the 10th to 12th centuries (Khoroshev and Sorokin 1992, 122–125), if not to the 15th century (Yanin 1992, 76). Generally speaking these properties fell within two types, smaller ones whose area did not exceed approximately 465 m^2 and larger ones that ranged from 750 to 1400 m^2. The smaller properties usually contained a main dwelling and two or three outbuildings, whereas the larger ones had a main dwelling together with other heated houses, plus outbuildings, mostly unheated. Traditional interpretations (*i.e.* from Soviet times) identified the larger structures as belonging to the landed boyars, whereas merchant houses were few and far between. For instance, extensive investigations in the 1950s at the Nerevsky site (c. 9,000 m^2) revealed some 18 properties, nine of which survived almost complete. Most of these were interpreted as dwellings of Novgorod boyars. Surprisingly perhaps, only one building was recognised as the house of a trader (dating from the 1130s). This was based solely on the finds found within it (Zasurtsev 1967, 116). This interpretation led to theories that the merchants' properties were separate from the places where the boyars lived, and were located, initially at least, outside the town's 'Ends' or main settlement areas, but this interpretation remains untested. Yanin (1992, 27) has argued that the boyar system, established from at least the 12th century, was based on hereditary holdings of both urban properties and far-flung hinterland estates, where the raw materials from the countryside were collected and brought into their extended family urban compounds where craft workers turned these into finished goods. Herein lies an important difference with medieval towns to the west, where craft production of a particular kind was often spatially concentrated in one part or street of the town (*e.g.* smiths, butchers, leather or textile workers) where craft guilds developed. In medieval Novgorod it seems likely that no such developments took place (Yanin 1992, 76–77), leaving the boyars to dominate both the economic and physical landscape of the town and countryside.

The dominant position of the boyars was further enhanced in other ways. Many of them obtained wealth by collecting taxes, a proportion of which went to them, as well as a tithe to the church, the bulk then going to the Prince. Evidence for this has come directly from archaeology not only in the form of birch-bark documents but also in the remarkable finds of wooden seals used to fasten the sacks of collected tribute (Yanin 2007). Some of this tribute came from great distances, such as one 12th-century wooden seal recording over 3,000 squirrel pelts collected along the River Pinega, more than 900 kms north-north-east of Novgorod. Other examples show tribute coming from territory near the Northern Dvina (Yanin 2007, 207).

This boyar dominance of the economic power in Novgorod (as opposed to in Kiev where the Prince held legitimacy due to military conquest) meant that the political power balance between boyars and Prince was more equally shared, yet still often contested. Nevertheless, it is this form of extended family power structure that has led some to identify 'the Novgorod Republic' with an early form of representative government (for example Fennell 1983; Yanin 1992, 93–96).

Merchants and trade

As discussed above, it seems more than likely that no, or very few, 'independent craftsmen' existed in medieval Novgorod. If this was the case, then could a true merchant class emerge as middlemen for the trade and exchange of surpluses? It is difficult to use archaeological evidence to answer this question.

For a start, as we have seen, it is difficult to distinguish merchants' houses from those of the boyars based solely on construction methods. Perhaps detailed artefact studies can help distinguish merchant houses and their properties. One of the leading proponents of this approach is Elena Rybina, who undertook an extensive study of merchants' properties in Novgorod. In various parts of the town she singled out a number of such properties, for instance in the Lubyanitsky, Slavensky, Duboshin and Troitsky excavations. For example, a prosperous merchant lived in a property excavated in the Lubyanitsky site, situated to the north of Yaroslav's Court near Buyany Street, in the second half of the 12th and the first third of the 13th centuries. Among the finds testifying to this are a number of birch-bark documents with content relating to trade, more than a thousand pieces of amber, many spindle whorls made of pink slate (from Ovruch, north-west of Kiev) and bronze folding-scales and weights (Rybina 1978; 2001).

In 2014, not far from the Lubyanitsky excavation, at the Rogatitsky II site (30, Bolshaya Moskovskaya Street) parts of two properties were excavated. On the basis of the finds it was possible to conclude that a craftsman had lived in one of them and in the other a merchant, who in the 14th century had been trading with Hanseatic merchants. In the second house several dozen West European lead seals were found, which had secured bales of textiles (mainly fine woollen cloth), the ends of oak barrels with various merchants' marks carved into them, and various items of jewellery of West European origin (Peter Gaidukov, pers. comm.).

For the identification of an excavated property in the town as belonging to a merchant, it would therefore seem appropriate to seek some or all of the following: (1) the presence of trading equipment (scales, weights, containers for goods); (2) finds of coins (especially foreign ones); (3) a concentration of imported pottery; (4) reserve stocks of raw materials or craft items; (5) birch-bark documents with contents relating to commodities; (6) the degree of frequency with which medieval Russian and West European commercial seals are found; (7) the location of the property (Perkhavko 2012, 133). It is worth noting that a younger generation of Russian archaeologists have been examining Novgorod properties for artisan and other activity based on artefact distributions, as well as the size and layout of properties (*e.g.* Petrov 2011; 2014).

It is perhaps equally difficult to say how the merchant traders obtained the commodities they sold: was it at the marketplace (Torg) or through middle-men? Archaeological data does not often provide this information, but it is possible to have occasional insights through some of the birch-bark documents when referring to trade. For instance, the late 12th-century birch-bark document 439 from the Lubyanitsky excavation records a merchant called Moses writing to his colleague Spirk, who appears to be a go-between, about sales of various commodities including items of tin, lead and what appear to be sheets of copper.

While we may speculate that there were probably a range of ways in which commodities were obtained (wholesale deliveries, through the Torg and using middle-men), there is little historical evidence for this, perhaps with one important exception. Namely in the 14th and 15th centuries several hundred German merchants used to visit Novgorod every year. It is likely that the number of Novgorod merchants who used to travel both to the West and to the East was also far from small.

By the latter half of the 15th century, things were changing. The boyar power base was being systematically destroyed by the rulers in Moscow, who saw it as a threat to their authority and militaristic rule. The first merchant guilds were emerging by then to fill the space once occupied by the boyars. One of the most important of these guilds was the Ivansky Merchants' Association (which built the Church of St John opposite Yaroslav's Court), and which brought together the privileged merchants who carried out commercial operations in the international markets of the Baltic region (Perkhavko 2012).

Novgorod Land: Byeloozero and sites along Lake Kubenskoye including Minino

In a ground-breaking project of international importance, Nikolai Makarov and his team from the Institute of Archaeology in Moscow extensively surveyed an area some 600 km to the north/north-east of Novgorod up to and along the southern fringes of Lake Kubenskoye. They followed this up with intensive excavation of selected settlement and burial sites from the period AD 900–1300. The results of the project are well published in three volumes (Makarov 2007; 2008; 2009).

The region known as Byeloozero (named after Lake Byeloye, literally the white lake, the largest lake in the area) lies on the edge of what would have been, briefly, the north-eastern limits of Novgorod Land before deeper penetration and colonisation took place to the north around Lake Onega and on to the White Sea, the Northern Dvina, and east eventually to the River Ob and the Urals (Figure 1.1). It was also an area contested between the two emerging medieval power bases of Novgorod and Rostov-Suzdal. The town of Byeloozero, now known as Byelozersk, on the south shore of Lake Byeloye is one of the earliest towns in the area with settlement evidence dating from the second half of the 10th century (Makarov *et al.* 2001). To the south-east of Lake Byeloye lies a long, thin body of water known as Lake Kubenskoye and it is along this lake edge and south to the River Mologa, a tributary of the Volga, that Makarov and his team conducted their survey of medieval settlements from 1996 to 2003 (Makarov 2012, 41).

The group of medieval sites near the village of Minino lies near the shore of the lake at the mouth of a small river (the Dmitrovka). The three sites that comprised the study sit on the main river terrace, the first one (Minino I, which is the one featured in the zooarchaeological study of this volume in Chapter 8) occupies a cape nearest of all to the lake. The site was found to have prehistoric antecedents, but the medieval settlement occupied an area of 1.4 ha and produced a significant number of artefacts, many of them dateable. It was therefore possible to say that the site began in the second half of the 10th century, and had uninterrupted occupation from the 11th to 13th centuries (Makarov 2012, 42, 54). The excavations revealed seven structures, including one jewellery workshop (Zakharov 2012, 58). All structures had substantial hearths or stoves and consisted of two rooms with an internal partition (the so-called 'five walled house' construction). Of interest for this volume is the fact that the excavators report finding clusters of calcined bones around the stoves (Zakharov 2012, 62).

In addition to the seven excavated structures, evidence for numerous other buildings scattered around the settlement consisted of the surviving remains of stoves just below the ground surface, and sometimes protruding from it. The excavators were able to identify and plan at least 40 of these, concluding that the site comprised a considerable cluster of buildings, possibly laid out along streets. Of course, without further excavation, the dating of the buildings that went with these stoves remains uncertain, but there was little in the way of material culture to indicate that the settlement lasted much beyond the 13th century.

This date range fits well with other sites excavated in the area, all of which show signs of long-distance trade orientated towards the Baltic to the west and the Volga river system to the east. This has led Makarov to see these settlements as stable and economically prosperous during the 11th and 12th centuries and representing the 'first wave of colonisation' that flowed out from the centres of Novgorod, Rostov and Suzdal in this expansionist period. Interestingly, these settlements were either abandoned or declined in size from the 13th century as new economic conditions prevailed (Makarov 2012, 57).

It is within this context that the animal bone studies from these sites and those of Novgorod should be seen. Sites such as Minino represent thriving rural settlements that were linked economically and culturally to centres despite what appears to be their precarious isolation as outposts in the forest. At least during the 11th and 12th centuries, these were significant places that contributed to the wider economic system that relied heavily on the exploitation of land, forest and natural resources.

– 2 –

THE HISTORY OF THE PROJECT: RETRIEVAL, RECORDING, ANALYSIS AND DISSEMINATION

Mark Maltby

1993

First impressions

It was in the small hours of a July morning in 1993 whilst sitting in the back of a former military truck somewhere on the road between St Petersburg and Novgorod that I began to appreciate more fully that we were embarking on an extraordinary project. The truck had been sent to transport us from the airport to our accommodation in Novgorod. Usually my involvement with an archaeological excavation began long after it had finished when boxes of animal bones were delivered to where I worked, hopefully accompanied by relevant phasing and context information. Thereafter in the comfortable working environments of various animal bone laboratories in the Universities of Sheffield, Southampton and Bournemouth, and even occasionally in my own dining room, I would identify and analyse the faunal material and write a report about the findings. Sometimes I had paid a site visit to the excavations from where the bones came and occasionally I had been asked to give advice on sampling. However, it would be fair to say that my role in excavation strategies was generally, at best, passive. This was, and still is, sometimes a source of frustration for all zooarchaeologists. I have never understood why many excavators embark on projects without having environmental and other specialists fully on board before and during fieldwork. Although, thankfully, there are an increasing number of exceptions to this, there is still much room for improvement (Maltby 2002; Baker and Worley 2019).

It was my colleague at Bournemouth University, Mark Brisbane, who had given me this opportunity to expand my horizons. He had first approached me the previous year and told me that subsequent to his recent visits to Russia, he planned to develop an ambitious collaborative project, which would involve archaeologists from institutions based both in Russia and Western Europe.

As Mark Brisbane has described in Chapter 1, it was abundantly clear that the archaeological discoveries from the waterlogged medieval deposits of the town of Novgorod are of exceptional quality and international importance, as acknowledged by its status as a World Heritage Site. In addition, archaeological investigations of the rural hinterland were beginning to produce potentially valuable comparative data from contemporary and slightly earlier medieval settlements. It did not take long to convince me that this was a worthwhile project with which to be associated.

It was clear from the published material (Thompson 1967; Brisbane 1992) that the potential of the resource was immense for environmental archaeologists. Many organic materials survived extremely well in the anaerobic deposits of the medieval town. A lot of attention had been drawn to the unique birch-bark documents and there was clearly an abundance of surviving wood, leather, plant remains and other materials that would not usually be preserved. Indeed, the abundance of finds from Novgorod itself seemed to be creating huge problems of retrieval, sampling, recording, conservation and storage. I was disappointed to learn that animal bones were not being kept. There were no samples being taken for environmental analysis and only the more complete or unusual leather and wooden objects were being retained for conservation and analysis.

It seemed that the main problems were the lack of financial, physical, technical and human resources. Everyone concerned with the excavations knew that they were recovering exciting and important finds. Excavations in various parts of the town over the past 50 years had provided a chronological overview of the layout of the town, the nature of its properties and an unprecedented range of artefacts. There had been many research projects into different categories of artefacts in particular. Many members of the project were also aware that recent scientific advances meant that the full potential of the archaeology was not being realised. However, appropriate equipment, storage and trained personnel were lacking. The attraction of the INTAS grant was that it was aimed at improving some of these resources, initially focusing on environmental archaeology.

We recognised that we would need to do some preliminary recording and analysis of any animal bones we looked at, in order to support any assertions we made about the potential of the material. We therefore asked our Russian colleagues to save some of the bones from the excavations that were to take place in 1993, so we could have a look at them when we arrived.

It was agreed that in the first year of the initial three-year project an assessment would be made of the potential of the animal bone available from the current excavations in the town along with an initial review of what other material was available from previous excavations. The visit was also intended to assess what facilities were available to carry out animal bone recording and analysis. Our primary aim was to advise on sampling and recording methods and to help to set up a system whereby a coherent programme of faunal analysis could be carried out in the future by members of the project team based in Russia.

One of the initial major challenges facing us was the lack of reference skeletal material. Mark was not aware of anything suitable in Novgorod. The nearest probable source was either in one of the museums or at the university in St Petersburg, which the truck was rapidly leaving behind, and is about 190 km from Novgorod, and, as I was about to discover, almost four hours away by road. I was not overly concerned about my own ability to identify most of the mammals and some of the birds without the aid of reference material. As far as I could gather, most of the bones were likely to belong to the domestic species commonly found on all European medieval sites. I hoped that I would be able to recognise any wild mammals without too much difficulty. However, I was not very confident about identifying some of the wild birds without assistance and I had only very rarely had to identify fish bones because I had always worked with someone who had greater expertise in this area.

Mark had agreed that we should invite someone with expertise in bird and fish identification to join us on the expedition. Sheila Hamilton-Dyer was an ideal choice. Sheila is a freelance zooarchaeologist based in Southampton. Both Mark and I knew of her work on bones from Saxon and medieval Southampton and, in addition to being extremely competent on the identification and analysis of mammal bones, we knew that Sheila was also proficient in the analysis of bird and fish bones. She had also been on other projects where bones were identified on site and she was therefore experienced in dealing with the practical problems of working in the field. Apparently this truck journey was a cakewalk compared with various minibus trips she had survived in the Egyptian desert. Sheila had brought with her annotated copies of various identification manuals for fish and bird, which were to prove extremely helpful.

Five nervous Bournemouth University undergraduates had also squeezed into the back of the truck. Hayley Bullock, Bob Edwards, Tamsin Goldring, Fiona Sharpe and Dave Tucker had all just completed the second year of their studies and were embarking on a five-week work experience placement. All of them had very little experience in animal bone analysis but after some initial training they, and students we were to bring to Russia in future years, made invaluable contributions to the project.

Apart from the driver, who spoke very little English, the other Russian in the truck was Lyuba Smirnova. She was to have a crucial role in the development of our project. In addition to her full-time fieldwork duties – she was a site supervisor on one of the excavations – she was also encumbered with being the main translator for a party of non-Russian speakers. As my own Russian vocabulary extended little beyond 'нет' (no) and 'рубль' (rouble), Lyuba was a vital communication link. I am still full of admiration for how she maintained her patience, politeness and cheerfulness in the face of all the demands we and others put upon her.

We survived our journey to Novgorod and the following morning we all gathered at the

Figure 2.1: *Novgorod Expedition Centre. Photograph by Sheila Hamilton-Dyer*

Figure 2.2: *Novgorod Kremlin wall. Photograph by Sheila Hamilton-Dyer*

Figure 2.3: *Troitsky Sites IX (background) and X (foreground) excavations. Photograph by Sheila Hamilton-Dyer*

expedition centre, which turned out to be in the grounds of a former monastery (Zamensky Sobor) within the eastern half of the medieval town (Figure 2.1). After introductions to Alexander Khoroshev (Moscow State University), the leader of the project and the man who organised our accommodation and food, we walked to the Troitsky excavations, which were located about 3 km away on the other side of the River Volkhov, to the south of the Kremlin (Figure 2.2).

My first reaction upon arrival at the excavations of Troitsky Sites IX and X was one of awe. Nothing really prepares you for the sight of these large-scale excavations. From the modern street level, we looked down upon an area of about 1,000 square metres, which was excavated down to a depth of between 4 and 5 m (Figure 2.3). Protruding out of the dark brown surface of the excavations were numerous large timbers that formed the footings of buildings and property boundaries. It was a scene of much activity. There were about a hundred people working on the site at that time. The excavation team consisted of site managers and site supervisors supported by students from Moscow and other universities and, for much of the three-month excavation season, local schoolchildren.

It transpired that the rate of excavation was surprisingly swift. Normal retrieval strategies in Novgorod involve excavation down through the deposits in 20 cm spits, demarcated at 2 m^2 intervals. Structures are carefully recorded, photographed and planned. Small finds are recorded by square and spit. Removal of the spits is carried out initially by shovels but all of the matrix is then disaggregated by hand in order to extract birch-bark documents and other finds. At that time these included any observed bones, pottery, metal, wooden and leather artefacts. However, not everything was destined to be kept and selection was made subsequently about what should be retained for recording. In previous years the bones, apart from those that were artefacts, were discarded at this stage. However, our request for bones to be retained from the current excavations had clearly been heeded because we spotted trays of bones amongst other finds.

Once any observed finds had been extracted, the spoil from the spits was piled on large collection trays. Once full, these were lifted by pairs of students or schoolchildren, who carried them up a steep, fairly precarious, wooden ramp to the top of the section and onwards to the base of the nearby very large spoil heap. There, the spoil was emptied onto a motorised open conveyor belt reminiscent of those at some former British collieries and occasionally used on British excavations (*e.g.* Winchester in the late 1960s and early 1970s). At the top of the slope where the belt disgorged its contents, other workers dispersed the spoil along the top of the spoil heap.

A brief examination of one of the former spoil heaps by the Bournemouth students soon revealed a treasure trove of discarded material. Within five minutes, large handfuls of pottery sherds, animal bones and even the wing case of a large beetle were discovered. It was clear that many categories of finds were not being retained. The enormity of the challenge began to be appreciated.

The sense of awe increased further when we were invited down onto the site to find ourselves about four metres below the top of the waterlogged deposits, in one place walking on a perfectly preserved 12th-century medieval wooden roadway (Figure 2.4). We carefully picked our way through a number of medieval properties, in which parts of the lower walls survived intact and where various features such as ovens were clearly evident.

Figure 2.4: *Medieval road surface, Troitsky excavations. Photograph by Sheila Hamilton-Dyer*

Assessing the potential

Initially, we had planned just one visit to Novgorod to assess the potential of the animal bones. During this initial site visit we held discussions with the site excavation directors, Alexander Sorokin (Department of Archaeology, Moscow State University) and Peter Gaidukov (Institute of Archaeology, Russian Academy of Sciences). It was agreed that we would first have a look at the bones they had both retained from recent spits from Troitsky sites IX and X. Thereafter, any animal bones observed during the excavation of the remaining spits from Peter's Troitsky X site would also be retained, so that these also could be examined by us at the Novgorod Expedition headquarters. This may appear to have been a simple request but for our Russian colleagues it represented a substantial logistical challenge. By road, the excavation site was probably no more than 3 km from the headquarters. However, at that time only the large ex-army truck that had carried us from the airport was available to transport equipment, finds and people around. A more difficult problem was the lack of suitable containers in which to store and transport bones. The expedition did not possess, nor could it afford to buy even if they had been available, large boxes or finds bags. Through a combination of ingenuity and goodwill, bones were delivered to us in a variety of containers, the most common of which, were fruit crates. Similarly, as there were no standard labels, the provenance of the bones in the crates was usually written on a scrap of paper. We had been warned of the paucity of resources and had brought some finds bags with us but the problems of the storage containers really brought into focus the severe economic difficulties that Russia faced at that time.

The next challenge was to find a room at the Expedition headquarters large enough to house seven people and the bones that were to be recorded. After some negotiation, a suitably large room, usually utilised as the drawing office, was commandeered. Although lighting was poor, even after the installation of some desk lamps, there were sufficient tables and space for our requirements.

Needless to say, bones could not be retained after inspection. There was simply no space or facilities at the Expedition or in the Museum. Most were put back into the crates and were returned to the excavations. However, because of the superb preservation of the bones and because many of them survived complete, it was possible to begin to build up a skeleton reference collection. Part of the first INTAS grant was used by Novgorod State Museum to purchase wooden boxes, in which the reference material could be retained. The reference material was expanded each year we visited whenever suitable bones were found. There are inevitably gaps in the reference material, particularly amongst the rarer wild mammals and the small bones of many of the smaller species. However, until modern skeletons can be acquired, the reference material is fit for the purpose of aiding identification of the major domestic and wild mammals and the most common species of birds.

Having got the agreement for any bones observed during excavation to be retained, we were then faced with the problem of what we should record. Our brief was initially to establish how well the bones had been preserved and what potential information could be obtained from them. No analytical work had been carried out on the mammal bones since the work of Tsalkin (1956) on material from the Nerevsky site. This was published as part of a comparative survey of northern Russian medieval material that focused mainly on metrical analysis. We also discovered that Sychevskaya (1965) had provided a list of fish remains identified from earlier excavations in Novgorod.

A number of factors in 1993 and subsequent years were involved in decisions about what to record and in how much detail. Our initial assessment confirmed that the preservation of the material was superb apart from the presence of a lot of gnawed bones. The bones had the dark brown colouration distinctive of waterlogged material. Most of the species identified in earlier published work were represented in the Troitsky assemblage. However, it was impossible to judge from a sample taken from part of one season of excavation whether there were any chronological or spatial variations in assemblage composition.

In addition, the assemblages had obvious limitations in relation to retrieval practices. There was clearly a preponderance of large bones and it seemed likely that the sample was biased heavily against the retrieval of smaller bones, particularly those of birds and fish. This is not unusual in any excavation that relies solely upon hand-retrieval. However, we had additional concerns that some of the inexperienced workforce may not have been very diligent in collecting smaller pieces of bone.

The question of retrieval standards was partially assessed during the wet-sieving experiments that were initiated in 1993. Peter Gaidukov was able to construct a sieve with approximately a 2mm-mesh for our use. Samples from Troitsky IX and X and from the Fedorovsky site (see below) were selected and sieved using the water of the nearby River Volkhov. Samples were processed by the Bournemouth students working in shifts on the edge of the river (Figure 2.5). Samples of approximately five litres were taken from a small number of squares. Sieving and sorting of each sample took approximately one hour. This slow rate of processing was due to the matrix consisting largely of birch twigs. Most of these did not pass through the mesh. This meant that a lot of the processing time was spent searching through the twigs in the sieve to find the bones.

Another serious problem facing the assessment of the animal bones was the lack of recording of the location of the bones recovered. Excavation in arbitrary 20 cm spits enabled some degree of vertical control, which could subsequently be translated into phasing information, although we expressed reservations about the validity of this

Figure 2.5: *Sieving samples from the Troitsky excavations in 1993 (Fiona Sharpe, Hayley Bullock and Tamsin Goldring). Photograph by Sheila Hamilton-Dyer*

approach. However, although collection of the bones from each square was carried out, limitations in the number of bags and other containers meant that bones from each of the two-metre squares were amalgamated into one large sample for each spit across the site. This clearly limited the contextual and chronological validity of the sample.

Given the limitations of the sampling strategies and the very short period (about four weeks) available to record the bones, it was decided that a limited record of the bones would be taken. There were no laptop computers available at the project and it had been decided beforehand that we would record bones on paper forms and subsequently transfer the records onto computer after we returned home.

Usually zooarchaeologists record bones individually. However, this would have been very time-consuming and would have provided more detail than was needed for the assessment. We therefore devised a species recording form for mammals. For each spit, we recorded the total number of specimens (NISP) for each element type, separating them by side where possible but not recording the parts or the amount of the bone present. All limb bones, cranial fragments and loose teeth were identified to species where possible. Ribs and vertebrae other than the atlas, axis and sacrum were not identified to species. Where possible, these and other unidentified fragments were assigned by size to large mammal (cattle, elk, horse) or medium-sized mammal (pig, sheep/goat, dog) categories.

The numbers of fused and unfused epiphyses and porous fragments were recorded for each element type. The numbers of gnawed, burnt and butchered specimens were recorded but no detailed records were made. No details of fragmentation or zones of the elements present were recorded. Metrical data and mandibular tooth ageing data were recorded on separate recording sheets.

The only categories for which individual records were made were for smaller mammals, birds and fish. Their numbers were small enough for this to be feasible and Sheila felt that more detailed records were necessary for her to make a considered assessment.

The Fedorovsky excavations

It so happened that a second substantial excavation was taking place in Novgorod in 1993. The Fedorovsky site was located in the eastern (Trade) side of the town (Figure 1.3, Site 30). This was also producing waterlogged deposits, although of less depth than at the Troitsky sites. The site director (Gena Dubrovin) showed us around the excavations, which were being undertaken employing the same methods as at the Troitsky sites. There were even greater pressures on the excavators of this site as the area was due for redevelopment in the near future and they needed to complete the excavations on a tight schedule, unlike Troitsky which is essentially a research excavation that first started in the 1970s and continues to this day (2019). It was tempting to ask for the bones to be retained from the Fedorovsky site as well. It would have been interesting to compare results from sites from different parts of the town. However, this was not realistic. We realised we already had more than enough material to get through from the Troitsky sites and we also did not want to add to the pressures already facing the excavation team to complete their work on schedule. We did take a small number of samples for sieving to compare results with those from the Troitsky sites.

In addition, a bone-working deposit was discovered on Fedorovsky. It consisted of the extremities of cattle metapodials discarded after initial processing. These bones provided good metrical data as well as evidence for production and we therefore asked for these bones to be collected.

Trips to the hinterland: Gorodishche and Georgii

Evgenij Nosov, from the Institute for the History of Material Culture in St Petersburg had been heavily involved in the development of the collaboration. As outlined in Chapter 1, he had been carrying out fieldwork since the 1970s on medieval sites in the vicinity of Novgorod (Nosov 1992). Evgenij organised two field visits to allow us to see the local environment including visits to the site of his previous excavations at Georgii and the current excavations at Gorodishche. Although this involved further bumpy rides in the back of heavy vehicles often over rough tracks, the trips were very illuminating.

The field visits allowed us to experience the landscape to the south of Novgorod. The topography is largely flat with few substantial hills. It was clear that the low-lying land on the marshy floodplain provides a good deal of pasture, although winter snow coverage and spring flooding restrict the seasons that this resource is available for grazing. It is (and was) a good area to keep cattle. Some of these zones, however, are less suitable for sheep because they are too wet (Figure 2.6). Woodland is much less abundant than in the early medieval period, when pollen analyses have confirmed that there was plenty available in which pigs could roam (Spiridonova and Aleshinskaya 2012). In most of the hinterland away from the lakeside, woodland was largely coniferous with spruce and pine being the dominant species. There were, however, also smaller areas of deciduous woodland, particularly in the Poozerie peninsula (Nosov 1992, 9; Alsleben 2001, 107–8; Hamilton-Dyer *et al.* 2017).

Figure 2.6: *Landscape of the Poozerie peninsula. Photograph by Sheila Hamilton-Dyer*

Arable farming was restricted mainly to areas not subject to flooding on the strip of higher ground between the north-western shore of Lake Ilmen and the River Veryazha – an area known as the Poozerie peninsula (Figure 1.2). This is the location for Georgii, which because it was above the floodplain did not produce any waterlogged deposits. Excavations had produced a small bone sample, which Evgenij had brought to Novgorod (Figure 2.7). We agreed we would look at this material during our stay. The results of this analysis are included in Chapters 3 and 7.

The excavations from Gorodishche were clearly producing much larger animal bone samples. Although also handicapped by lack of resources and equipment, sieving of deposits was being routinely carried out in a systematic manner in support of normal excavation methods. The excavations at that time were concentrated in the areas on either side of the Sivers Canal. We were ferried to the south bank in a rubber dinghy to inspect the sieving process (Figure 2.8). Animal bones along with other macroscopic finds were being retained, although no specialist had been employed to analyse these. Collaboration with several European partners had by 1993 already led to some sampling for pollen and macroscopic plant remains to be carried out prior to the inception of the INTAS projects. However, there was clearly further work to be done, which led to the inclusion of Gorodishche and other sites in the hinterland in this project. Although preservation of the archaeological deposits was not as good as at Novgorod, there were still substantial waterlogged deposits and it was clear that bone preservation was very good. It was also observed that large numbers of fish bones were being retrieved from the sieving. Material from previous excavations in 1979 had been retained and was available in Novgorod. However, there had not been any sieving at that time. We agreed we would examine the 1979 assemblage during this visit and that, should

Figure 2.7: *Georgii excavation site. Photograph by Sheila Hamilton-Dyer*

Figure 2.8: *Mark Maltby being transported across the Sivers canal by Evgenij Nosov to visit the South bank excavations at Gorodishche. Photograph by Sheila Hamilton-Dyer*

opportunities become available, we would strongly recommend that the material from the well-dated deposits from the recent excavations should also be examined in detail.

Progress by August 1993

A lot had been achieved by the end of our stay in 1993. Mark and I had to depart after about 10 days, during which time the recording methods had been set up for the Troitsky material and all the bones from Georgii and the 1979 excavations at Gorodishche had been recorded. Sheila and the five students were left to record as many bones as possible from the Troitsky sites. They made remarkable progress and had cleared most of the backlog by the time they returned to England. Over 9,000 mammal bone elements had been recorded from Sites IX and X together with substantial numbers of bird bones and a few fish bones. The material from the sieved samples had also been examined, identified and recorded.

Any complete bones were retained to form the basis of a reference collection and already a fair collection had been gathered, particularly for cattle and pig. Sheila and the students, with the aid of Lyuba's interpretation skills, were also able to deliver a presentation of their work to members of the Expedition. We felt this was an important thing to do, as we were keen that the Russian students in particular were aware of the questions we were asking and the potential of the data we had gathered. The presentation was well received, although we were disappointed that we had not found any member of the Expedition who was willing and able to continue our work.

1994

Zooarchaeology and environmental archaeology

The initial assessment of the results was made as part of the annual report required by INTAS. We were able to convey the positive results of the previous year as well as noting that there were many problems to address. It transpired that Mark was able to obtain additional European funding as well as a grant from the Society of Antiquaries of London, which meant that Sheila and I could accompany Mark on our second visit to Novgorod in July 1994. We were joined by Mick Monk from University College, Cork (UCC), an archaeobotanist who had much experience and expertise in the analysis of plant macrofossils from archaeological sites including medieval towns.

His brief was similar to ours, in that he was to assess the potential for the study of waterlogged and carbonised remains of seeds, fruits and grains in Novgorod and provide advice on sampling. His assessment would involve the selection, processing, recording and analysis of specific samples from the Troitsky sites. Inevitably, he had to restrict the programme to a few samples, as all the processing, identification and recording had to take place during the two weeks or so he initially spent in Novgorod.

He was spoilt for choice! However, he also expressed concerns that the traditional excavation methods in Novgorod were not the best for obtaining suitable samples.

Mick's presence with us in this and subsequent years was greatly beneficial. We were able to compare results and discuss ideas relevant to both our specialist areas. One of the more memorable incidents of this visit was Mick and Sheila's discovery of a deposit of horse dung amidst some hay in a Troitsky X 11th-century building. The texture and indeed the aroma were unmistakable. We were discovering horse bones amongst the faunal assemblage from the site indicating their presence, but here was direct evidence that horses had at one time been stabled within at least one of the properties. It was a cogent and pungent example of how integrated approaches to environmental archaeology can produce a more complete and coherent picture of what life was like in a town such as Novgorod.

Recording Gorodishche

We had decided that our priority during our stay was for myself and Sheila to record the animal bones recovered from the recent excavations at Gorodishche. We already knew that preservation of the bones was generally very good and retrieval rates were of a high standard thanks to the extensive sieving programme. We also wanted to demonstrate that such sampling would provide better results than were possible using the retrieval methods practised in Novgorod. We had no hesitation in making individual records for every element recovered, although at times we regretted that decision when faced with the large number of small fragments accruing from sieved samples. Unfortunately, the sieved material had usually been re-amalgamated with bones recovered by hand, so it was not possible to compare the contents of the unsieved and sieved assemblages. Again, the lack of suitable bags was the reason for this. Labels were written on paper and there were some problems in deciphering provenance if these had become wet. Evgenij Nosov, who was based at Gorodishche, was a frequent visitor to the expedition and was very helpful in providing information about the provenance of the assemblages. We also visited him to have another look at the excavations and to discuss the project (Figure 2.9). This

Figure 2.9: *Evgenij Nosov inspecting a section of the Gorodishche excavations in 2004. Photograph by Sheila Hamilton-Dyer*

concluded with a convivial evening at his campsite after which Evgenij demonstrated his rowing prowess during a nerve-wracking crossing of a backwater of the River Volkhov on the way back to our hotel in Novgorod (I again regretted I had never learned to swim).

'The boys from Voronezh'

At the excavations, Peter Gaidukov was nearing the bottom of Troitsky X. He continued to collect animal bones, which provided us with the opportunity to assess whether there were any significant differences in the assemblages from a broader range of spits.

The sieving programme initiated in 1993 was continued by students from Voronezh University, Russia. They processed further samples from the lowest spits of Site X. It continued to be slow and painstaking work, which they undertook very effectively and with good humour under Sheila's guidance. It became clear, however, that extensive sampling of the deposits would not be feasible as processing of the samples would not be able to keep up with the pace of excavations of the large area involved.

Mark had brought over a set of sieves from Bournemouth University to use for a pilot sampling programme of plant remains, which was organised and supervised by Mick Monk. This was carried out using the water supply at the Expedition. Although the priority was to obtain plant macrofossils, the residues inevitably also included animal bones, which were kept for our perusal.

Although the experiment indicated that a lot of bones, particularly those of fish, were being overlooked, it proved impracticable for routine sieving to be carried out in subsequent seasons of the Troitsky excavations. This was partly because of time restraints and the availability of suitable and willing labour, and partly because of the limitations of the contextual information available.

Progress by August 1994

Sheila and I had largely completed the identification and recording of the bones from Gorodishche. There were a few mystery bird and fish bones, which we could not identify. We obtained permission to bring a few bones back to England to confirm identifications against more extensive reference material. We also brought back to England all the paper records we had made for the material from Gorodishche and Novgorod, in order to transfer them to computer. Before we left, however, we photocopied the Troitsky record sheets, so that a copy of the paper archive remained in Novgorod. We did this as a matter of routine in subsequent years. Although we had time to record a small amount of the material from the sieved and hand-collected samples from the lower levels of Troitsky X, there remained a backlog of material to record from these. We were also aware that more bones would be collected during the remainder of the excavation season after we had left.

1995

Russian assistance

Sheila returned to Novgorod with Mick and his assistant Eamonn Cotter, a graduate of UCC, for two weeks at the end of July. On this occasion Sheila's student assistance came from Andy Miller, who had just completed his BSc Archaeology degree in Bournemouth. Most of their work focused upon the completion of the identification and recording of bones that had been retained from the lowest spits of Troitsky X in 1994.

This year we began to benefit from the assistance of Natasha Efimova, of the Novgorod State Museum, where her work involved the cataloguing and curation of some of the archaeological collections. She had been seconded to the Novgorod expedition for the duration of the excavation season but was keen to develop further post-excavation experience, which would be beneficial for her work in the Museum. She did not have any experience in zooarchaeology but was very willing to learn. Sheila invited her to join the bone team and Natasha was provided with an excellent introduction to the principles of identification and recording (Figure 2.10). It was agreed that on our return in 1996, Natasha would rejoin the team to gain further training and experience.

Figure 2.10: *Bone identification in 1995 (Natasha Efimova and Sheila Hamilton-Dyer). Photograph by Mick Monk*

Excavations and sampling in Novgorod

In 1995, Peter Gaidukov began excavations of Troitsky XI. He agreed that the bones from the site would be separated by property as well as by spit. Although one property (G) covered most of the site, it has been possible to compare assemblages from there with parts of two other properties. Although this was a step in the right direction, it still proved impracticable to carry out more detailed intra-site variability studies within and between individual properties. We were not able to obtain samples by individual two-metre squares mainly due to logistical problems on site and the continuing lack of storage space and containers. Bones from Site XI were delivered in a range of containers including large packing cases and sacks.

Gena Dubrovin had been charged with directing another excavation at the Fedorovsky site. These again involved rapid excavations in the wake of property development, allowing no time with the resources available for a coherent zooarchaeological sampling programme. However, one deposit from a late medieval context was collected for assessment, to see if it differed from the earlier medieval assemblages that were currently available from the Troitsky sites.

Mick and Eamonn meanwhile were taking further samples to obtain plant macrofossils from specific contexts on the Troitsky sites. They were retaining animal bones from these.

Progress by August 1995

Without doubt the most important development during 1995 was the recruitment of Natasha to join the zooarchaeological team. In addition to assessing the potential of the bones from the sites of Novgorod and its hinterland, our brief was to attempt to set in place resources that would enable research to be carried out on future excavations, particularly in Novgorod itself. We had enhanced the reference collection annually by extracting suitable bones recovered from the excavations but had not found anyone based in Novgorod available or willing to be trained in zooarchaeology. This problem had now been addressed. In addition to completing the recording of material from the lowest spits from Troitsky X and from Spits 1–2 from Troitsky XI, a significant amount of training had been achieved. Sheila also carried out further recording of sieved material from samples that had been retrieved in 1994.

During the previous spring, Sheila and I had transferred data from Gorodishche onto a computer in preparation for writing up a report on the assemblage. However, the site was still producing material from excavations from the North Bank, which we had not examined.

1996

All hands to the tea chests

Sheila had reported back that there was a significant backlog of material building up from Troitsky XI. At that time, there was no guarantee that any further significant grants for travel and training would be forthcoming. We therefore decided to take over another group of students in order to process as much of the material as possible in the summer of 1996. By doing so, we would have a substantial sample from the later medieval period from the upper spits from Troitsky XI with which to compare with the earlier material we had recorded from the lower spits of Troitsky IX and X.

Matty Britten, John Steane and Jeremy Bell, three second-year archaeology students from Bournemouth, volunteered to go to Novgorod with us as part of their placement experience. Clare Hamilton-Dyer, who had already had some experience in identification working alongside her mother, also joined us. We spent a week in Bournemouth in preparation for the trip, during which time we provided the students with training in the basics of identification and recording. The party which assembled at Gatwick on July 13th also included myself, Sheila, Mark, Mick and Eamonn.

Figure 2.11: *Bone identification in 1996 (Matty Britten and Natasha Efimova). Photograph by Sheila Hamilton-Dyer*

The six of us working on the bones were joined by Natasha and we clearly needed to find larger accommodation than the room we had used in 1994 and 1995. Alexander Khoroshev arranged for one of the storerooms to be cleared and we moved into this room along with the reference collection and the Expedition's large, cast iron safe!

Although space was tight and the lighting barely adequate, we developed an efficient system of identification and recording (Figure 2.11). Bones collected on Troitsky XI were by this time usually being packed in large tea chests, of the type once commonly used by removal companies to pack crockery and other household items. In some instances, bones from more than one spit/property were stored in the same crate separated only by a layer of paper. This meant that we had to be careful to prevent smaller bones packed in the top half of the crate from slipping into an adjacent assemblage from another spit. We all participated in the initial sorting and identification of an assemblage, which commonly consisted of several hundred bones. We set aside tables or trays for each species and for the more abundant species, placed the bones in separate piles of different elements. This initial communal sorting phase enabled the less experienced workers to check identifications with Sheila and myself as we proceeded and they were also encouraged to use the ever-expanding reference collection.

After the initial sorting, we delegated personnel to record different species. At least two people were assigned to the cattle assemblage, which was always the largest. Recording was carried out using the same methods we devised in 1993. Everyone was given the opportunity to take measurements and record tooth ageing data. We continued to create paper records as it was not practicable to use laptop computers and it was initially quicker to record on paper, although I was concerned that considerable time needed to be found in the future to transfer a large amount of data onto computer. As noted previously, the Novgorod assemblage provided an excellent source of training material. Although quality control was an issue with several different people recording material, Sheila and I were on hand to supervise and ensure that standards of recording were as efficient and consistent as possible. Sheila also continued to concentrate on the more detailed recording of bird, fish and smaller mammal bones.

More sieving

Mark and I departed after a fortnight leaving Sheila to supervise progress for a further week. In addition to recording more hand-collected material, the Bournemouth students were given the opportunity to excavate on Troitsky XI. Under Mick and Peter's guidance they excavated in Spit 8, Square 1237 and obtained two samples for sieving, one from within a building and one from an outdoor context. We continued to be keen to investigate the probability that there was substantial intra-site variability in faunal assemblages within properties. Although this was very much a token gesture, it was hoped we could again demonstrate the potential of such an approach. The students processed the soil samples themselves under Eamonn's supervision and the material was made available for Sheila to record.

Reference material

In addition to the mammal and bird bones that were added incrementally to the reference collection from the material we looked at, Sheila had been gradually building up reference material for fish. Her analyses were beginning to demonstrate that most of the fish eaten in Novgorod and Gorodishche were the same species as those eaten in the town today. Most of us had eaten meals based on zander, perch, pike and cyprinid species readily obtainable from Lake Ilmen and the River Volkhov. Sheila needed to enhance her reference material to include a greater range of these species. Fortunately these fish were readily available in a local market situated close to our hotel. Sheila processed several of these in 1996 to add to her reference material.

Progress by August 1996

A substantial number of bones from Troitsky XI Spits 2 to 7 had been processed involving the sorting, identification and recording of around 8,000 bones. This meant we had material from early and late phases of occupation in the Troitsky area, albeit from different trenches. We also had recorded material from two different properties. Natasha had gained further valuable experience and training. However, a substantial amount of data needed to be transferred to digital databases and spreadsheets.

1997

Zooarchaeological training

INTAS funding included money for travel, accommodation and subsistence for various Russian colleagues to visit Britain to develop and share their specialist skills.

During the course of all three INTAS Projects over a dozen Russians visited the UK, some coming more than once. Natasha Efimova was duly included in these visits and stayed in Southampton with Sheila during early 1997. Under Sheila's guidance, Natasha was given intensive training in zooarchaeological methodology and she significantly developed her recording skills. She worked mainly on British material to gain this experience but also transferred some of the paper records of the animal bone data from Novgorod and Gorodishche onto a computer. We were keen to train at least one of our Russian colleagues in zooarchaeology, so that such work could become an established part of the Novgorod post-excavation process. Natasha certainly benefited from this experience and became more confident in her zooarchaeological skills.

Walrus skulls

Lyuba Smirnova had continued to act as an interpreter during our annual visits. At the beginning of our 12-day visit to Novgorod in July 1997, she kindly arranged for Sheila, the recently graduated Matty Britten and myself to visit the Zoological Museum in St Petersburg to look at some of the skeleton collections. There were reports that objects of walrus ivory had occasionally been found in Novgorod. As we were not familiar with walrus skeletons, we welcomed the opportunity to look at walrus skulls and the skeletons of some other species, for which we did not have reference material. I was amazed by how heavy walrus skulls are, although this is not surprising when one reflects their need to support the weight of their tusks. It was clear that if walrus ivory was traded to Novgorod as a raw material for manufacture, it would be very unlikely that it would have been transported with the skulls and therefore it would be very surprising if we found any walrus remains apart from tusks in Novgorod.

The project expands

The award of the second INTAS grant, allowed the project to expand. This involved collaboration and financial support to develop the studies of ceramics, dendrochronology and wooden artefacts and brought in new partners from University College London, the British Museum, The National Museum of Denmark and the University of Lund. At various stages of the visit we were joined by David Gaimster (at that time based at the British Museum), Clive Orton and Jon Hather (University College, London) along with specialists from Denmark and Sweden and various other visitors from England and Ireland. Clive and Jon went on to develop assessment and analytical research on the ceramics and wooden artefacts respectively and, incorporating contributions from Russian colleagues, edited two of the monographs already published in this series (Brisbane and Hather 2007; Orton 2006). David Gaimster, whose expertise includes the analysis of medieval and post-medieval European ceramics, Mark and Clive went on various visits both locally and further afield to other early medieval towns such as Pskov and Stara Russa to look at ceramic collections and meet other ceramic specialists.

More tea chests

New working accommodation had been arranged and the British and Irish contingent were allotted two rooms at the side of the monastery building itself. The bone room along with its reference collection boxes became an environmental archaeology room, which we shared with Mick Monk and Penny Johnston, a former student of Mick's at UCC and now based in Dublin. Penny had replaced Eamonn to process more samples from the Troitsky excavations and identify plant remains from them. Conditions were cramped and lighting far from ideal, particularly for Sheila, who again concentrated on recording fish and bird remains from Gorodishche and the Troitsky sites. Matty, Natasha and I concentrated on making further headway processing the stack of tea chests of bones collected from Peter Guidukov's Troitsky XI.

The adjacent room was where Mark and Clive set up their base to begin studying the ceramics. It was fascinating and instructive to discuss aspects of their and Mick and Penny's work. They faced many of the same questions and challenges that we had encountered with the faunal remains, such as issues to do with sampling and intra-site variability. These discussions and other observations further convinced me of the value of having integrated post-excavation projects.

Progress by August 1997

Significant further progress had been made on the summary recording of animal bones from Troitsky XI. However, there was still much more to do and Peter was still excavating the lower levels and producing further material. However, this is where we hoped our training of Natasha was to come to fruition. She was to continue with further recording of the bones after we departed until the end of the excavation season, when she would return to her normal duties at Novgorod Museum. With the completion of the identification of the fish bones from Gorodishche, Sheila had already acquired enough data to present a paper of her work to the ICAZ fish conference in Panama in March 1997.

1998

The UCL excavations

This was a very busy year for Mark in co-ordinating the INTAS 3 project, which now included dendrochronologists from the National Museum of Denmark, who were collaborating with the Novgorod tree-ring specialist, Olga Tarabardina. In addition, not only were Clive and David continuing their work on ceramics, but Jon Hather was beginning his analysis of woodworking technology and Penny also returned to do further identification of the plant remains. On top of all this, Mark facilitated the arrangements for the four weeks of excavations of specific buildings within

Troitsky XI and XII that were carried out under the direction of Andrew Reynolds (then) of UCL. Alexander Khoroshev and the sites' supervisors Peter Gaidukov and Alexander Sorokin kindly gave permission for these excavations to take place. Using a workforce of students drawn from UCL and Bournemouth, the excavations were designed to assess whether single-context recording was feasible on such sites. The results were encouraging and it proved possible to recognise fine details of wall and floor construction, which added further understanding to the broader information that had been obtained from the traditional methods of excavations (Reynolds and Suggs 2001).

With all this activity going on, we had decided not to add to the logistical complexities by sending any zooarchaeologists to Novgorod in 1998. We hoped that Natasha would be available to carry on recording the Troitsky XI material during the excavation season. However, as Natasha was soon to go on maternity leave, this understandably proved not to be possible.

Progress by August 1998

Although progress on the Troitsky material had stalled and the backlog was increasing, Sheila and I did complete the first detailed analysis of the Gorodishche bones and sent the report to the excavator.

1999

Natasha returns

Natasha returned from maternity leave in time to make a second training visit to England in April. On this occasion she spent some time in Bournemouth, where Lyuba Smirnova was now embarked upon a PhD on combs under Mark's supervision (Smirnova 2005). Natasha gained further experience in analytical skills, computer-recording and English!

Yet more tea chests

The complexities of the annual Novgorod visits continued in June and July with various colleagues from Great Britain and Scandinavia descending on the town at various times. On this occasion the zooarchaeology contingent consisted of Sheila, her younger daughter Joyce, Natasha and Mel Bavington, who was one of the Bournemouth students who was in the excavation team in 1998. Mel has subsequently become a graduate of Bournemouth's MSc Osteoarchaeology programme.

During the first two weeks of July they carried out further identification and summary recording of the Troitsky XI backlog. They were housed in the same room

Figure 2.12: Bone identification in 1999 (Mel Bavington, Sheila Hamilton-Dyer and Natasha Efimova). Photograph by Mick Monk

as in 1997, again sharing facilities with Mick Monk and Penny Johnston. Altogether they recorded around 5,000 mammal bones from the three spits, with Sheila carrying out more detailed recording of the bird and fish bone from these spits and from those set aside by Natasha at the end of the 1997 season (Figure 2.12).

The EAA conference

Bournemouth University hosted the fifth annual meeting of the European Association of Archaeologists in September 1999. Mark Brisbane and David Gaimster organised a session on Novgorod and its hinterland. This session gave us a platform to disseminate not only our own work but also the research of our Russian colleagues, many of whom were able to come to the conference and deliver their papers personally. I delivered a paper on behalf of Sheila and myself on the zooarchaeological work. All the proceedings of this session were subsequently published as a monograph of the British Museum (Brisbane and Gaimster 2001).

Progress by the end of 1999

The summary recording of bones from Troitsky XI had continued after a year's delay but we had reached only Spit 13 out of the eventual 24 by the middle of July. Although Natasha continued to record material during the later summer, there was clearly much more to be done.

2000

No progress!

As there was no funding for travel for any zooarchaeologists to Novgorod this year, we hoped that Natasha would be able to continue the recording of the Troitsky XI material. Clive Orton, Jon Hather and various others accompanied Mark Brisbane on the visit in July. However, Natasha for very sound reasons had decided to leave Novgorod Museum and start a new life in Moscow. She went with our best wishes but of course this still left us with the problem of the animal bone backlog. There was nobody else trained in zooarchaeology at the expedition and this meant that we had to set contingency plans into place for the following year.

2001

Assessing the assessment

It was clear that if we were ever to complete the Troitsky XI material, either we had to find someone else to complete the work, which would be difficult given that they had not been involved with the earlier work and our methods of recording, or complete the work ourselves. It was decided that I would pay a short visit in 2001 to assess how much time was needed to complete the identification and recording. I visited Novgorod for ten days at the end of July, accompanying Mark, Jon Hather, Clive Orton, Mick and Penny. We were joined by Kathy Judelson, who acted as our interpreter since Lyuba had now also left the expedition. Kathy brought a vast amount of experience and expertise and she has for many years translated Russian archaeological texts into English, including Novgorod text for our publications since 1991. She has, for example, translated Mikhail Sablin's report on animal bones from later seasons of excavation at Gorodishche for this volume (see sections of Chapters 3, 6, 7 and 9). We also welcomed Eileen Reilly, from Dublin, an expert on insect remains, who was brought into the project to assess the potential of such studies in the waterlogged deposits at Troitsky. My suspicions that a lot remained to be done were confirmed – the tea chest mountain had reached Alpine proportions. At least Peter had now completed the Troitsky XI excavations, which meant that it was possible to have a finite view of what remained. Evgenij Nosov had also begun to deposit further boxes of bones from Gorodishche from excavations in 1998, which had returned to the investigation of well stratified 9th- and 10th-century deposits. He hoped optimistically that we might find time to look at them. I was also able to complete the recording of a couple of spits from Troitsky XI including the detailed recording of the domestic bird bones. I also began to record butchered bones in detail. Although there would not be a complete record of such observations throughout the deposits, I felt that it would be worthwhile to assess the potential of such work on a sample of the material. I recorded butchery using protocols I had devised and have used subsequently on prehistoric and Roman assemblages in Britain (Maltby 1989; 2007).

Progress by August 2001

At the completion of the assessment, I estimated that to record the remainder of the Troitsky bones would take about a month for one person to complete. A small amount of extra time would be needed for the remaining bird and fish bones to be identified and recorded.

2002

Moscow

Mark and I visited Moscow in April for a variety of meetings. For INTAS3, Mark had brought Nikolai Makarov, then Deputy Director, now Director of the Institute of Archaeology, Russian Academy of Sciences into the collaboration. Several seasons of his excavations of a group of medieval sites at Minino had been producing many animal bones. Minino is located on Lake Kubenskoye some 500 km to the east-north-east of Novgorod on the borders of its territory (Figure 1.1). Its inhabitants were heavily involved with the fur trade and it was rumoured that the faunal assemblage included a substantial proportion of beaver bones. Novgorod's involvement in the fur trade is well known but few bones of beaver, squirrel or other fur-bearing species had been found in the samples we had examined. The opportunity to add information about faunal material from a site in an area where hunting such species for their pelts was being carried out was one we were keen to include. This mapped closely on to our general integrated approach to the analysis of production and consumption of materials in Novgorod and its region, a major theme running through all of the INTAS Projects.

The visit to Moscow provided me with the opportunity to visit Arkady Savinetsky at the Institute of Ecology and Evolution, Russian Academy of Sciences to see the mammal bones he was investigating from the Minino sites and to discuss our project with him. Arkady has wide-ranging zooarchaeological experience being involved in projects as far afield as the Aleutian Islands and the Negev Desert as well as in Russia. We were very pleased that he could disseminate the results of some of his and Olga Krylovich's work in this volume (Chapter 8).

Arkady, however, does not analyse fish bones and these were being kept separately at the Institute of Archaeology, where I was allowed access to them to assess their value. Sieved sampling had been taking place at the excavations and a substantial number of fish bones had been collected. The most spectacular group consisted of thousands of scales of fish that had been found at the bottom of a pit. The overall sample was too large to be analysed in full given the time restraints. It was decided therefore that a representative sample would be selected and taken to Novgorod for inspection, recording and analysis by Sheila when she next visited.

On one of our visits around Moscow we were taken to see excavations taking place at the site of the Moscow Mint. The director was Nikolai Krenke, who we knew from his work on the analysis of the ploughsoil at the base of the Troitsky X excavations and from his contributions to the EAA conference in Bournemouth in 1999. We were shown the finds that included animal bones. One of the finds team, Anna Glazunova was working with the animal bones and clearly interested in learning more about how to identify and analyse them. We invited her to join us in Novgorod in the summer, if she was available.

Scaling the mountain

It was agreed that the completion of the bone recording would be in two stages. The first involved the completion of the summary recording of the material from Troitsky XI. The second stage would involve the recording of fish and bird bones extracted from this material and the recording of the fish bones from Minino. If time and resources allowed, the bones from the more recent excavations from Gorodishche would also be recorded. The stages would take place in successive seasons in 2002 and 2003.

Ellen Hambleton and I carried out the first stage of the work at the end of July 2002. Ellie was at that time a Research Fellow at Bournemouth University and is now a Principal Academic in Zooarchaeology. The most constructive use of Sheila's time would be during the second stage when the fish and bird bones had been extracted and were ready for recording, so we arranged for her to come over in 2003.

I was pleased that Anna was able to join us for that fortnight and although she was not able to come back to the project in subsequent years, there is no doubt that she gained a lot of knowledge about zooarchaeology in a short space of time.

Against all my pessimistic expectations we managed to complete the recording of the mammal bones from all the remaining spits, although decreasing the mountain to a molehill was probably at the expense of our sanity (Figure 2.13). We looked at and recorded something in excess of 10,000 bone fragments during that time. I somehow even found time to continue detailed recording of butchery marks. Luckily for Mick, Penny and Eileen, who also all returned in 2002 to complete their identifications on plant remains and insects respectively, they were able to find refuge in the other room allocated to us along with Mark and Clive. Again the presence of specialists from different fields was extremely stimulating in terms of comparing ideas, approaches and results (Figure 2.14).

Figure 2.13: *Bones awaiting identification and recording in 2002. Photograph by Ellie Hambleton*

Figure 2.14: *The 2002 research team. From left to right: Mark Brisbane; Clive Orton; Eileen Reilly; Penny Johnston; Mick Monk; Ellie Hambleton; Mark Maltby. Photograph by Ellie Hambleton*

Progress by August 2002

By the end of the visit, I felt that we were back on course for the completion of the recording stages of what had become so much more than the initial assessment planned in 1993. We had completed recording all the mammal bones we were going to look at from Novgorod. There remained the final challenge of recording the fish and bird from the earliest levels at Troitsky XI and the fish from Minino. We also still hoped to record the remaining bones from Gorodishche.

2003

The final recording push

This was the last year for data collection. Unfortunately I was absent on sick leave, which left the burden of recording the bones in the hands of Sheila and Ellie, supported by Krish Seetah, Juliet Mant and Vickie Jamieson. Krish was a graduate of the MSc Osteoarchaeology programme in Bournemouth, who had embarked on a PhD at the University of Cambridge on butchery studies. Juliet was another graduate of the MSc Osteoarchaeology programme. Vickie was at that time an undergraduate at Bournemouth and has since become yet another graduate of the MSc Osteoarchaeology programme. The party accompanied Mark Brisbane and Clive Orton, and two other colleagues from UCL, Dean Sully and Thilo Rehren, who are respectively experts in the identification and analysis of leatherwork and metallurgy. Kathy again acted as our interpreter. The party worked in Novgorod for a fortnight in July.

Ellie and Krish carried out the remaining identification of the mammal bones from Gorodishche. For the first time bones were recorded directly onto a database. Sheila completed the recording of the bird and fish bones from Troitsky XI and also identified the bird and fish bones from Gorodishche. In addition, she recorded the fish bones from Minino that had been brought from Moscow after their initial assessment the previous year.

Monkey business

In addition to the routine recording of the remaining Gorodishche material, continuing excavations of the site produced a partial skull of a monkey, which had been discovered from unstratified levels. This skull was eventually brought to Britain (after the required permissions had been granted) for further identification. After Ellie compared it with skulls in the British Museum (Natural History) collections, it was confirmed to be from a Barbary macaque (*Macata sylvana*), a remarkable find of a species normally associated with Gibraltar and North Africa. We had to sacrifice one of its teeth for radiocarbon dating but the results showed that it most likely dated to the 12th century. This led us to carry out a review of similar finds in Europe (Brisbane *et al.* 2007a; 2007b).

Progress by August 2003

The last stages of recording were pressurised because of the large amount of material that remained to be recorded, exacerbated by my unexpected absence. It is testament to the hard work and organisational skills of Sheila and Ellie that our targets for completion of the recording stages were achieved. Over the previous decade, we had gathered substantial amounts of data from three of the Troitsky sites (IX–XI) and small samples from Troitsky VIII and XII and two other sites in Novgorod (Fedorovsky and the Kremlin). We had also recorded large well-excavated samples from Gorodishche and small samples from Georgii and other early medieval sites in the hinterland. We also recorded a substantial sample of fish bones from Minino to supplement the mammal bone analysis being carried out by Arkady Savinetsky in Moscow. Although excavations at all the major sites were planned to continue in 2004 and beyond, it was clearly high time we stopped to evaluate the data we had already collected. It is this analysis which forms the basis for the rest of this volume.

2004–2008

Data transformation and analysis

It is one thing to have collected a large amount of data, it is another to get it into a form that can be used effectively. By the end of the recording phase, we undoubtedly had substantial amounts of data. Unfortunately, it was in various formats. Some of the data were still on paper records; other information had been transferred to spreadsheets; yet other information was stored on two different types of database. Much of the time preparing this volume has been taken getting these data into a usable form for analysis and archival storage. All the data from Gorodishche has now been transferred to database files (Microsoft Access). The bird and fish bone data from the Troitsky site has also been put into a database. Summaries of the mammal data from the Troitsky sites have been transferred to spreadsheets (Microsoft Excel).

Changing publication strategies

At the outset of the project in 1993, a complete monograph dedicated to zooarchaeological studies was not part of the agenda. Apart from our requirement to report progress and results to INTAS, it was not envisaged that anything beyond a short article outlining the potential of the assemblages would be published. Indeed, this was essentially the paper that Sheila and I wrote subsequent to the EAA conference in 1999 (Maltby and Hamilton-Dyer 2001a). A slightly amended version of this was also published in German (Maltby and Hamilton-Dyer 2001b). However, it had become clear much earlier that further publications were necessary. The initial work at Gorodishche (pre-2000) had already been written up and submitted to Evgenij Nosov, for publication in Russian. In addition, the amount of data we had collected from the Troitsky excavations had become too large to be incorporated within summary articles.

The expansion of the INTAS collaboration to incorporate other environmental, artefact and technology specialists gave rise to the development of the much more ambitious publication programme, of which this volume forms a part. Volumes of papers on ceramic studies (Orton 2006) and wood technology (Brisbane and Hather 2007) have already been published. It was originally envisaged that the third specialist volume would be a more broadly-based environmental volume that included the work of Mick Monk and Penny Johnston on the macroscopic plant remains from Novgorod, Almuth Alsleben on the plant remains from Gorodishche and Minino, and Eileen Reilly on the insect remains from Novgorod. However, with the onset of the fourth integrated volume of papers (Brisbane *et al.* 2012), it seemed more appropriate to place the other environmental archaeology papers in that volume. This provided more space in this volume for the animal bone data to be covered more comprehensively.

2008–2019

Delays and distractions

It was originally intended to complete this book in 2010 and it was meant to precede the fourth book in the Novgorod series (Brisbane *et al.* 2012). I did contribute a chapter in that volume entitled 'From *Alces* to zander: a summary of the zooarchaeological evidence from Novgorod, Gorodishche and Minino' (Maltby 2012). This provided a fairly detailed summary of the conclusions, which have since been expanded for this volume, for which it has been possible to include more of the data.

It is regrettable that it has taken a decade to complete the monograph. There are many reasons for the delays including personal challenges resulting from various family crises. With regards to the project itself, the final digital database took longer to complete than was originally envisaged because of the complexities and inconsistencies in the original paper records, which Sheila and I had to sort out in our limited spare time.

I confess that my attention was also often diverted to other projects with which I was fortunate to be involved. I was quite heavily involved in the Ecology of Crusading Project led by Aleks Pluskowski at the University of Reading. From 2014, a significant portion of my time has been devoted to being one of the leaders of a major AHRC-funded project focused on the complex history and relationships of humans with chickens. Within Britain, I made contributions to several conferences and produced a number of publications on the zooarchaeology of Roman Britain including producing a monograph on the animal bones from Winchester (Maltby 2010). I also returned to Russia with Mark in 2011 to look at medieval sites in the Suzdal region, the neighbouring territory to Novgorod, that were being investigated by Nikolai Makarov.

Since 2013, accompanied by Mark and/or Ellie, I have returned to St Petersburg four times to analyse bones from Chalcolithic sites in the North Caucasus excavated by colleagues from Saint Petersburg State University and the Hermitage Museum. These projects may have distracted me from completing this book, but, as will become apparent below, there have been opportunities taken during several of these visits to develop aspects of the analysis.

Dialogues and dissemination

Fortunately it has been possible to disseminate information about our research through lectures, seminars and conference presentations and there have been several publications since 2008 that have presented some of the results to a variety of audiences.

I have given lectures to undergraduate archaeology students and postgraduate osteoarchaeology students at Bournemouth University annually over the past two decades on aspects of our work in Novgorod, updating the presentations as the research developed. In 2013 Mick Monk and I were invited by Tim Williams of UCL to give a seminar to his MA Urban Archaeology students. We have done this on three occasions since and we were joined by Mark for this session in 2017.

I have also presented aspects of the research to the Archaeology and Anthropology research seminar series in Bournemouth and was invited to give similar seminars to Reading and Sheffield University students and staff. It was good to get positive feedback and constructive criticisms of our project, which we have taken on board.

Wider dissemination has been achieved via conference presentations and their proceedings and through other publications. In 2008, I was invited by Alice Choyke of the Central European University to present a paper to the 'Fauna and Medieval Space Conference' in Budapest. Unfortunately I was not able to attend in person because of a family bereavement but I was subsequently delighted to be asked to provide a chapter for the conference proceedings. I contributed a paper on aspects of the beaver and horse evidence (Maltby 2017a).

In 2009, I presented a paper at the Association of Environmental Archaeologists conference held in York. It was entitled 'Environmental Archaeology in Novgorod: towards an Integrated Approach'. As the title indicates, I took the opportunity to discuss how working with other environmental archaeologists such as Mick, Penny and Eileen, had benefited our zooarchaeological studies.

In 2010, Mark and I were invited by the Institute of Archaeology of the Russian Academy of Sciences to participate in a conference on 'The Archaeology of the Baltic Region: new Investigation and Discoveries'. The conference was held in Kaliningrad in November. This gave us the opportunity to meet up again with Nikolai Makarov, who was the chairman of the organising committee, along with other scholars from Russia, Germany, Poland and Scandinavia to share information about our research. Amongst the conference delegates was Andrei Zinoviev of Tver University. In addition to being an excellent interpreter, Andrei had also been working on animal bones from sites in Novgorod. We invited him to contribute to this volume and are delighted that he was able to accept. His contributions can be found mainly in Chapters 4 and 7.

Kaliningrad is a Russian enclave that few British archaeologists have visited in recent times and it was fascinating to be shown some of the sites within the city and the region despite the Stygian gloom of the November weather. At the conference, I gave a presentation entitled 'The exploitation of animals in towns in the medieval Baltic trading network: a case study from Novgorod'. This is the only time that I have delivered a presentation that was simultaneously translated. It took me a while to get accustomed to the delayed reaction of the audience, particularly when I tried to insert unscripted humorous or other comments! It did of course enable us to convey our results to non-English speaking Russian colleagues. The conference proceedings were subsequently published and I contributed a paper based on the presentation (Maltby 2013). This expanded previous papers by incorporating more comparisons of the Novgorod species representation with those from other medieval towns in northern Europe. Mark also presented a paper entitled 'Early trading centres of the Baltic', which was published in the conference proceedings (Brisbane 2013a).

Mark presented a paper in 2010 at the Society of Post-medieval Archaeology conference held at Memorial University, St John's, Newfoundland. The focus of the paper concerned the motivations that lay behind people moving into new territories. He used Novgorod and the north of Novgorod Land as a case study focusing on the trading of furs and other forest products by incomers who exploited the natural resources of these remote locations. His paper was published in the conference proceedings (Brisbane 2013b). Some of this discussion was also incorporated within a paper written by Mark, Sheila, Ellie and myself for a monograph honouring the career of Nikolai Makarov, edited by Peter Gaidukov (Brisbane *et al.* 2015). In this paper we discussed the development of the settlements at Minino and the effects on the local environment. We also compared and contrasted the zooarchaeological evidence from Minino and Novgorod, focusing on the evidence for the exploitation of beavers. We drew attention to some innovative research Mark was co-ordinating to create models mapping the impact that Novgorod had on the forests of its hinterland.

Our visit to Russia in August 2011 at the invitation of Nikolai Makarov allowed Mark and I to visit sites Nikolai and others were excavating in the Suzdal region (Figure 1.1). Suzdal was an important regional centre to the south-east of Novgorod and at times the two states came into conflict, notably when Novgorod was besieged by a Rostov-Suzdalian army led by the Prince of Vladimir in AD 1169. The visit gave us insights into the landscape and history of this adjacent polity. I also carried out an assessment of a small assemblage of animal bones from the site of Shekshovo 2, near Suzdal, which was being excavated by Nikolai and his team. The species data has been included in the comparative analysis in Chapter 11 (Table 11.1).

Subsequently we were driven from Suzdal to Novgorod to catch up with colleagues and get updates on the latest developments. On the way we picked up Peter Gaidukov and ceramics specialist Peter Malygin in Tver. I do not remember much about that car trip, because I shut my eyes much of the time as we narrowly avoided head-on collisions on the motorway! On arrival we found that several excavations were taking place that summer in Novgorod. Investigations at Troitsky had been extended to new areas; Mikhail Petrov was excavating the Nutny site on the eastern (market) side of the Volkhov and rapid rescue excavations were taking place on sites in the north-east sector of the town. I was disappointed to discover that, despite meticulous recording of buildings and other features, no routine sieved sampling was taking place and the excavation and recording methods continued to be the same as had traditionally been carried out. On a positive note, hand-collected bones from Mikhail's excavations at the Nutny site were being retained, although no one had been assigned to examine them. Mikhail kindly allowed me to do a scan of the bones that had been excavated up until that date. I only had time to record species and areas of the body represented and make notes of any unusual finds. The counts of the domestic species have been included in the comparative analysis in Chapter 11 (Table 11.1). The scan revealed that there continued to be biases towards the retrieval and retention of large bones. We also revisited Gorodishche with Evgenij Nosov and it was pleasing to see a photograph of the Barbary macaque on one of the display boards at the excavations. Apparently its discovery had been mentioned to President Putin when he visited Novgorod!

Our visit, and in particular discussions with Evgenij, convinced us that despite the potential of investigating further sites from Novgorod, it was not worthwhile simply to expand the dataset, given our reservations about the sampling and retrieval methods plus logistical considerations. We already had a sufficiently large database to serve the needs of the original aims of the project.

Earlier in 2011, I paid the first of three visits to Latvia at the invitation of Aleks Pluskowski of the University of Reading. Aleks was leading the EU-funded Ecology of Crusading Project, which was researching multiple aspects of environmental change in the medieval eastern Baltic. I had previously worked with Aleks and Krish Seetah in the analysis of animal bones from the Teutonic Order Castle at Malbork (Marienburg), Poland, which formed part of the preliminary groundwork for the major funding application (Maltby *et al.* 2009). The project gave me another opportunity to work with a multidisciplinary team composed of scholars from Britain and eastern Europe. What was particularly attractive was the chance to examine material from Latvia, with which to make comparisons with Novgorod.

Aleks, Krish, Rose Calis, a graduate from the University of Reading, and I began recording bones from several medieval settlements from Latvia in January 2011. We returned in April 2013 accompanied by Sheila Hamilton-Dyer and Richard Madgwick of the University of Cardiff. We worked in a rather cramped and poorly-lit basement of Riga Museum, which was very reminiscent of working conditions in Novgorod! We recorded bones from a total of eight sites examining material ranging from the 11th to 18th century in date. Sheila was also able to directly compare the fish and bird bones with those she had studied from Novgorod, Gorodishche and Minino. The results of our work, which also included data from sites in Estonia (Rannamäe and Lõugas 2019) and Aleks' detailed analysis of the documentary evidence, was published as one of the chapters in the project's two volumes (Maltby *et al.* 2019).

Comparisons of our data from Latvia and Estonia with results of work on Teutonic order sites in Poland and Germany have also been published (Pluskowski *et al.* 2019). Data from our Livonian research have been included in comparisons with Novgorod in Chapter 11 of this volume.

Aleks organised the Ecology of Crusading Project's concluding conference held in Riga and Cēsis Castle in September 2014. This brought together nearly all of the people who had worked on the project alongside archaeologists and other interested parties from the Baltic States. I presented a paper on the zooarchaeology of Novgorod in order to compare the results with those we had obtained from Livonia. This was subsequently published in the second of the Ecology of Crusading volumes (Maltby 2019).

In June 2013, Mark and I flew to St Petersburg to visit Evgenij Nosov and some of his colleagues. The main purpose of the visit was for Mark to seek a formal partnership agreement between Bournemouth University and Saint Petersburg State University (SPSU), building upon the good working relationship we had established with the Novgorod project. My role was to explore the potential of further zooarchaeological collaborations. We held fruitful discussions with two of the SPSU archaeology lecturers, Anton Murashkin and Evgenij Cherlenok. This resulted in us subsequently being given the opportunity to work on bones from a Chalcolithic cave site at Meshoko in the northern Caucasus, which was being excavated by Evgenij Cherlenok.

We also took the opportunity to revisit the Zoological Museum in St Petersburg with Evgenij Nosov and catch up with Mikhail Sablin, who had worked on the Gorodishche animal bones collected from 2000 onwards. Mikhail had already sent us a report on this material and understandably enquired when it might be published. We have incorporated his work in Chapters 3, 7 and 9 of this volume. It was great to see the Barbary macaque skull again after its safe return from Britain. Mikhail also showed us the complete horse skulls that had been discovered in a ditch at Gorodishche and we discussed with him and Evgenij how best these could be interpreted.

In June 2014 I went back to St Petersburg with Mark, Ellie Hambleton and four Bournemouth University students to record the bones from Meshoko Cave. During the visit Mark took the opportunity to revisit Novgorod and Gorodishche. In June 2016 Ellie, Mark and myself returned to St Petersburg to carry out further recording of material from Meshoko Cave that had been excavated since our last visit. We also recorded the bones from the largely contemporary Meshoko settlement site that were assembled from excavations directed by Sergei Ostashinskii from the State Hermitage Museum. Mark again visited Novgorod to pursue his research, funded this time by a Leverhulme Fellowship, into the impact that the inhabitants of the medieval city of Novgorod had on the surrounding forest and ecology. Ellie and I paid our final joint visit to St Petersburg in 2017 to complete the bone recording and prepare our final reports on the material from the two Chalcolithic sites.

Although Ellie and I were focusing on material from a completely different period and region of Russia, each visit gave us further insights into Russian archaeology and its approaches to the study of animals. Many of our Russian colleagues expressed surprise and were clearly mystified by the fact that the animal bones were being studied by archaeologists rather than by zoologists. The concept of social zooarchaeology is one which needs to be developed further in Russia and we hope that this book and other publications that have emanated from this research will stimulate further interest and foster greater understanding into our approach. Every year we went to Novgorod, we met Evgenij Nosov and he always enquired about the progress of the book and was clearly frustrated by the delayed publication. I have a great deal of sadness and guilt about my inability to complete the work before his death in February 2019. We have dedicated the book to Evgenij and the late Eileen Reilly and we are preparing a paper for Evgenij's memorial volume. However, that is little compensation for the loss of a friend and colleague. Without his foresight, knowledge and negotiating skills, none of our work in Novgorod would have been possible.

In May 2015, Sheila was invited to present a paper to the Conference of Environmental Archaeology of European Cities held in Brussels, Belgium. The paper was entitled 'Fish, fur and feather: exploitation of wild animals in medieval Novgorod'. She updated the delegates on the results of our work on the birds, fish and mammals from Novgorod and its territory. Subsequently she was the first author of a conference paper based on her presentation, which also incorporated further consideration of Mark's research into the effects of Novgorod on the local forests (Hamilton-Dyer *et al.* 2017).

Finally, I was invited by Umberto Alberella of the University of Sheffield, to provide a chapter on Novgorod for the Oxford Handbook of Zooarchaeology, for which he was lead editor (Maltby 2017b). This chapter incorporated a lot of the material we had previously published but its inclusion in that wide-ranging volume has meant that the research in Novgorod will be widely consulted by current and future generations of zooarchaeologists.

Although there have been several publications that have reported upon the results of this project, space has precluded the inclusion of all of the relevant details. Most of the earlier publications have been based on incomplete datasets, which have subsequently been expanded or modified. This monograph provides the opportunity to consider all the zooarchaeological evidence in much greater depth than has previously been possible. For example, it is the first time that we have published the ageing, butchery and metrical data in detail. We have also taken the opportunity to expand our comparisons with other sites taking into account more recent work both in Novgorod and elsewhere. Following on from the integrated approach, which I highlighted in a previous volume in the Novgorod publication series (Maltby 2012), the final chapter (Chapter 11) will incorporate further considerations of other archaeological and documentary evidence.

The structure of this volume

The first two chapters have considered the history of the INTAS projects and their funding (Chapter 1) and how the zooarchaeological studies have developed (this chapter). Chapter 1 has also set the scene with regards to some aspects of the history and archaeology of medieval Novgorod and its territory. The following eight chapters will consider the zooarchaeological evidence in detail. We will begin in Chapter 3 with the evidence for the exploitation of domestic animals at Gorodishche and other hinterland sites. This will mainly be based on evidence from assemblages collected from excavations in 1979 and the 1990s but we will also incorporate Mikhail Sablin's report on domestic mammals from the 2000–2004 excavations at Gorodishche.

Chapter 4 will focus on the exploitation of domestic mammals from the town of Novgorod itself. Evidence will mainly be derived from the three Troitsky sites (Troitsky IX–XI) but we will also consider data from a small sample derived from excavations in the Kremlin and from a larger assemblage from the Desyatinny-1 site, the latter having been studied by Andrei Zinoviev. The evidence considered in Chapters 3–4 includes species representation, body part representation and ageing evidence based on both tooth ageing and epiphyseal fusion data.

Chapter 5 focuses on the evidence for carcass processing of domestic mammals from Novgorod and the hinterland sites. Chapter 6 provides a detailed review of the metrical data of the domestic mammals from these sites. In Chapter 7, we turn our attention to the zooarchaeological evidence for the exploitation of wild mammals in the assemblages from Novgorod and the hinterland sites. Here we consider all aspects of the evidence including species and body part representation, butchery, ageing and metrical data. We include evidence provided by Mikhail Sablin and Andrei Zinoviev from their studies as well as our own. Chapter 8 is devoted to a report on the mammal bones from Minino by Olga Krylovich and Arkady Savinetsky, which focuses mainly on the wild species from medieval deposits, although it also includes some information about earlier periods and domestic species.

Chapter 9 considers the evidence for the exploitation of birds and is largely derived from the detailed analysis carried out by Sheila Hamilton-Dyer on the bones from the Troitsky sites and Gorodishche. It also includes a summary of the bird bones recovered from the 2000–2004 excavations from Gorodishche based on identifications made by A.N. Panteleyeva, which forms a section of Mikhail Sablin's report. Chapter 10 focuses on the fish bone evidence from Novgorod, Gorodishche and Minino based on Sheila's work.

Chapter 11 has two main sections. It will first consider all the zooarchaeological evidence discussed in Chapters 3–10. In addition to comparing results between the different sites studied in this research, it will draw on comparisons of zooarchaeological evidence from other sites in Novgorod and towns in north-east Europe. The discussion will also consider information about animal exploitation derived from documentary and other forms of archaeological evidence. The second section will critically evaluate the current state of the evidence, and will particularly focus on sampling and methodological issues. It will highlight what we have achieved and assess the current state of our knowledge. But it will also acknowledge its limitations and will discuss how we can enhance our knowledge in future investigations.

-3-

THE EXPLOITATION OF DOMESTIC MAMMALS AT THE 9TH- AND 10TH-CENTURY SITES IN THE HINTERLAND OF NOVGOROD

Mark Maltby, Sheila Hamilton-Dyer, Ellen Hambleton and Mikhail Sablin

INTRODUCTION (Mark Maltby)

This chapter will examine the zooarchaeological evidence from settlements occupied in the immediate hinterland of Novgorod in the century prior to its foundation. The most important of these sites is the fortified settlement on the eastern bank of the River Volkhov at Ryurik Gorodishche (Nosov 1992; 2007). This site will henceforth be referred to as Gorodishche. Over 10,700 mammal bone fragments were examined from the 9th- and 10th-century deposits at Gorodishche, mostly from deposits excavated in the 1990s. These samples include over 4,300 identified domestic mammal bones. The assemblages are of sufficient size to enable detailed analysis of species and element representation, and, for some species, sufficient tooth ageing, epiphyseal fusion and sexing data to provide scope for a detailed analysis of stock exploitation by the inhabitants of this non-urban but high status, fortified settlement in the period immediately prior to, and overlapping with, the earliest deposits at Novgorod. The extensive wet-sieving retrieval policy has also enabled more detailed analysis of taphonomic and body part representation to be undertaken than was possible at Novgorod (Chapter 4).

A further sample of animal bones from the 2000–2004 excavations from Gorodishche has been examined by Mikhail Sablin of the Zoology Institute of the Russian Academy of Sciences. This assemblage included further material of 9th- and 10th-century date but also some later material of 11th- and 12th-century date. His report on this assemblage appears at the end of this chapter.

Contemporary settlements situated on Poozerie on the north-western side of Lake Ilmen have also been excavated (Nosov 1992). The sites of Georgii, Prost and Vasilievskoye (Figure 1.2) have all produced small samples of domestic mammal bones that are also included in this discussion.

THE DOMESTIC MAMMAL BONES FROM GORODISHCHE 1979 AND 1993–1998 (Mark Maltby, Sheila Hamilton-Dyer and Ellen Hambleton)

Introduction

Gorodishche was founded in the mid-9th century. It lies just 2 km to the south-east of Novgorod just to the north of where the River Volkhov emerges from Lake Ilmen. It lies on a small hill and promontory overlooking the east bank of the river and the settlement was effectively situated on an island bounded by the Volkhov and two of its tributaries (Figure 1.2). It was situated just above the floodplain that is inundated each spring. It is estimated that the defended settlement covered up to 7 ha during the 9th and 10th centuries (Nosov 1992). It was at that time a royal administrative centre for the region with evidence for long-distance trading and a significant military and diverse ethnic presence (Nosov 2001). It was gradually replaced as a regional administrative, and trading centre by Novgorod from the mid- to late 10th century but maintained a high status by being the residence of the princes for a substantial part of the medieval period after the royal household departed from Novgorod and returned to Gorodishche in the late 11th century.

Parts of the site have suffered substantially from subsequent disturbance, not only from building and agricultural damage but also from the construction of the Sivers canal in the post-medieval period (Figure 3.1). This cut east–west through the southern part of the site and has destroyed a section across the site of about 100 m in that area (Nosov 1992). The preservation of the site also suffered during World War II when the area was heavily used by the entrenched Soviet army (Figure 3.2).

Gorodishche has had a long history of excavations but the most extensive and scientific have been those that have taken place since 1975 (Nosov 1992; 2007). Many of these investigations have focused on the lower promontory area of the settlement on either side of the canal, where archaeological layers are less disturbed and substantial amounts of organic deposits have been preserved. Initially animal bones were not systematically retained apart from the 1979 season. This assemblage, which like the samples subsequently retained for analysis in this report, dates to the 9th and 10th centuries. However, unlike the samples collected in later years, there was no sieving programme, which means that many of the smaller bones, particularly of fish, were not recovered.

As outlined in Chapter 2, from the early 1990s, a substantial sieving programme was established for material being excavated from both sides of the canal (Figure 3.3). The animal bones from those excavations were recorded during 1993 to 1998 and summary results subsequently published (Maltby and Hamilton-Dyer 2001). Further material from the North bank was recorded in 2003 and this sample has been added to the earlier assemblages for this analysis.

The animal bones were identified and recorded at the Novgorod expedition headquarters. A comparative reference skeleton collection has been built up during the

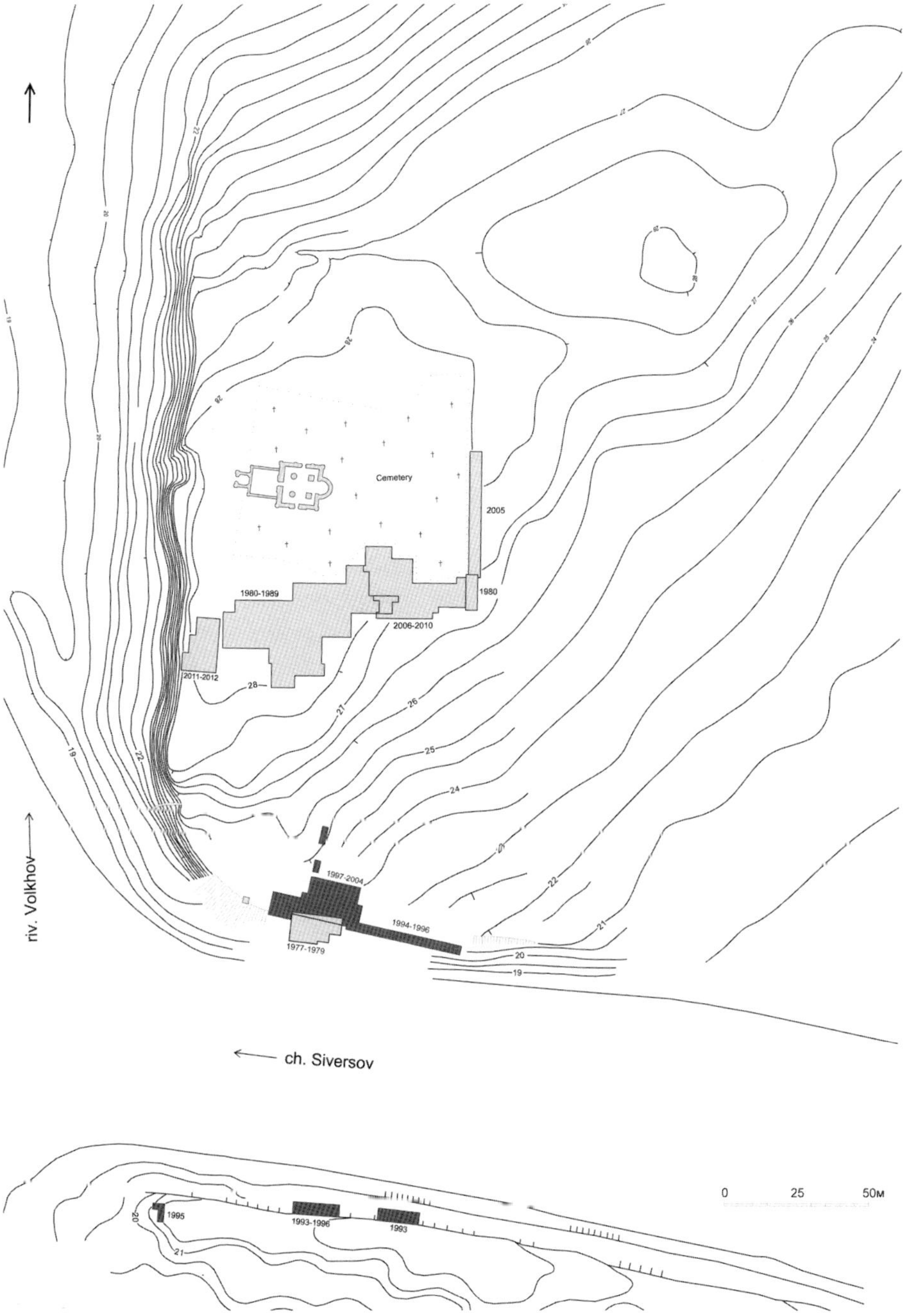

Figure 3.1: *Plan of Gorodishche showing location of trenches and years they were excavated. The animal bones were studied from the trenches shown in black. Supplied by N. Khvoshchinskaya*

Figure 3.2: *Gorodishche: view looking north towards Novgorod; the bomb craters can be seen in the foreground. Photograph by Sheila Hamilton-Dyer*

Figure 3.3: *Sieving at Gorodishche. Photograph by Sheila Hamilton-Dyer*

INTAS project, to aid identification. Reference was also made, where necessary, to identification manuals. Identification, recording and analysis of the fish, bird and smaller mammal bones was carried out by Sheila Hamilton-Dyer. Mark Maltby was responsible for the recording and analysis of the larger mammal bones, including the domestic mammal bone discussed in this chapter.

All fragments from the excavations were recorded individually, initially onto a paper archive for subsequent digital transfer. In 2003, Ellen Hambleton recorded the mammal bones from the 1998 excavations directly onto a Microsoft database.

The provenance of each fragment was recorded by area (North or South bank), excavation square (where known) and context (*e.g.* brown layer; black layer; pit). All mammal bones were recorded to species where possible, apart from ribs and most vertebrae, except when these formed associated groups. However, the atlas (1st cervical vertebra), axis (2nd cervical vertebra) and sacrum were identified to species where possible. It is often impossible to differentiate between sheep and goat bones, particularly if they are fragmented. Although all of their bones have been recorded as 'sheep/goat' in the archive, some specimens could be further identified to species and a note was made of these.

Inevitably, given the high degree of fragmentation in the sample, many mammal bones could not be identified to species. Where possible, such fragments were assigned by size to unidentified large mammal (cattle; horse; elk) or unidentified medium-sized mammal (pig; sheep/goat; dog) categories. Some specimens were too fragmented or too poorly preserved to be assigned even to these size categories and were designated as 'unidentified mammal'.

In addition to species, anatomy and context, the following data were recorded for each specimen where appropriate onto the database:

a) side of the body
b) part of the bone represented (*e.g.* proximal; shaft; distal)
c) percentage of the bone represented (<25%; 25–50%; 50–75%; 75–95%; 95–100%)
d) greatest length of the specimen (<1 cm; 1–5 cm; 5–10 cm; 10–15 cm; >15 cm)
e) gnawing damage
f) evidence for surface erosion and burning
g) epiphyseal fusion evidence
h) tooth eruption and wear evidence
i) butchery marks
j) evidence of pathology
k) cross-reference to associated bones from the same skeleton
l) measurements
m) any other observations

Tooth eruption and tooth wear data from mandibles of cattle, pig and sheep/goat were recorded using the method of Grant (1982). The details are also stored in the database archive.

A grand total of 13,433 bone fragments were recorded in this way. In order to investigate possible variations within the assemblage, these were assigned to several groups for comparison. In the 1993–97 assemblage, bones from the brown layer found across the site and the subsequent black layer were counted separately. Bones from pit fills were assigned to a third group. A fourth group consisted of miscellaneous contexts that could not be assigned to the above categories (*e.g.* stones; post holes). In addition, for each of these categories, bones from the excavations to the north of the Sivers Canal were counted separately from those from the South bank, in order to investigate whether there were variations in the assemblages from different areas of the settlement. A small number of bones could not be assigned to the north or south areas either because the information was not recorded on the label or it had become illegible.

Similarly, bones from the 1998 excavations were subdivided into those from brown layers, black layers and other contexts. Bones recovered from the 1979 excavations of the North bank were also counted separately. There was no sieving during that season and the assemblage partly represents a judgemental collection of 'interesting bones' rather than a sample that was systematically collected.

Domestic mammal representation at Gorodishche (1979 and 1993–1998)

Species of domestic mammal identified were cattle, pig, sheep, goat, horse, dog and cat. Counts were based initially on the number of individual specimens (NISP) identified to each species. From all contexts studied, 4,351 fragments were identified to these species (Table 3.1). In addition, probably nearly all of the 6,295 fragments assigned to the unidentified categories would have belonged to these species (Table 3.2).

Cattle (2,076 fragments) and pig (1,580 fragments) were the best represented of these species in all the excavation seasons (Table 3.1; Figure 3.4). However, there

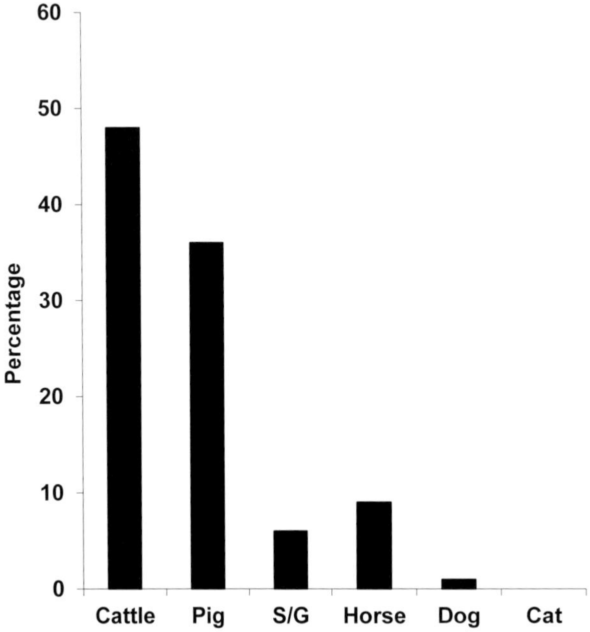

Figure 3.4: *Gorodishche 1979–1998 NISP percentages of domestic mammal bones all seasons combined*

were significant variations in the relative abundance of these two species (Table 3.2). Pig fragments were the most common (53%) in the 1993–97 North bank assemblage. They contributed between 44% and 70% of the identified fragments of domestic mammals in the different samples from that area. Pig fragments were most common in the pit fills in this area (70%), although their total was swollen by the recovery of 27 bones of two partial skeletons of immature pigs found together in one sector of a pit. However, even if these bones are excluded, pig fragments still contributed 67% of the identified larger mammal fragments in these pit fills.

In contrast, the 1993–97 assemblage from the South bank was dominated by cattle bones in all types of deposit. Percentages based on the total fragment counts (Table 3.2) were again slightly biased by the presence of several small groups of associated bones of horse and cattle. Cattle fragments provided 66% of the identified domestic mammal fragments from this area compared with only 32% from the North bank (Table 3.2; Figure 3.5). Pig provided only 25% of the fragments from the South bank. Pit fills provided the highest percentage of cattle (79%) in this area. Other layers and features produced figures of between 61% and 62%. Pig fragments were less well represented in the assemblages from the brown layer on the South bank (Table 3.2).

The 1998 sample from the North bank produced equal percentages of cattle and pig fragments overall (39%). Whereas the cattle percentages remained consistent in the different layers, pig elements were again less well represented in the brown layer (33%) than in the black layer (43%). There is a possibility therefore that there was an increase in the relative abundance of pig bones being deposited in the later black layer, perhaps reflecting a modest increase in the importance of pork and bacon in the diet in comparison to beef in the 10th century.

The sample from the 1979 excavations from the North bank produced 50% cattle and only 32% pig fragments (Table 3.2; Figure 3.5). Cattle elements were therefore relatively more frequently retrieved than in subsequent excavations in this area. However, this mainly reflects the different retrieval strategies. Larger, well preserved bones of cattle are more likely to have been observed and retained in 1979, whereas

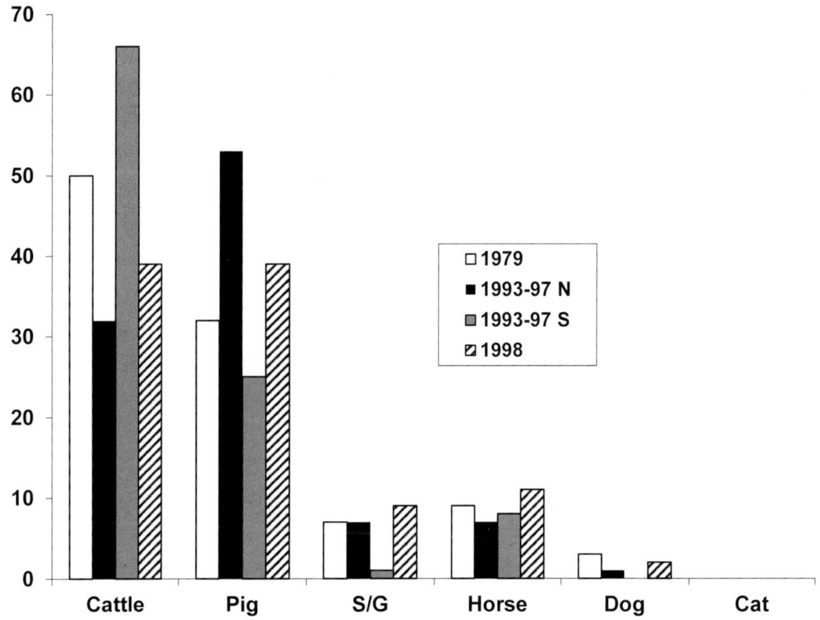

Figure 3.5: *Gorodishche 1979–1998 NISP percentages of domestic mammal bones by season*

every attempt was made to collect all bone fragments during the 1993–97 and 1998 excavations. Consequently, higher percentages of bones of pig and other smaller species are to be expected from the latter excavations.

In comparison with cattle and pig, other mammal species were poorly represented at Gorodishche. Sheep/goat fragments accounted for only 250 (6%) of the identified domestic mammal bones (Figure 3.4). They were particularly uncommon in the 1993–97 excavations of the South bank. No bones were identified in the pits from that area and sheep/goat contributed only 1% overall. Sheep/goat fragments were better represented (7%) in the 1993–97 assemblage from the North bank (Figure 3.5). They were again poorly represented in pit fills (3%) and in the small sample from the brown layer (1%) but were better represented in the black layers and other contexts (9%) (Table 3.2).

Sheep/goat were slightly better represented in the assemblage collected from the 1998 excavations of the North bank, providing 9% of the domestic mammal NISP counts overall (Figure 3.5) and as high as 11% in the black layers. Therefore sheep/goat elements are consistently better represented in the area to the north of the canal and in the later black layers.

Horse bones (379 fragments) were found more frequently than those of sheep/goat, although this total includes 87 bones from several associated groups. Twenty-five of the associated bones were found in the brown layer from the South Bank in the 1993–97 excavations, which partially accounts for the high percentage (19%) of horse bones in this sample. Thirty-five further associated horse bones were recovered from brown layers excavated in 1998, also contributing to the percentage of horse (17%) in that assemblage (Table 3.2). The largest group consisted of 23 lumbar and thoracic vertebrae and ribs of an adult animal from the brown layer in Square 39. Including bones in associated groups, horse bones contributed 9% of the total of identified domestic mammal bones (Table 3.1; Figure 3.4). This figure varied comparatively little between the north and south areas excavated in 1993–97 (Figure 3.5). Horse bones were well represented in the South Bank brown layer (13%) but were poorly represented in the pits from the same area (1%). The 1998 excavations produced slightly higher percentages of horse overall (11%) (Figure 3.5).

Only 59 bones of dog were identified (Table 3.1), 18 of which are from the 1979 excavations including 10 bones from one partial skeleton. Four further associated bones were found in the 1998 excavations. Even including associated bones, dog never formed more than 3% of the identified larger mammal bones in any of the assemblage groups and contributed only 1% of the overall total of domestic mammal bones (Table 3.2; Figure 3.4). Bones that had been gnawed by dogs were, however, quite frequently encountered. Gnawing damage was recorded on nearly 400 of the domestic mammal elements, forming 11% of their total assemblage (Table 3.3). Dogs clearly had access to many carcasses processed at Gorodishche, which implies they were present in substantial numbers.

Only a dozen cat bones were recorded from all the Gorodishche excavations under consideration, forming just 0.3% of the identified domestic mammal bones (Table 3.1).

A large number (6,290 fragments) of unidentified mammal bones were also recorded (Table 3.2). Nearly all of these would have belonged to the domestic mammals. Comparisons of the identified and unidentified portions of the mammal assemblage show that the highest percentage of identified material (79%) was recorded in the 1979 assemblage, again reflecting the selectivity of the retrieval policy and the lack of sieving. Unidentified fragments were much more common in the sieved deposits from later seasons of excavation. Identified fragments formed only 42% of the total domestic and unidentified mammal bones in the 1993–97 assemblage from the South bank and only 31% in the sample from the North bank (Table 3.2). The highest percentages of identified bones were found in the pits, where about half the mammal bones were identified. This reflects better preservation of material in those features where deeper burial protected them from subsequent trampling and other disturbance and less exposure to acidic soils.

The black layer from both banks produced the lowest percentages of identified bones from the 1993–97 excavations (25%–33%). This reflects the higher levels of fragmentation of the bones, particularly in the sample from the North bank. Many bones in this layer also had evidence of surface erosion, which also indicates poorer preservation and explains the high incidence of unidentified fragments. The brown layer produced higher percentages of identified larger mammal bones (38%–48%). Here, bones tended to be slightly less fragmented and suffered less from surface erosion. The waterlogged conditions in this layer would account for the better preservation of the bones, which in turn would increase the chances of a bone's identification. The same contrast was observed in the material from the 1998 excavations, where the brown layers also produced higher percentages of identified elements (47%) than the black layers (36%).

The relative proportions of unidentified large mammal (ULM) and unidentified medium-sized mammal (UMM) fragments show interesting similarities to those of cattle and pig. Given the dominance of these two species amongst the identified material, it is likely that they were also the main contributors to the large and medium-sized categories of unidentified mammal bones respectively.

Overall, large mammal fragments were slightly more common, providing 57% of the total fragments of the two categories (Table 3.2). However, in the 1993–97 assemblage this figure varied between 43% in the total assemblage from the North bank and 69% in the assemblage from the South bank. This contrast mirrors the higher percentage of cattle in the southern area. Indeed, there is a good correlation between the relative percentages of cattle and unidentified large mammal fragments in all of the assemblage groups (Figure 3.6). This lends support to the impression that there was a genuine difference in the relative abundance of cattle and pig bones deposited in the two areas. The 1979 sample produced the highest percentage (75%) of ULM fragments, again indicative of the selection bias towards larger bones. The 1998 assemblage produced values of 60%–64% ULM fragments within the three sub-samples, with an overall percentage of 62%.

Figure 3.6 compares the percentages of cattle expressed as a total of cattle and pig NISP counts against those of ULM expressed as a total of large and medium-sized mammal (UMM) fragments in all 13 sub-samples from Gorodishche. To avoid repetition, the figure excludes the overall percentages for the assemblages from the 1993–97 North bank (38% cattle; 43% ULM), the 1993–97 South bank (73% cattle; 69% ULM), the 1993–97 overall counts (59% cattle; 54% ULM), the 1998 overall counts (50% cattle; 62% ULM), and the combined counts from all the excavations (57% cattle; 57% ULM). The results show a strong positive linear correlation (r = 0.87) confirming that assemblages with higher percentages of cattle also had higher percentages of ULM fragments.

Counts of individual fragments do not take into account biases produced by differential fragmentation of bones of different species. Bones from cattle carcasses are likely to be broken more frequently and into more pieces than those of pig, for example. In addition, the generally denser bones of cattle are more likely to survive the destructive effects of carnivores and erosion than those of pig. Alternative methods for quantifying species representation are therefore needed to test whether the results from NISP counts can be a reliable indicator of the relative frequencies of these species.

One way to decrease the problems of differential preservation is to estimate whole bone equivalents (WBE), using in this case only the major limb bones, scapula and mandible. Fragments were given scores of 0.1, 0.25, 0.5, 0.75 and 1.0 on the basis of the approximate amount of the complete bone that was present. The accumulated scores are given in Table 3.4.

The WBE method tended to decrease the percentage of cattle represented. Overall, pig (39%) WBEs are slightly higher than cattle (38% – Figure 3.7). However, differences can still be observed between assemblages obtained from the 1993–97 North and South

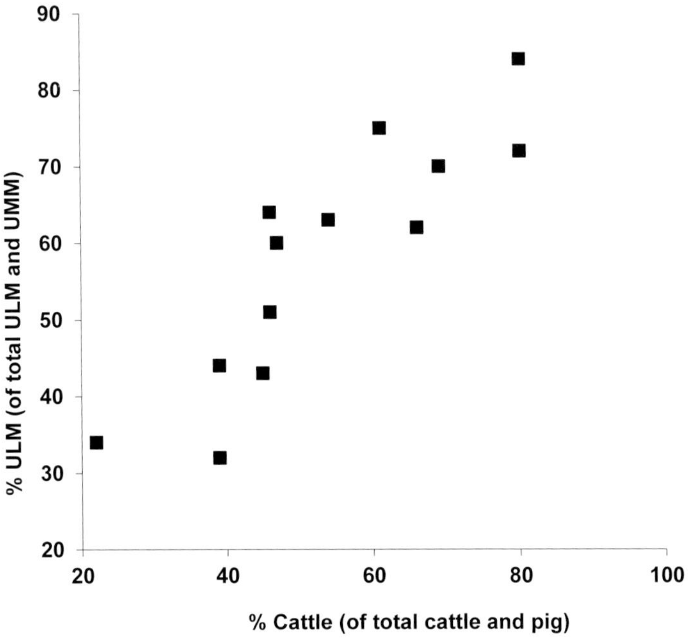

Figure 3.6: *Gorodishche 1979–1998 comparison of cattle and unidentified large mammal (ULM) NISP counts in sub-samples. UMM – unidentified medium-sized mammal*

banks. Pig provides 56% of the total WBEs in the former and only 27% in the latter, whereas cattle increase from 28% to 60%. The very high percentage of pig WBEs in the North pit fills (78%) reflects the presence of several complete bones in the associated groups of bones. Similarly, the high percentage of horse in the South brown layers (23%) reflects the disposal of several associated bones in this deposit (Table 3.4). The results confirm the dominance of cattle and pig and the variations between the North and South banks. They also suggest that pigs may have been more common than indicated by simple fragment counts. Pig WBEs are also the most common in the 1998 assemblage (39%), although the gap between pig and cattle (29%) is much closer and there is also a significantly higher percentage of sheep/goat (17%) than found in earlier seasons. Both dog and cat are also better represented in the 1998 assemblage (Figure 3.8). These variations are similar to those obtained via NISP counts (Figure 3.5).

The cattle assemblage from Gorodishche (1979 and 1993–1998)

Cattle associated bone groups

Only two small groups of associated cattle bones were recorded, both in the 1993–97 South bank assemblage: the first consisted of the distal portion of the tibia and four tarsals found in Pit 2; the second was comprised of five thoracic vertebrae of an adult animal from the brown layer. In addition, six lumbar vertebrae recorded as unidentified large mammal in the brown layer may have belonged to the same adult cow. Although a further 16 cases of associated groups of between two and four cattle bones

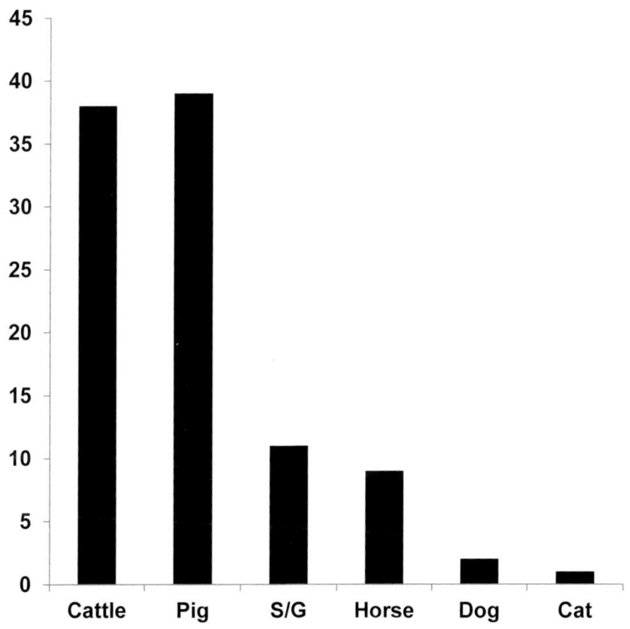

Figure 3.7: *Gorodishche 1979–1998 whole bone equivalent (WBE) percentages of domestic mammals all seasons combined*

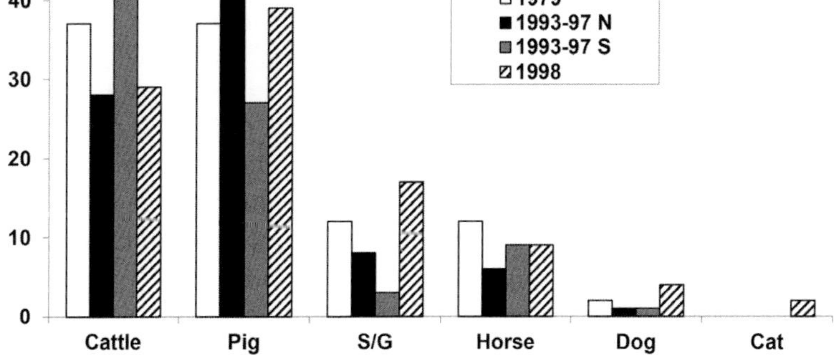

Figure 3.8: *Gorodishche 1979–1998 whole bone equivalent (WBE) percentages by season*

were recorded, several of these included butchered bones and most of the others were from the limb extremities. There is nothing to suggest that any of these groups were not dumped as part of the normal butchering process.

Cattle anatomical representation

The NISP counts of each anatomical element for cattle were recorded for each context. These have been summarised by assemblage group in Table 3.5. The small size of some of the samples makes detailed comparisons of these assemblage groups of limited value. However, comparisons of the total assemblages from the 1993–97 North bank, the 1993–97 South bank, and from the 1998 and 1979 excavations are viable and do reveal some interesting patterns.

The cattle assemblage from these four assemblages was compared by assigning anatomies into the following eight areas of the body:

1) head (horn cores; skull fragments; mandible; hyoid)
2) loose teeth
3) upper limb bones (humerus; radius; ulna; femur; tibia; patella)
4) girdle bones (scapula; pelvis)
5) carpals and tarsals
6) metapodials
7) phalanges and sesamoids
8) vertebrae (atlas; axis; sacrum only)

Amalgamation of the anatomies in this way enables the investigation of broad patterns in the disposal of different parts of the body in different areas of the settlement. However, results derived from this study should be interpreted with care. The eight areas of the body comprise different numbers of bones. There are more carpals and tarsals than metapodials, for example. Therefore equal numbers of bones are not expected in these categories. In addition, the analysis does not take into account differential fragmentation. However, the relative numbers of bone fragments in these categories can still be compared.

The results (Table 3.6) show that there are significant variations in the cattle assemblage between the four broad assemblage groups. The 1993–97 South bank assemblage includes higher percentages of head, loose teeth and phalanges and correspondingly fewer upper limb and girdle bones than the 1993–97 assemblage from the North bank (Figure 3.9). Higher percentages of upper limb bones were found consistently in all types of deposit on the North bank (Table 3.6). The results indicate that the area around the South bank was used for the disposal of a greater proportion of cattle bones of low meat utility than in the North. This in turn may suggest that more primary butchery of cattle carcasses was carried out in this area. In addition, the brown layer from the South bank included significantly more cattle horn cores than any of the other deposits, perhaps indicating the presence of waste from skinning or possibly horn-working (Table 3.5).

The excavations of the North bank in 1998 produced a cattle assemblage similar to that recovered from that area in previous seasons. Indeed, there are even higher percentages of upper limb bones, particularly from the brown layer (Table 3.6). Girdle bones are also better represented than in the other samples. Despite the extensive sieving programme, phalanges, carpals and tarsals are poorly represented in comparison with the other assemblages (Figure 3.9).

Comparisons between the 1979 cattle assemblage with the three sieved assemblages show that loose teeth are less well represented (Table 3.6; Figure 3.9). This reflects retrieval bias. Apart from this, the assemblage is quite similar to the other North bank samples, although it did include a higher percentage of phalanges.

Comparisons of the whole bone equivalents of the major cattle bones also clearly demonstrated differences between the assemblages obtained from the 1993–97 assemblages from the North and South banks. This method decreases the biases of differential fragmentation of these bones. The whole bone equivalent scores obtained from the eight different anatomies involved were fairly consistent in the North bank assemblage, ranging between 1.95 (humerus) to 4.15 (mandible, femur and metatarsal). In contrast, whole bone equivalent scores from the South bank assemblage varied markedly. Scores were high for the mandible (17.00), metacarpal (15.15) and metatarsal (15.60), whereas scores for the upper limb and girdle bones were low, ranging between 2.30 (humerus) and 5.15 (radius). Again these trends were found consistently in different layers and features within the two areas (Table 3.7). Figure 3.10 clearly shows the marked contrast in the relative abundance of these major bones in the two areas. It also shows that the 1998 cattle assemblage is very similar to that from the 1993–97 North bank.

The most likely explanation for these variations is that the southern area was used more often for the disposal of primary butchery waste from cattle carcasses. Head and foot bones are areas of the body associated with relatively little meat and are often discarded at an early stage of carcass segmentation. Although these types of bone were also found in the North bank assemblage, there they were outnumbered by good meat-bearing bones. This in turn suggests that the northern area was where

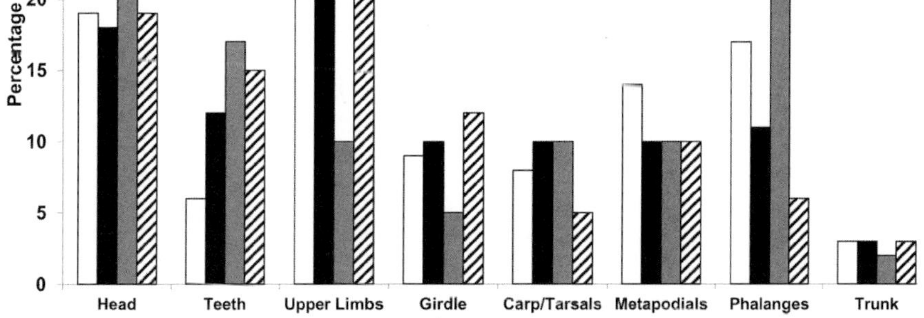

Figure 3.9: *Gorodishche 1979–1998 NISP percentages of cattle body areas by season*

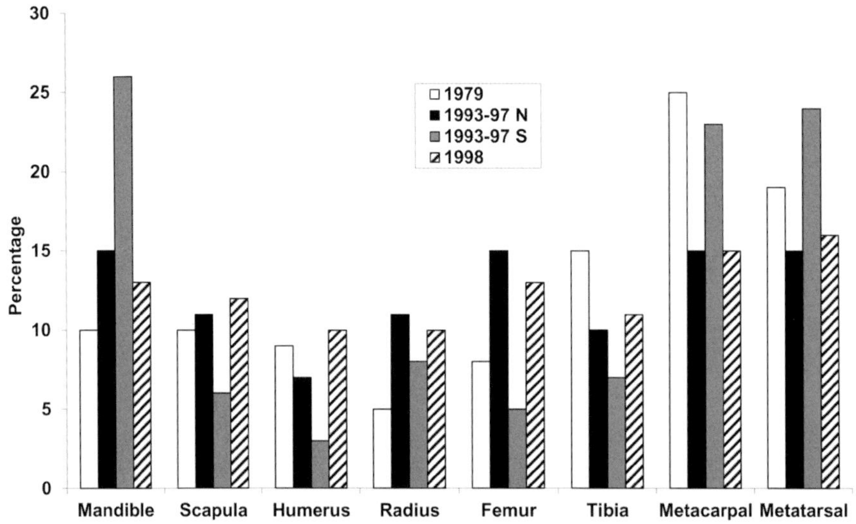

Figure 3.10: *Gorodishche 1979–1998 cattle whole bone equivalent (WBE) percentages of major bones*

more secondary processing of cattle carcasses took place and perhaps where more beef was consumed.

Cattle ageing and sexing evidence

Tooth eruption and tooth wear data from mandibles of cattle were recorded using the method of Grant (1982). The detailed records are provided in Table 3.8. The cattle mandibles were then divided into the following seven stages of eruption and wear:

Stage 1 = 4th deciduous premolars (dp4) not in wear;
Stage 2 = dp4 in wear; 1st molar (M1) not in wear;
Stage 3 = M1 in wear; 2nd molar (M2) not in wear;
Stage 4 = M2 in wear; 3rd molar (M3) and permanent premolars not in wear;
Stage 5 = M3 in wear; 4th permanent premolar (P4) not in wear;
Stage 6 = P4 in wear; Grant mandible wear stage (MWS) <46;
Stage 7 = MWS >45.

The results are summarised in Table 3.9 and illustrated in Figure 3.11. Where sufficient teeth survived, numerical values were assigned or estimated, again using the system of Grant (1982). The results showed that at least 60% of the 50 mandibles analysed had completed their tooth eruption sequence (Stage 6) and belonged to adult animals of over four–five years of age (Figure 3.11). A few of these specimens possessed very heavily worn teeth (Stage 7) and belonged to elderly animals. However, a larger proportion of the specimens had teeth that were less worn and appear to have belonged to young or mature adult animals killed in their prime rather

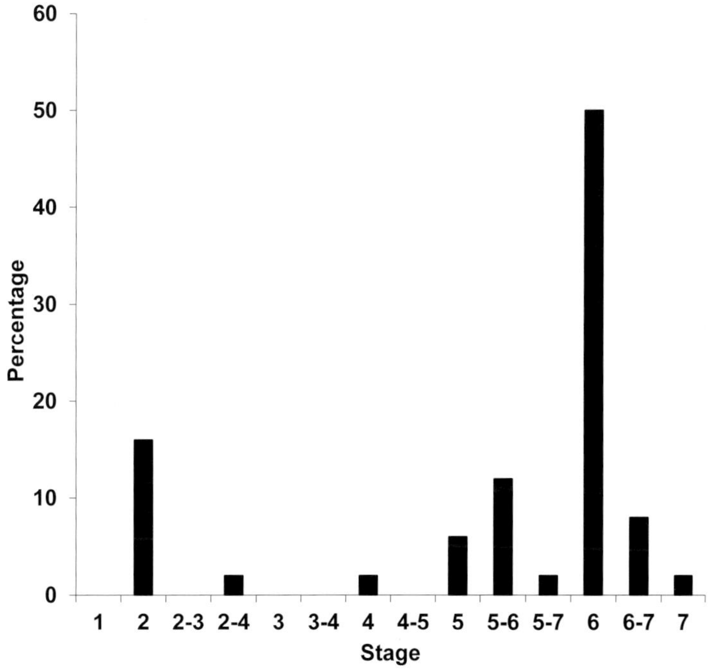

Figure 3.11: *Gorodishche 1979–1998 cattle mandible age stages percentages*

than in their old age. This suggests that the preferred age of slaughter of cattle may have been between four and perhaps seven or eight years of age (Jones and Sadler 2012). Eight (16%) specimens belonged to much younger animals. Most of these only had slight wear on their deciduous molars whilst the first molars were not erupted (Stage 2). They belonged to calves that died under three months of age. These may be interpreted as natural mortalities amongst breeding cattle herds kept at and around Gorodishche. Alternatively they may represent the deliberate culling of some young calves for their meat, perhaps as a by-product of dairy production. There were hardly any mandibles of older immature cattle recovered.

Epiphyses of limb bones of all mammal species were recorded as unfused, fused or just fusing. The last category was restricted to specimens whose epiphysis had fused but where there was still a clear epiphyseal fusion line visible on all sides of the join. For cattle, 17 different fusion points provided data. These were divided into three broad age groups (early- later- and latest), following estimates provided by Silver (1969) and Grigson (1982).

The cattle epiphyseal fusion data (Table 3.10) broadly support the results of the tooth ageing analysis. The results from the three sets of excavations were initially kept separate, in case there had been differential selection of fused or unfused specimens in the 1979 assemblage. In fact, the analysis revealed no significant differences in the early and later-fusing groups between the three samples. Very few unfused specimens (2%) were found amongst the early-fusing group of cattle epiphyses, indicating negligible kill-off of cattle under a year old (Figure 3.12). The discrepancy between the tooth eruption (16% of mandibles at Stage 2 – Figure 3.11) and the epiphyseal fusion evidence could be the result of differential preservation with fewer limb bones than mandibles of young calves surviving the destructive effects of erosion, gnawing and trampling. However, porous bones from young calves were also noted and a total of 40 such specimens were recorded, again representing only 2% of the cattle assemblage excluding loose teeth.

Cattle distal tibia and metapodials usually fuse sometime between 18 months and 30 months of age (Silver 1969), although estimating accurate ages of epiphyseal fusion is problematic. This sample also contained a high percentage (90%) of fused specimens (Table 3.10; Figure 3.12). This supported the tooth eruption evidence that indicated that, provided cattle survived their first few months, they could expect to live for several years.

Overall, 59% of the latest-fusing epiphyses were fused and belonged to cattle aged over about four years of age and probably substantially older in some cases. The 1998 assemblage included rather higher percentages of fused specimens than the other two samples (Figure 3.12). The results, however, generally support the tooth eruption evidence in that they indicate that over half of the cattle represented were fully mature adults. The increase in the percentage of unfused specimens in the latest-fusing sample compared with the distal metapodials and tibia, however, does suggest that a substantial proportion of the animals represented (around 30%) were aged between three and perhaps five years old, indicating the slaughter of prime beef cattle.

Fusion data were also recorded on the bodies of large mammal vertebrae. Of the 130 specimens that definitely or probably belonged to cattle, 56 (43%) were fused and another 10 (8%) were fusing. These epiphyses fuse at a relatively late age in cattle. Silver (1969) suggests about six years, although there is probably a lot of variation involved. It does support the conclusion, however, that a substantial number of cattle were culled between four and six years of age, although over 40% of the cattle represented may have been older animals.

The high percentage of adult cattle represented could indicate that a substantial number were used for traction and/or as dairy animals prior to being slaughtered. However, the main peak of slaughter appears to be of quite young adults, which suggests that the supply of prime beef was important to the inhabitants of the settlement. Older animals may have been mainly kept as breeding stock prior to slaughter. This, of course, makes the assumption that the cattle consumed at Gorodishche were representative of the cattle herds reared in the region. Given its high status, it is conceivable that the inhabitants were able to acquire cattle of preferred ages for slaughter.

Metacarpals of male cattle tend to be broader than females in relation to their length (Grigson 1982). It has been suggested that the relationship between these two measurements can indicate the sex of the animal, a contention recently supported by DNA analysis (Davis *et al.* 2012). Greatest lengths (GL) of 11 complete metacarpals were compared with their greatest distal breadth (Bd) (Figure 3.13). Eight with Bd/

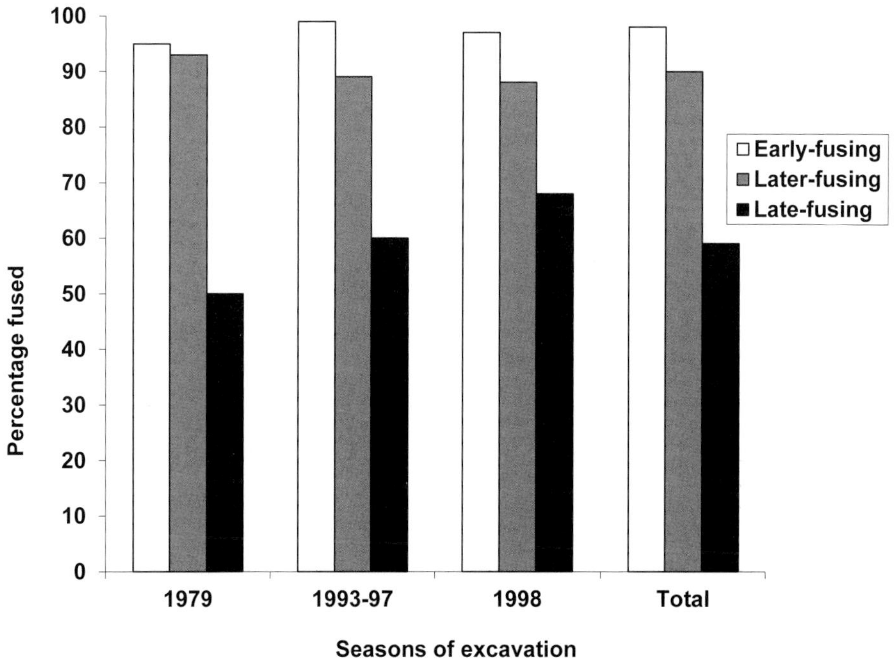

Figure 3.12: *Gorodishche 1979–1998 cattle epiphyseal fusion data*

GL indices of less than 0.30 probably belonged to females; the other three broader specimens probably belonged to males, with the two longest of these more likely to have been from oxen rather than bulls. Eight of these provisionally sexed specimens are included in the scatter diagrams of the metacarpal distal breadth measurements in Figure 3.14, which also includes three other metacarpals with only distal ends measurable. Six assigned as 'female' all fall in the cluster to the left of the distribution. The three broadest specimens scattered to the right of these were all provisionally categorised as 'males'.

If these attributions are largely correct, it suggests that about 73% of the fused distal metacarpals belonged to females. As only fused metacarpals over the age of 24–30 months were measurable, it implies that the majority of the sub-adult and adult cattle found at Gorodishche were females. Given that the epiphyseal fusion analysis showed that nearly all the distal metacarpals were fused (Table 3.10), it would appear that there was a bias towards the acquisition of prime and mature female cattle for slaughter. It could also imply that the majority of young calves represented by the Stage 2 mandibles (Figure 3.11) were males.

Most cattle limb bone breadth measurements are known to vary between sexes with male specimens within a closed herd tending to be larger than those of females (Grigson 1982). The analysis of metrical data of all domestic mammal species from Gorodishche will be further explored in Chapter 6.

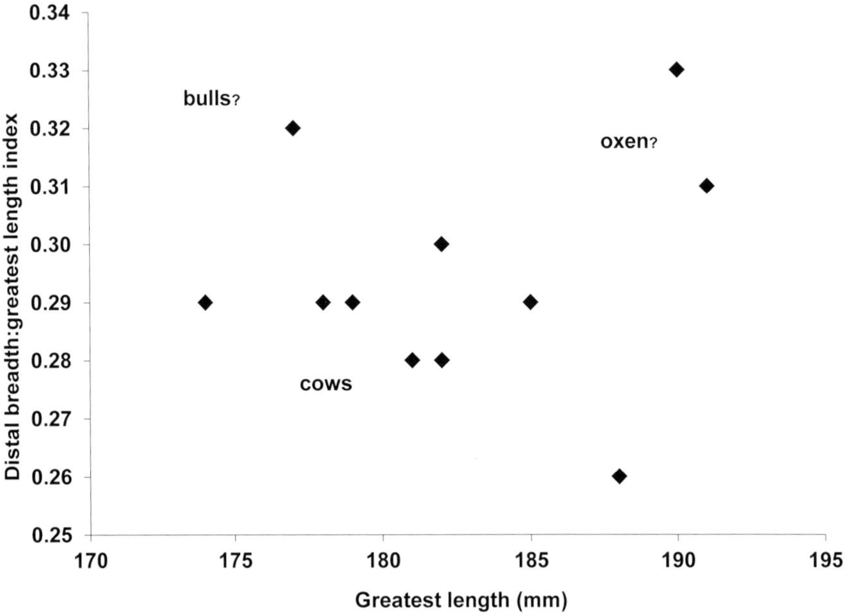

Figure 3.13: *Gorodishche 1979–1998 cattle metacarpal lengths and distal breadth indices*

The pig assemblage from Gorodishche (1979 and 1993–1998)

Pig associated bone groups

The only substantial accumulation of associated pig bones came from the 1993–1997 assemblage and was found in a pit located in Squares 10/11 of the North bank. A minimum of 29 bones from two skeletons were recovered. In addition, 38 ribs and 5 vertebrae categorised as unidentified medium-sized mammals belonged to these animals. Both pigs were very young. Only 8 bones were assigned to the younger animal (mandible; 2 scapulae; 2 femora; 2 tibiae and a fibula), although 13 of the medium-sized mammal ribs are also likely to have been from this skeleton. All these bones are very porous and the mandible has deciduous premolars that were only just coming into wear. This skeleton is therefore from a neonatal mortality probably only a few days old. The second pig is represented by 21 bones. Both forelimbs (scapulae, humeri, radii, ulnae) are present plus two of the metacarpals. Both sides of the pelvis, and both femora and tibiae are also present plus a fibula, an astragalus, two metatarsals, one of the mandibles, part of the skull and the atlas. Twenty-five medium-sized mammal ribs and five vertebrae also probably belonged to this skeleton. It was a little older than the first specimen. Its bones are a little less porous and there is some wear on all of the mandibular deciduous premolars. The first molar has not erupted, however, and the pig is not likely to have been more than a few weeks old (Bull and Payne 1982). No butchery marks were found on any of the bones and the presence of bones from all areas of the body apart from the limb extremities suggests that both

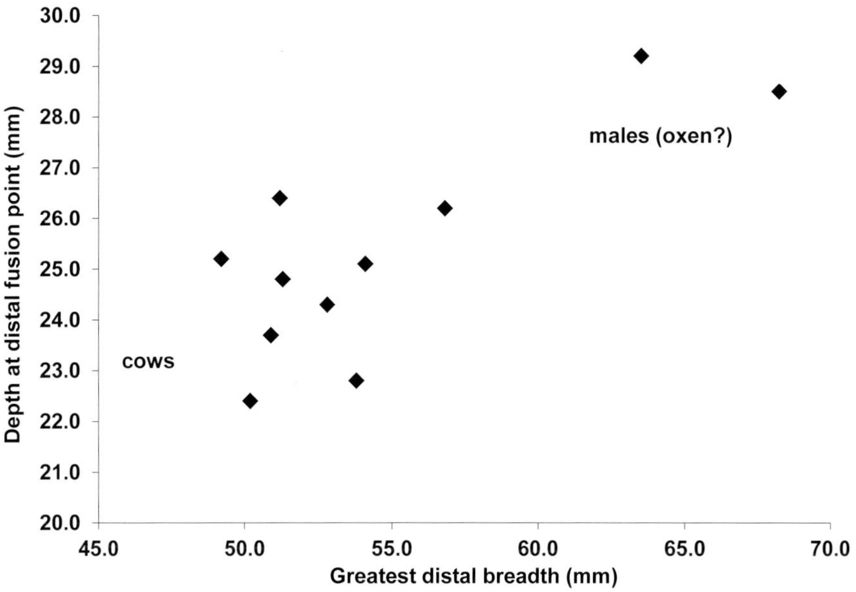

Figure 3.14: *Gorodishche 1979–1998 cattle metacarpals distal measurements*

pigs may have been deposited as complete carcasses. The presence of pigs of this age strongly suggests that they were kept and bred at Gorodishche. These may have been natural mortalities, which were dumped in the pit, although it is impossible to rule out the possibility that they were the remains of sacrificial offerings.

The associated pig bones from this pit account for less than 2% of the total pig assemblage recovered from the excavations. There is no doubt that pigs were heavily exploited for their meat.

Pig anatomical representation

The pig assemblages from the North and South banks differed less markedly than cattle (Table 3.12). The assemblage from the 1993–97 South bank included slightly higher percentages of cranial elements and phalanges and fewer upper limb bones than from the North bank but the differences are not statistically significant. The 1998 assemblage included higher percentages of head and upper limb bone fragments and lower percentages of loose teeth and foot bones. There were also significant differences between the 1993–97 samples and the 1979 assemblage, which also included much fewer loose teeth and phalanges (Table 3.13; Figure 3.15). The most likely explanation for these variations is that retrieval of the smaller elements was more efficient in the seasons where extensive sieving programmes took place. There is no evidence that joints of bacon or pork were imported to Gorodishche and it seems likely, given the representation of bones from all parts of the body, that most of the pigs were slaughtered at the settlement.

Pig ageing and sexing evidence

The pig assemblage includes 78 mandibles and 42 maxillae with tooth eruption data (Table 3.14). The pig mandibles were divided into the following seven stages of eruption and wear (Table 3.15):

Stage 1 = 4th deciduous premolars (dp4) not in wear;
Stage 2 = dp4 in wear; 1st molar (M1) not in wear;
Stage 3 = M1 in wear; 2nd molar (M2) not in wear;
Stage 4 = M2 in wear; 3rd molar (M3) and permanent premolars not in wear;
Stage 5 = P4 in wear; M3 not in wear;
Stage 6 = M3 at Grant (1982) wear stages a–b;
Stage 7 = M3 at Grant (1982) wear stages c–h.

Again all tooth eruption stages were represented including at least two (3%) neonatal (Stage 1) and 12 (15%) young juvenile mortalities (Stage 2). The evidence suggests that substantial numbers of pigs were probably being kept at Gorodishche. The killing of pigs during their first and second years was quite common, as indicated by the presence of at least 29 (37%) mandibles assigned between Stages 3 and 5 (Table 3.14; Figure 3.16). This is to be expected, as pig populations can tolerate high rates of slaughter of young animals. However, the sample also included at least 24 (31%) mandibles with the third molar in wear (Stages 6–7), all probably belonging to pigs over two years old and those at Stage 7 probably over three years old (Bull and Payne 1982). This percentage is higher than commonly observed in samples from most European sites from historic periods including Britain. It implies that many of the pigs consumed at Gorodishche were nearly fully grown before being slaughtered to supply fresh pork or bacon. However, no very old pigs were represented by mandibles.

Epiphyseal fusion data for pigs are shown in Table 3.16. These were again divided into three broad age groups (early- later- and latest), following estimates provided by Bull and Payne (1982). The results indicate that about a quarter of the pigs represented by unfused early-fusing epiphyses died under a year old and two-thirds were under two years of age. There were very few fused epiphyses in the latest-fusing group (Table 3.16; Figure 3.17), suggesting that very few of the pigs represented had survived much beyond three years of age. About 6% of the pig elements excluding loose teeth are porous and from very young piglets.

A limited amount of sexing data was derived from loose canine teeth and from jaws in which they were embedded. Domestic boar tusks grow substantially larger than those of sows. Unsurprisingly, male specimens form a high percentage (79%) of the 28 loose canines where sexing data were recorded (Table 3.17). Even with extensive sieving, it is likely that samples are going to be biased towards the larger male specimens. However, male specimens are also more common (63%) amongst the 30 jaws that were recorded with canine teeth present. This may still be the result of retrieval bias but it is also possible that there was some preference for the acquisition of males for consumption. In later medieval Britain, high status sites tend to have higher percentages of male

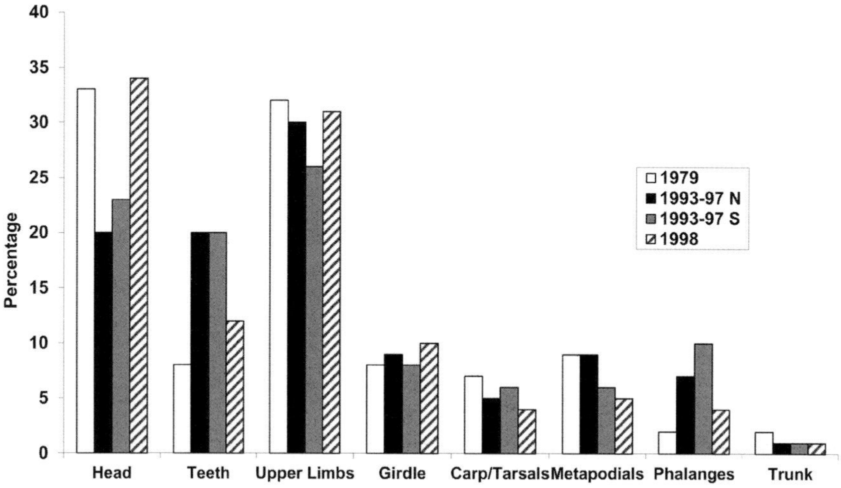

Figure 3.15: *Gorodishche 1979–1998 NISP percentages of pig body areas by season*

animals (Albarella 2005). Unfortunately, very few of the mandibles with extant canines possessed cheek teeth with ageing data. However, four of the six male mandibles were at Stage 5 of the tooth eruption sequence, which suggests perhaps that boars were commonly slaughtered between 18–24 months of age, perhaps in order to supply bacon and other preserved meat during the winter months.

The sheep/goat assemblage from Gorodishche (1979 and 1993–1998)

Relative abundance of sheep and goat

Of the sheep/goat bones that could be identified to species, 20 were identified as sheep and 15 as goat. Eight bones of each species were identified in the 1979 sample. The fact that a high proportion of the sheep/goat assemblage (43%) could be identified to species in that assemblage again reflects the nature of the recovery methods. It is easier to distinguish species from the more complete bones that tended to be recovered and retained, whereas in the sample from the 1993–1997 excavations, only 10 (12%) of the sheep/goat bones could be assigned to a particular species. Five of the sheep bones were found on the North bank and two from the South; two goat bones were found in the North and one in the South. The 1998 assemblage includes at least five sheep bones and four goat horn cores. Therefore, both sheep and goat were represented, with perhaps the former being slightly more common. However, neither species appears to have been a very important component of the meat diet.

Sheep/goat associated bone groups

No large groups of associated sheep/goat bones were recorded. The only associated bones consisted of two cervical vertebrae and the proximal portions of a goat's radius and ulna, both in the 1979 sample.

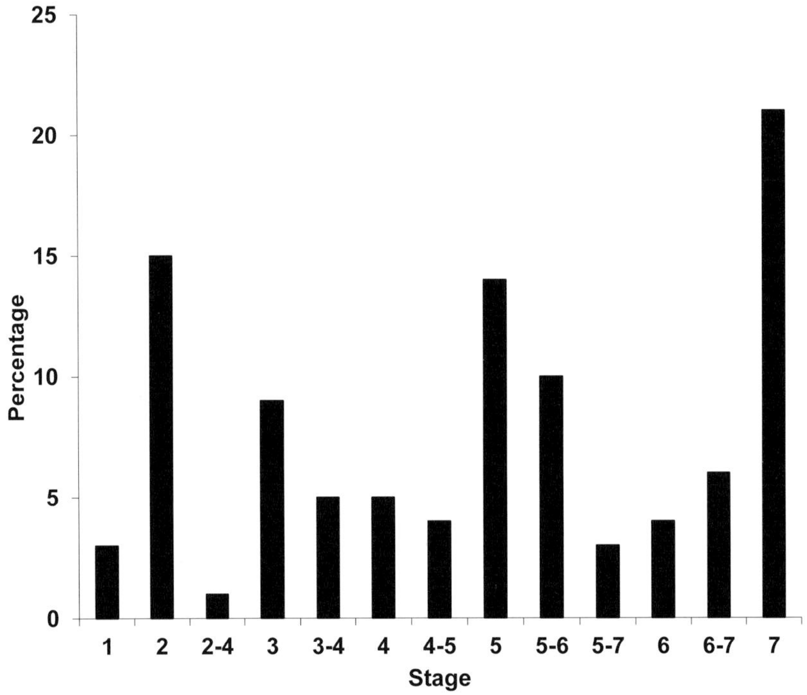

Figure 3.16: *Gorodishche 1979–1998 pig mandible age stages percentages*

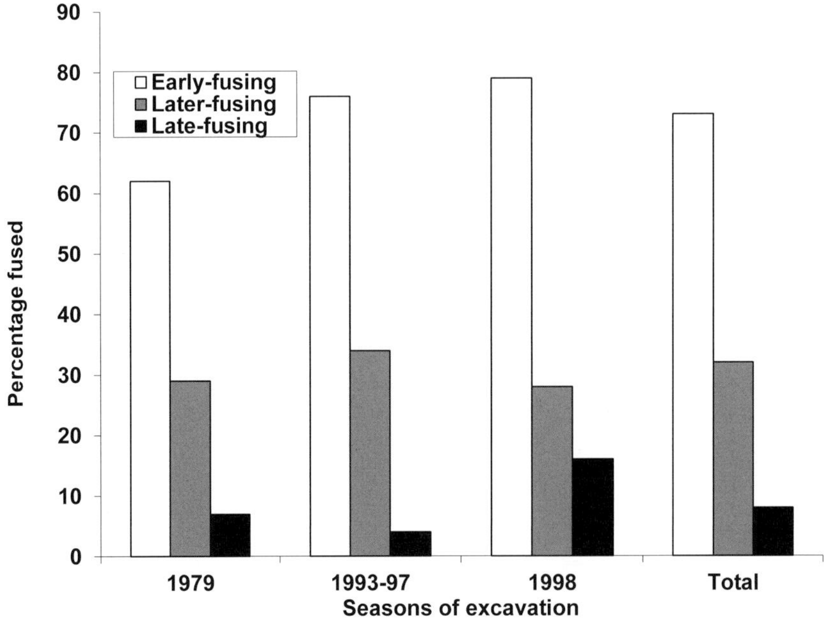

Figure 3.17: *Gorodishche 1979–1998 pig epiphyseal fusion data*

Sheep/goat anatomical representation

Sheep/goat element NISP counts are provided in Table 3.18 and the body areas represented in Table 3.19. The assemblage is too small for detailed conclusions to be gained regarding variations between assemblages. As expected, small bones and loose teeth are poorly represented in the 1979 sample indicating substantial retrieval bias. This sample also contained significantly more vertebrae than the other samples (Table 3.18; Figure 3.18). The sieved 1993–97 assemblage contained substantially more loose teeth, and also more carpals, tarsals and phalanges, reflecting more efficient retrieval. Variations between assemblages from the North and South banks are not statistically significant, as the sample from the South bank is so small. The 1998 assemblage has some similarities with the 1979 sample, particularly with regard to the low percentages of carpals, tarsals and phalanges and it also contains lower percentages of teeth than in the 1993–97 assemblage (Figure 3.18), perhaps indicating that retrieval was not as efficient. Head and upper limb bones are both well represented in the 1998 sheep/goat assemblage, which in many respects is similar to the pig assemblage (Figure 3.15). Overall, head and upper limb bones were recovered in roughly equal numbers and loose teeth were frequently recovered, particularly in sieved samples (Table 3.19). Generally, differential preservation and fragmentation largely account for the variations in the sheep/goat assemblage. Loose teeth, mandibles and tibiae are the best represented elements (Table 3.18), all of which are robust elements that survive well in most archaeological assemblages. There are fewer metapodials represented than perhaps would be expected given their relative robustness, possibly indicating that the assemblages were biased in favour of better quality meat-bearing limb bones.

Seven of the 15 elements identified as goat are horn cores, some of which have evidence of processing (see Chapter 5). Four of these were found in quite close proximity to each other in the brown layer excavated in 1998 and could indicate that some horns were brought to the settlement for processing. Apart from these, there are no significant differences in the types of bone identified specifically as sheep or goat.

Sheep/goat ageing and sexing evidence

The sheep/goat assemblage has only 20 mandibles and nine maxillae with tooth eruption data (Table 3.20). The mandibles were divided into the following seven stages of eruption and wear:

Stage 1 = 4th deciduous premolars (dp4) not in wear;
Stage 2 = dp4 in wear; 1st molar (M1) not in wear;
Stage 3 = M1 in wear; 2nd molar (M2) not in wear;
Stage 4 = M2 in wear; 3rd molar (M3) and permanent premolars not in wear;
Stage 5 = M3 in wear; M1 at Grant (1982) wear stage g;
Stage 6 = M1 at wear stages h–m; M2 at stage g;
Stage 7 = M1 and M2 at wear stages h–m.

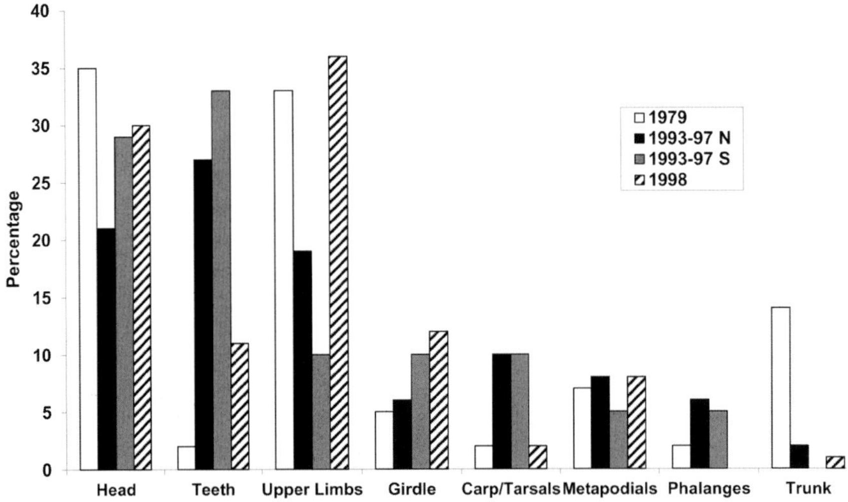

Figure 3.18: *Gorodishche 1979–1998 NISP percentages of sheep/goat body areas by season*

The results demonstrated that the animals represented ranged in age from young lambs or kids (Stage 2) to quite old animals (Stage 7), with no evidence for any particular peaks in the mortality profile (Table 3.21). Over half of the mandibles belonged to immature animals of under approximately two years of age (Stages 2–4). There was insufficient evidence to suggest whether sheep and goats had different life expectancies. Four sheep mandibles, however, were at Stage 4, indicating they were probably slaughtered in their second year.

The epiphyseal fusion data were also very limited (Table 3.22) and indicated that only 41% of the 17 latest-fusing limb bones epiphyses were fused and belonged to adult animals probably over three years old. Only 12% of the 17 early-fusing epiphyses are unfused and only eight (4%) of the surviving bones were porous. About 30% of the later-fusing epiphyses belonging to animals under about two years of age are unfused. Although there may well be significant bias against the survival of the bones of young animals, the evidence would suggest that sheep and goats were quite commonly slaughtered for meat between two and four years old.

Sexing data are also very limited. Three pelves, probably all from adult sheep, were recorded as females. On the other hand, most of the horn cores, particularly of goat, belonged to males. Larger male horns may have been acquired preferentially for processing.

The horse assemblage from Gorodishche (1979 and 1993–1998)

Horse associated bone groups

Groups of horse bones were more frequently recovered than for other domestic mammals. Nine groups were recorded comprising of a total of 87 bones. One of the largest groups consisted of six cervical and eight thoracic vertebrae and six ribs of an adult horse. These were found in the 1993–97 South bank brown layer in Squares 12 and 13. Most of the thoracic vertebrae bore evidence of pathology in the form of swelling of the dorsal process and extra growth of bone. This was probably the result of infection of the bones in the area above the shoulder; a condition sometimes referred to as fistulous withers.

A second group of pathological vertebrae from an adult horse were recovered from the North bank during the 1979 excavations. Three thoracic vertebrae, six lumbar vertebrae and the sacrum were found in association. All the lumbar vertebrae had fused together pathologically. Extra growth of bone from around the vertebral bodies had resulted in this fusion. The thoracic vertebrae and sacrum also bore evidence of extra growth of bone. This pathology was probably the result of the condition known as ankylosing spondylitis, an arthritic condition of the spine. This condition in the lumbar region of the spine sometimes develops on horses that have been ridden (Pluskowski *et al.* 2010).

Another group of six cervical vertebrae of an adult horse was also recovered from the 1979 excavations. The 1993–97 excavations produced the radius, ulna and three of the carpals of the left forelimb of an adult horse from a pit on the North bank and a third phalanx and four sesamoids were found in the brown layer in the same area.

Four further groups of horse bones were found during the 1998 season. Three were from the brown layer. Square 33 produced a complete radius, ulna, two carpals, the third metacarpal and a peripheral metacarpal from the right leg of an adult horse. Square 35 produced a peripheral metacarpal and the three phalanges from the right foot of another adult. Finally, Square 29 produced six thoracic vertebrae, six lumbar vertebrae, the sacrum and 11 ribs of an adult horse. This group included a rib with a healed fracture and most of the vertebrae have evidence for lipping, bridging and ankylosing spondilitis. The group therefore bears similar pathology to the group of thoracic and lumbar vertebrae found in 1979 and again there is the possibility that the condition was associated with stress created by riding (Pluskowski *et al.* 2010). A test pit in Square 43 produced most of a radius, the third metacarpal, a peripheral metacarpal and the three phalanges of a left foot of another adult horse.

Associated groups consisting of five bones or more accounted for 23% of the total horse assemblage. This is a much higher percentage than for cattle (0.5%) and reflects the fact that horse carcasses generally were not processed for meat and marrow as intensively as cattle. None of the associated horse bones bore evidence of butchery. There are a further 12 cases where groups of two to four associated horse bones were recorded. In some cases these may represent processing waste. For example, a set of three phalanges recovered from the black layer on the South bank (Square 13) bore evidence of skinning marks in the form of knife cuts on the first phalanx. However, a complete humerus, radius and ulna and a complete femur and tibia were also recovered from the 1979 excavations. No butchery marks were observed on either group. These good meat bones appear to have been abandoned without processing.

Horse anatomical representation

The horse element counts (Table 3.23) include bones from the associated groups. This accounts for the higher percentages of trunk bones than other species in many of the sub-samples (Tables 3.23–3.24; Figure 3.19). The 1993–97 horse assemblage showed similar trends to cattle in that the assemblage from the South bank included slightly higher percentages of head bones and phalanges (Table 3.24). However, the contrast between the relative proportion of upper limb bones and bones from areas of poorer meat utility was less marked than for cattle. Phalanges were well represented in both areas. The 1979 sample from the North bank did contain a high percentage of upper limb bones but this may again be the result of selection bias and small sample size. The percentage of loose teeth, carpals and tarsals recovered and retained is much smaller than in subsequent seasons. The 1998 assemblage had significant variations between the brown and black layers, partly resulting from the inclusion of associated groups of vertebrae and ribs in the counts from the brown layers. Metapodials are better represented in this assemblage, due mainly to the recovery of 20 peripheral metapodials (Table 3.23).

Horse ageing and sexing evidence

Only four horse mandibles and two maxillae provided tooth eruption evidence. In each case all the permanent cheek teeth were erupted and in wear. Three of the mandibles possessed well developed canines and belonged to males, one of which had heavy wear on all its cheek teeth, indicating it was from an elderly animal. Only one of the 117 surviving horse limb bone epiphyses is unfused (Table 3.25) and there is therefore little evidence that horses were slaughtered for meat at a young age. The use of horses for riding and as beasts of burden meant that they could expect relatively long lives, although, as will be discussed in Chapter 5, meat was sometimes butchered

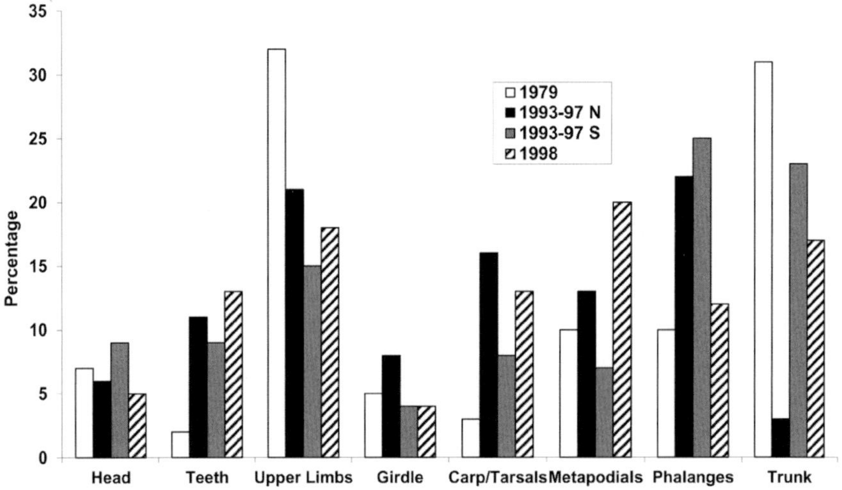

Figure 3.19: *Gorodishche 1979–1998 NISP percentages of horse body areas by season*

from their carcasses. The prevalence of pathology amongst the associated groups of bones also indicates that horses were kept on as working animals despite developing these pathological conditions.

The dog assemblage from Gorodishche (1979 and 1993–1998)

The larger of two groups of associated dog bones consisted of 10 bones recovered from the 1979 excavations. A thoracic, a cervical and six lumbar vertebrae, a rib and left pelvis probably all belonged to the same adult animal. The brown layer in Square 34 of the 1998 excavations produced the distal part of the right ulna and three of the metacarpals of another adult dog. The bones in these groups comprised 24% of the total dog assemblage. No evidence of butchery was observed on either group. Although no other groups of dog bones were recorded, there is little evidence that their carcasses were commonly butchered. Gnawing damage was only observed on 2% of the dog elements (Table 3.3), indicating that their carcasses were probably usually buried.

Apart from the partial skeletons discussed above, the dog assemblage consisted of scattered bones from all parts of the skeleton (Tables 3.26–3.27). The high percentage of trunk bones reflects the presence of the associated group of vertebrae from the 1979 excavations (Figure 3.20).

Ageing data are limited. The mandible and associated maxilla of a young puppy were recovered from the black layer during the 1998 excavations. Only some of the deciduous teeth have fully erupted and the first mandibular molar has not erupted, indicating the jaws belonged to a puppy that lived for only a few weeks. This suggests that dogs were being bred at Gorodishche. Another immature dog was represented by an unfused distal radius. On the other hand, 12 of the limb bones have

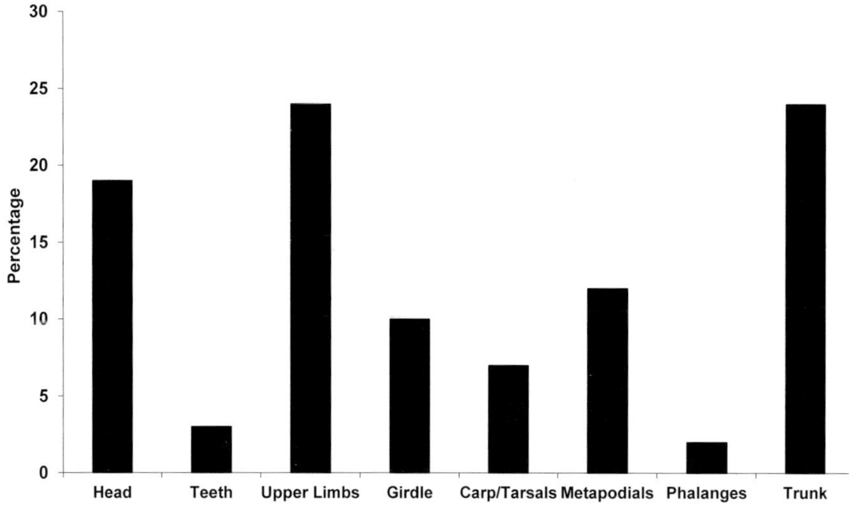

Figure 3.20: Gorodishche 1979–1998 NISP percentages of dog body areas all seasons combined

fused epiphyses and belonged to adult dogs. The high prevalence of gnawing marks (Table 3.3) indicated that dogs had access to discarded carcass processing waste and domestic refuse.

The cat assemblage from Gorodishche (1979 and 1993–1998)

The 12 cat bones were found in both areas of the 1993–97 excavations and in the 1979 and 1998 material (Table 3.1). Although the bones could be from the native European wildcat (*Felis sylvestris*), the small size of the bones match those of domestic cat, although there is some size overlap in modern populations. Several of the bird bones have tooth marks made by cats, which attest to the presence of domestic animals at Gorodishche, as wild cats are unlikely to have lived in the settlement. No evidence of butchery was observed on any of the bones and no records of gnawing were recorded.

Summary of domestic mammal exploitation at Gorodishche (1979 and 1993–1998)

Analysis on the animal bones from the different seasons of excavation and in different areas and layers has revealed significant differences in the faunal assemblages. Some of these differences seem to be the result of differential retrieval standards. Of greater interest, however, are the marked differences in species and anatomical representation between the areas excavated between 1993 and 1997 on the North and South banks of the Sivers Canal. The southern area seems to have been where the greater proportion of cattle bones were dumped, particularly bones of little meat value. The area may have been frequently used to dispose of butchery waste of cattle. Pigs were better represented in the northern area, which also produced a greater proportion of cattle upper limb bones, indicating that this area was where cooking and table waste was more commonly deposited.

Analyses of ageing data have provided some insights into the nature and intensity of domestic animal exploitation at Gorodishche. These results have indicated that the bulk of the meat supply was provided by cattle. Many cattle were killed between four and six years of age for meat and the majority of adult cattle were females. Older cows may have provided milk for human consumption but their main value was probably as breeding animals.

Pork and bacon were also important commodities. Most pigs were killed immature for meat, although over a third of the pigs represented had survived to over three years old. Pigs were probably reared at Gorodishche in substantial numbers, judging by the presence of bones from neonatal animals.

Sheep and goats were consumed in much smaller numbers. Horse meat appears to have been eaten occasionally but many horse carcasses were not heavily exploited for meat and their main use was as beasts of burden.

Tooth marks attested to the presence of domestic cats and dogs in Gorodishche, although their numbers are probably not fully reflected in these assemblages, which consist mainly of carcass processing waste and household rubbish.

THE DOMESTIC MAMMAL BONES FROM OTHER HINTERLAND SITES (Mark Maltby)

Georgii

Introduction

The Poozerie, to the south-west of Novgorod, is a 5 km-wide strip of land running north–south bordered by the northern part of Lake Ilmen to the east and by the River Veryazha to the west (Figure 1.2). The land is rich in floodplain meadows and includes substantial areas of good soils for agricultural production. A substantial number of early medieval sites have been discovered along the line of the River Veryazha (Nosov 1992).

The fortified site of Georgii (Figures 1.2 and 2.7) is located above the floodplain of the River Veryazha. It began as an unfortified settlement probably in the 8th century and the subsequently defended site was occupied until the 10th century (Nosov 1992, 16–18). Mammal bones from excavations that took place between 1989 and 1993 were examined by Mark Maltby in 1993. The bones were obtained from hand excavation only. The bones were bagged by the year of excavation and by square(s) and sometimes further contextual details were provided. Unfortunately, some of the writing on the paper labels had been obliterated prior to examination and there is therefore an incomplete contextual record. All the bones were recorded individually initially onto a paper record, which was eventually transcribed to a Microsoft Access database, using the same fields as described for Gorodishche. Where possible, the year, squares and other information were recorded and all this information is stored in the animal bone archive. However, given the limitations of the surviving contextual information, it was decided not to carry out intra-site analysis and the assemblage has been treated as a single entity dating mainly from the 9th and early 10th centuries. Unlike Gorodishche, none of the bones were recovered from waterlogged deposits.

The overall assemblage consists of 828 animal bone fragments, of which 533 were identified as domestic mammals and a further 187, which were split into various categories of unidentified mammal fragments, were probably nearly all also from domestic mammals. NISP counts of elements of the identified domestic species plus those assigned to unidentified size categories are given in Table 3.28.

Georgii: cattle

Cattle form 38% of the identified domestic mammal fragments (Figure 3.21). Loose teeth provide 35% of the identified elements (Tables 3.28–3.29), a figure significantly higher than at Gorodishche (Table 3.6; Figure 3.22). This reflects poorer preservation of the bones at Georgii. Fifty (25%) of the cattle elements were recorded as eroded and at least 13 have been damaged by gnawing. Percentages of loose teeth tend to increase in more poorly preserved samples. Upper limb bone elements (15%) are, however, better represented than cranial elements (10%), as was the case in the assemblages from the northern sector of Gorodishche (Figure 3.22). Phalanges are quite well represented forming 11% of the cattle NISP counts, with the larger first phalanges being better represented than the others (Table 3.28). Five of the tarsals form an associated group from the right ankle of an immature animal (the group includes an unfused calcaneus). Although there is no evidence of butchery on any of these bones, it is likely that this represents bones discarded during an early stage of processing. The presence of both good and poor quality meat bones indicates that most cattle were probably slaughtered at the settlement.

Only three cattle mandibles with tooth ageing evidence survived. All three were at Stage 4 of the tooth eruption sequence (as defined above for Gorodishche), indicating that they belonged to sub-adult animals probably killed in their third year. Animals killed at this stage of development formed only a small proportion of the Gorodishche mandibles (Figure 3.11). This contrast may also be reflected in the epiphyseal fusion data. Whereas around 90% of the distal metapodials and tibiae were fused in the samples from Gorodishche (Table 3.10; Figure 3.12), only six of the 11 specimens from Georgii (55%) had fused (Table 3.30). Although this is based on very limited samples, it is possible that Georgii received more cattle selected for culling for meat at a younger age than at Gorodishche. Only one specimen from the early-fusing group was unfused, again indicating that few of the cattle elements recovered were from calves.

Georgii: pig

Pig is the most common species represented by NISP counts, forming 47% of the domestic mammal assemblage (Table 3.28; Figure 3.21). The 249 pig elements include six (two femora; two humeri; an ulna and a rib) that are all probably from the same young piglet discovered in the 1992 season (further context details are unavailable). The pig assemblage also contained a large number (44%) of loose teeth (Tables 3.28–3.29), again forming a significantly higher proportion of the assemblage than at Gorodishche, which is indicative of poorer preservation. Other cranial and upper limb bone elements are both quite well represented. Pig vertebrae and ribs were recorded more frequently than at Gorodishche (Figure 3.23). Sufficient quantities of foot bones were recovered to suggest that most, if not all of the pigs were butchered at the settlement.

Although the partial piglet skeleton could indicate the presence of animals bred at the site, only one other pig bone (a first phalanx) came from a very young animal and none of the 18 mandibles with tooth eruption data were from very young animals (Table 3.31). Similarly, none of the ten maxillae with tooth ageing data were from very young animals. This contrasts with the results from Gorodishche where at least 16% of the mandibles were from neonatal or young juvenile pigs (Table 3.15; Figure 3.16). It is probable that this simply reflects the better preservation conditions and sampling strategies at Gorodishche, which were more favourable to the retrieval of small, porous bones from neonatal mortalities. It may also indicate that, despite their overall abundance, relatively fewer pigs were actually farrowed at Georgii.

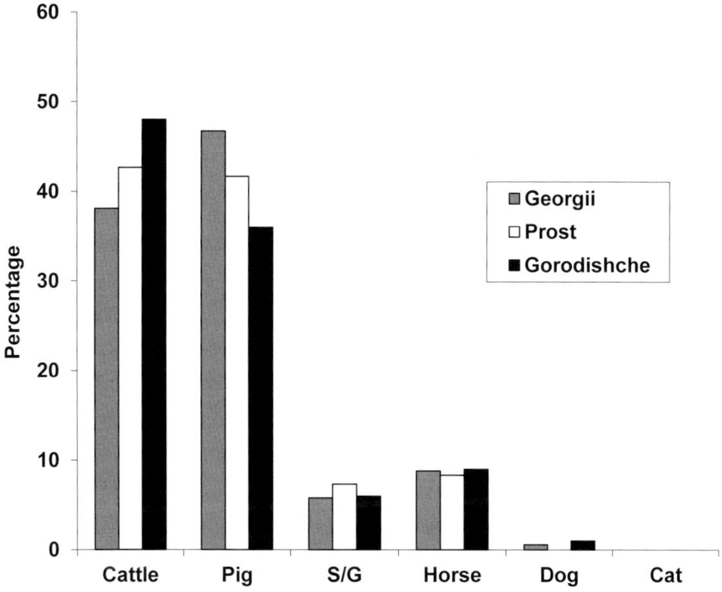

Figure 3.21: *Georgii, Prost and Gorodishche domestic mammal NISP percentages*

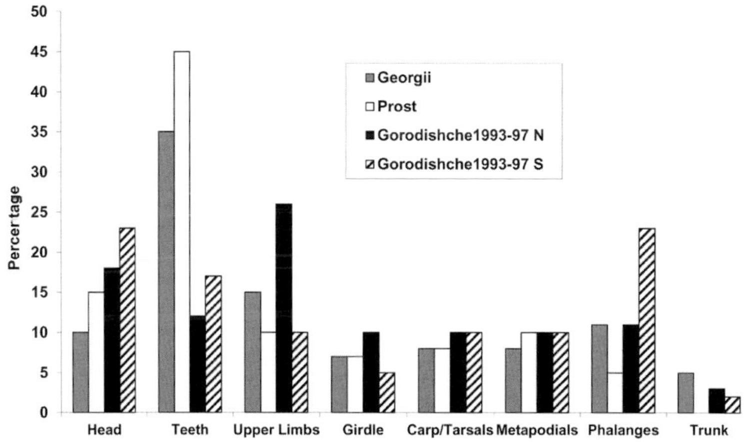

Figure 3.22: *Georgii, Prost and Gorodishche NISP percentages of cattle body areas*

Five (28%) of the pig mandibles from Georgii are from animals that only had the first molars in wear (Stage 3) and are likely to be mainly from pigs culled in the autumn and winter of their first year (Bull and Payne 1982). At least two, and possibly as many as four, of the ten maxillae belonged to pigs of a similar age. There are relatively few Stage 3 specimens in the Gorodishche sample, suggesting that more pigs were selected for slaughter at this age at Georgii. At least five of the 18 mandibles belonged to pigs culled in their second year (Stages 4–5). Six mandibles have all three molars in wear, although five of these have only slight wear on the third molar

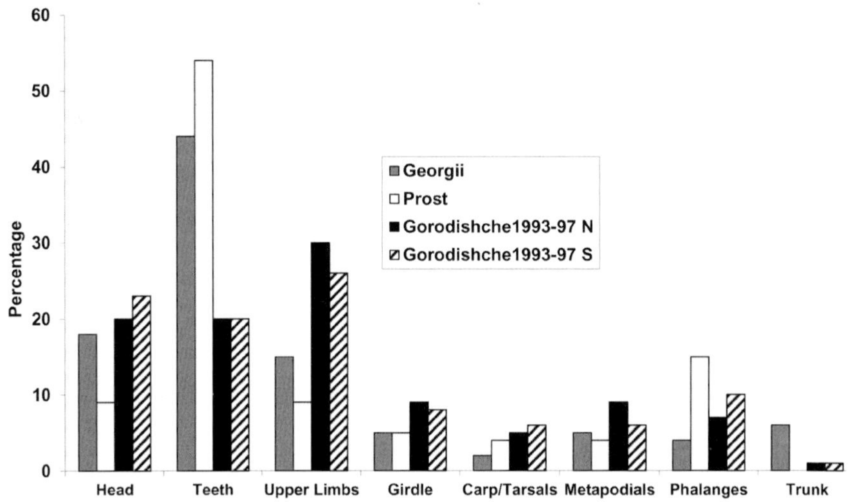

Figure 3.23: *Georgii, Prost and Gorodishche NISP percentages of pig body areas*

(Stage 6), which probably means that most of these were from pigs slaughtered in their third year (Table 3.31). Relatively fewer pig mandibles had reached Stage 7 of the tooth eruption sequence than at Gorodishche (Table 3.15), although very few pig mandibles from either site have very heavy wear on the molars. The epiphyseal fusion evidence shows that there are no fused specimens amongst the ten surviving epiphyses in the latest-fusing group, indicating that no fully mature pigs are represented (Table 3.30). This is similar to the results from Gorodishche (Table 3.16; Figure 3.17). Therefore, although there is an emphasis on the culling of juvenile and immature pigs at both sites, fewer very young and fully adult pigs are represented at Georgii, perhaps suggesting that there was greater emphasis on the procurement of pigs aged between 9 and 24 months of age for consumption at Georgii. However, it needs to be borne in mind that the Georgii results are based on a small sample and ideally further samples are needed to test whether the observed variations in mandible ages are statistically significant.

Georgii: sheep/goat

Sheep/goat elements are poorly represented, providing less than 6% of the identified mammal fragments (Table 3.28; Figure 3.21). Four bones were identified specifically as sheep but no bones of goat were positively identified. Loose teeth provided 35% of the assemblage, a further indication of the relatively poor preservation of the bones. No phalanges, ribs or vertebrae were identified but bones from all other parts of the body were recorded (Table 3.28).

Ageing evidence is limited to a sheep mandible at Stage 3 of the tooth eruption sequence (6–12 months), a maxilla of another immature animal and a handful of limb bones with epiphyseal fusion evidence (Table 3.30).

Georgii: horse

Although providing only 9% of the identified domestic mammal fragments, horse elements outnumber those of sheep/goat (Table 3.28; Figure 3.21). However, the horse totals include two associated bone groups. The larger group consists of 12 bones (4 carpals; metacarpals 2–4; phalanges 1–3; 2 proximal sesamoids) from the left fore-foot of a foal. Cut marks on the first phalanx are indicative of skinning. The proximal epiphysis of the first phalanx is unfused and that of the second phalanx is just fusing, indicating the animal was probably about 9–12 months old (Silver 1969). The second group consists of three phalanges of an older horse. Excluding bones in these associated groups, there are approximately equal numbers of sheep/goat and horse elements. These groups also partly account for the relatively high number of phalanges in the horse assemblage. In contrast to the other domestic species, loose teeth only account for 6% of the horse assemblage. Other horse cranial elements are even scarcer and it seems that few skulls and mandibles were deposited in the areas of the site excavated. Most upper limb bones are well represented and at least four different radii and ulnae are present. Scapulae and pelves are also quite well represented but vertebrae and ribs were rarely recorded (Tables 3.28–3.29).

Ageing evidence is restricted to a small amount of epiphyseal fusion data (Table 3.30). The only unfused specimens belonged to the horse skeleton described above. Most of the horses represented are therefore probably from adults, as at Gorodishche.

Georgii: dog and cat

Only three bones of dog were identified, although their presence at the settlement is also indicated by gnawing marks recorded on 11% of the identified mammal elements (excluding loose teeth). No bones of cat were identified.

THE DOMESTIC MAMMAL BONES FROM PROST (Mark Maltby and Ellen Hambleton)

Prost

Introduction

Prost was quite a large undefended settlement situated at the northern end of the Poozerie just to the west of the head of the River Volkhov (Figure 1.2). Various excavations have indicated that the site was occupied in the 9th and 10th centuries and the features excavated include pits and buildings (Nosov 1992, 13). Domestic mammal bones from excavations in 1997–98 were identified and recorded by Ellen Hambleton in 2003. Sampling included an extensive sieving programme. The bones were subdivided by square and over 30 sub-samples from these produced animal bones. Detailed contextual information has not been obtained but it is believed that all the bones date to the 9th and 10th centuries. All the bones were recorded individually onto a Microsoft Access database. The overall sample consisted of over 3,000 bone fragments, of which only around 200 were positively identified as belonging to domestic mammal species. Large numbers of unidentified mammal fragments, some of them burnt, (Table 3.32) were recovered from the sieving.

Although animal bone fragments are recorded by squares in the archive, and some variability was observed between different sub-samples, the context information was insufficiently detailed to merit intra-site analysis of the domestic mammal bones. NISP counts of elements of all the domestic species plus those in the unidentified size categories (with the exception of small mammals) are given in Table 3.32.

Prost: cattle

Of the 87 cattle fragments, 45% consist of loose teeth, many of them fragmentary (Table 3.32; Figure 3.22). The high numbers of loose teeth reflect both poor preservation and efficient retrieval of fragmentary material. The rest of the cattle sample is too small and too poorly preserved to extract any significant variations in body part representation. Three of the scapula fragments probably belonged to the same bone. Similarly, four heavily eroded horn core fragments were from the same square and it is probable that they were from the same core. Nearly all of the larger cattle elements survived only as small fragments.

Only one mandible survived with any teeth and this consisted of a worn third molar from an adult animal (Stage 6–7). The very limited epiphyseal fusion data did not reveal the presence of any calves, although given the poor preservation, this is not totally surprising. Two proximal femora belonged to cattle that were not fully mature (Table 3.33).

Prost: pig

A very high percentage of the 85 pig elements consists of loose teeth (54%), again indicative of poor preservation. There are also quite a large number of smaller foot bones, indicating both high standards of retrieval and that these elements have tended to survive in a less fragmented state, making them more readily identifiable (Table 3.32; Figure 3.23). However, the three sesamoids and two of the phalanges, all of which are calcined, came from the same squares. These bones could have belonged to the same trotter.

Ageing evidence was forthcoming from one of the five mandible fragments and both maxillae. All three are from immature pigs. The mandible has the first molar in a fairly early stage of wear (Stage 3–4). One of the maxillae was from a younger animal with only the deciduous premolars in wear (Stage 2). The other maxilla possessed a fourth permanent premolar which had erupted but was not in wear (Stage 4). Epiphyseal fusion evidence revealed the presence of other immature animals (Table 3.33). Four porous bones are from juvenile animals. The loose teeth do include two third molars in wear indicating the presence of older pigs but the majority of the bones with ageing data are from juvenile and immature animals killed in their first or second year. Five loose canines provided sexing data; three were from males and two from females.

Prost: sheep/goat

Sheep/goat elements are poorly represented and two-thirds of these elements are loose teeth (Table 3.32). Only one element, a sheep metacarpal, was more closely identifiable. No teeth survived within jaws. The loose teeth include specimens from both sub-adult and adult animals. A porous metatarsal fragment belonged to a juvenile mortality. Only two limb bones provided fusion data (Table 3.33).

Prost: horse

Although horse elements outnumber those of sheep/goat, they include seven from one square (83). Three of these consist of a set of three metacarpals, which probably belonged to the same horse as the first and third phalanges, and possibly also to the pelvis and mandible, found in the same square. Three loose teeth from square 77 also could have belonged to one mandible. As will be discussed more fully in Chapter 5, another horse first phalanx has evidence of skinning marks.

Two horse mandibles provided tooth eruption evidence. Both of these possessed fully erupted tooth rows from adult animals. One of these has a well developed canine and belonged to a male. The crown height of the second premolar suggests it belonged to a horse aged between five–seven years old (Levine 1982). Crown heights taken from five teeth of the second specimen indicated that this was from a much older animal, probably well over ten years old. The limited epiphyseal fusion evidence also failed to provide evidence for the presence of immature horses (Table 3.33).

Prost: discussion

Although handicapped by the small sample size and by the fragmented nature of the assemblage, analysis of the Prost bones does show that, as at Gorodishche and Georgii, cattle and pigs were the most common species exploited, providing 43% and 42% respectively of the NISP counts of identified domestic mammals. Sheep/goat (7%) and horse (8%) are again much less well represented, with the horse percentage perhaps inflated by the probable presence of associated bones. No bones of domestic dogs and cats were identified (Figure 3.21).

Ageing data are limited beyond observations that sub-adult and adult cattle, juvenile, sub-adult and adult sheep/goat, and juvenile, immature and sub-adult pigs are represented. Only bones and teeth of adult horses were recovered.

THE DOMESTIC MAMMAL BONES FROM VASILIEVSKOYE (Mark Maltby and Sheila Hamilton-Dyer)

Vasilievskoye

Introduction

The third site from the Poozerie that has produced a small assemblage of animal bones is that of Vasilievskoye. This was a small unfortified settlement only 300 m from Georgii

(Figure 1.2). Excavations in 1987 produced evidence for rubbish pits and houses sunk into the natural. The settlement appears also to have been occupied in the later 9th and 10th centuries (Nosov 1992, 19). Bones were recovered from six layers or groups of layers. All the bones were recorded individually initially onto recording sheets and subsequently transferred to a Microsoft Access database. The total animal bone assemblage, which was originally recorded by Sheila Hamilton-Dyer and Mark Maltby, consists of only 156 fragments, of which just 52 were identified to domestic mammals. A further 28 mammal fragments were unidentified to species (Table 3.34). Some sieving was carried out but exact details of the sampling programme are uncertain. The assemblage was quite well preserved apart from the presence of 13 burnt fragments, all but one of which was found in layer 2. Eleven of these were unidentified fragments.

Vasilievskoye: cattle

Ten fragments were identified as cattle and most of the nine unidentified large mammal fragments probably also belonged to cattle. Three pelvis fragments were recovered from at least two animals. One of these belonged to a calf (unfused acetabulum); the other to an older animal (fused acetabulum). The four cattle phalanges from the same layer group could conceivably have belonged to one animal (Table 3.35). Both the first phalanges have fused proximal ends.

Vasilievskoye: pig

Nine fragments of pig include two tibiae from different animals (Table 3.35). One porous specimen with an unfused distal epiphysis was from a neonatal mortality. The other tibia was from an older, albeit still immature pig, with an unfused proximal end. One of the two mandibles provided tooth ageing evidence. Only the deciduous premolars and the first molar were in wear with the second molar nearly fully erupted but unworn (Stage 3 of the tooth eruption sequence). This pig was therefore probably slaughtered late in its first, or early in its second year. A first phalanx with an unfused proximal epiphysis also belonged to an immature pig.

Vasilievskoye: sheep/goat and horse

Only two sheep/goat teeth fragments from layer 2 were identified. Both teeth were burnt. Two horse limb bone shaft fragments were identified (Table 3.35). One of the rib fragments in the unidentified large mammal category was also similar to horse.

Vasilievskoye: discussion

The faunal sample from Vasilievskoye is too tiny to provide much additional information regarding the exploitation of domestic stock to that obtained from the other sites discussed in this Chapter. As in the other assemblages, cattle and pig were the most common of the domestic species with sheep/goat and horse poorly represented. The absence of cat and dog is not necessarily significant given the small sample size. Dog bones are stated to have been present in material from other excavations of the site (Nosov 1992, 19). The pig ageing evidence did reveal the presence of a neonatal mortality, implying some pigs were farrowed at the settlement. The immature pig mandible was from an animal similar in age to the most common age category encountered at the neighbouring settlement at Georgii (Table 3.31).

DOMESTIC MAMMALS FROM GORODISHCHE 2000–2004 (Mikhail Sablin)

Introduction

In this section attention is directed to the osteological material from Gorodishche obtained during the excavations carried out in 2000–2004, which is held in the Zoology Institute of the Russian Academy of Sciences. The osteological collection (8,425 identifiable fragments) is not homogenous and can be divided into eight 'blocks', which had arisen in different ways: red sandy loam; floor of the ditch; slope between the natural soil and the ditch; in-fill of the ditch; in-fill of the oak chambers; 10th-century organic material; grey loam; 12th-century organic material. The faunal remains from Gorodishche were divided by the excavators into three parts relating to three different periods: to the 9th century (floor of the ditch, slope between the natural soil and the ditch, in-fill of the ditch and infill of the oak chambers); to the 10th century (red sandy loam and 10th-century organic material) and to the 11th–12th centuries (grey sandy loam, 12th-century organic material). Identifiable bone remains were in the main food waste, which had accumulated naturally in the habitation levels. The osteological material was well preserved, particularly fragments found on the floor of the ditch, which would have enabled them to remain for a long time in a wet environment. Taken together the unidentified remains make up 56.8% of the total amount of bones and most of them were rib and vertebra fragments. Overall the number of unidentified fragments of medium-sized mammals was slightly higher than the number of large mammals, although this varied between phases (Table 3.36). Traces of burning were only visible on a few fragments.

It is held that the number of bones in a sample must be in excess of 300–400 items in order to provide a minimally reliable reflection of an archaeozoological situation (Grayson, 1984). This condition is met by the bones from the 9th-century levels (598 identified fragments), the 10th-century levels (6,548 identified fragments) and the 11th–12th-century levels (1,279 identified fragments). In general the faunal remains from Gorodishche which have been studied are of a variety common for Novgorod and the surrounding area (Tsalkin 1956; Maltby and Hamilton-Dyer 1995; 2001), but the list of identified species of wild mammals and birds contains several new species not noted by previous researchers (See Chapters 7 and 9). In addition it proved possible for the first time to trace the evolution in the composition of the faunal assemblage over a period of almost four centuries.

Domestic mammals

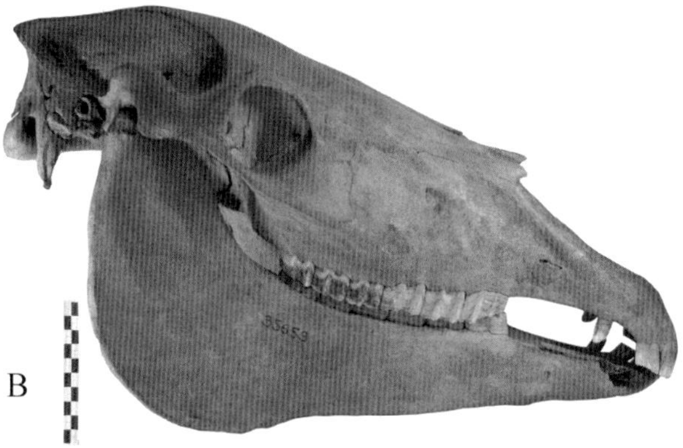
B

Figure 3.24: *Horse skull and mandibles from Gorodishche ditch. Photograph by Mikhail Sablin*

Comparison of the mammal bone samples from the 9th-, 10th- and 11th–12th-century deposits reveals the following trend: while in the 10th-century levels the number of pig bones was greater than that of cattle bones, in the 11th/12th-century levels the share of pig bones decreased from 38% to 18% of the NISP counts (Table 3.36). A similar decrease can be found in the MNI calculations (Table 3.37). Moreover, the relative quantity of horse bones in the 11th–12th-century levels also decreased to half the levels of earlier assemblages. It should be noted that bones of domesticated cats only first appeared in the 10th-century levels. It is also worth pointing out that there is a lack of reliably definable fragments of goat bones in the 11th–12th-century levels, although in the 9th and 10th centuries goats had constituted 14.3% (as regards numbers of individuals) of the total sheep/goat assemblage (Table 3.37). The carcasses of cows, pigs, goats and sheep were butchered on the spot in the settlement, since, in the kitchen waste there have been found not only body parts of high meat value, but also those with little meat value (phalanges, including hooves, caudal vertebrae, horn cores, teeth etc). Significantly more cranial fragments of pigs have been found than those of other mammals, because heads of domestic pigs were a far richer source of meat than the heads of cattle or sheep/goat. It is worth noting that virtually all the pig skulls had been cut along their sagittal axis. Detailed analysis of the anatomical composition of the cattle, pig and sheep/goat remains was not undertaken. Eight horse skulls were found on the floor of the ditch (9th century). Attention should be drawn to the fact that all eight skulls were those of large stallions of the same age, namely eight or nine years old. It can be assumed that the horses were in their prime and that they had been deliberately selected for sacrifice (Figure 3.24).

Half the pig bones, a third of the sheep bones and a quarter of the cattle bones belonged to young or semi-adult individuals, which had been slaughtered for meat. The remains of horses, cats and dogs belonged to adult animals and do not bare traces of their having been cut up for cooking purposes.

– 4 –

THE EXPLOITATION OF DOMESTIC MAMMALS IN NOVGOROD: THE EVIDENCE FROM TROITSKY IX, X AND XI AND OTHER SITES IN NOVGOROD

Mark Maltby with Sheila Hamilton-Dyer, Ellen Hambleton and Andrei V. Zinoviev

INTRODUCTION

Most of the data presented in this and subsequent chapters are derived from the assemblages examined from excavations on three of the Troitsky sites in the Lyudin End of the town (Figure 1.3). The data derive from three adjoining sites (Troitsky IX, X and XI) excavated in the late 1980s and 1990s (Figure 4.1) Samples of animal bones were examined from the lower fills of Troitsky IX and X and the whole sequence from Troitsky XI. These were identified and recorded either directly by Mark Maltby, Sheila Hamilton-Dyer and Ellen Hambleton or by students and others under their supervision. In addition, a small assemblage was examined by Mark Maltby from the 1996 excavations in the Kremlin. A large assemblage from the Desyatinny 1 site also from the Lyudin End (Figure 4.2) has been analysed by Andrei Zinoviev and has been included for comparison.

DOMESTIC MAMMAL BONES FROM TROITSKY IX–XI (Mark Maltby with Sheila Hamilton-Dyer and Ellen Hambleton)

Introduction

Irrespective of the problems and challenges discussed in Chapter 2, over 50,000 mammal bone fragments including material from sieved samples were recorded from the Troitsky sites investigated. These included over 34,000 identified bones of domestic mammals (Table 4.1). This chapter will discuss much of the information obtained from these bones. Major themes include the relative contribution of domestic mammals (mainly cattle, pig, goat and sheep) to the meat diet. The mortality profiles of all domestic species will be evaluated from evidence derived from

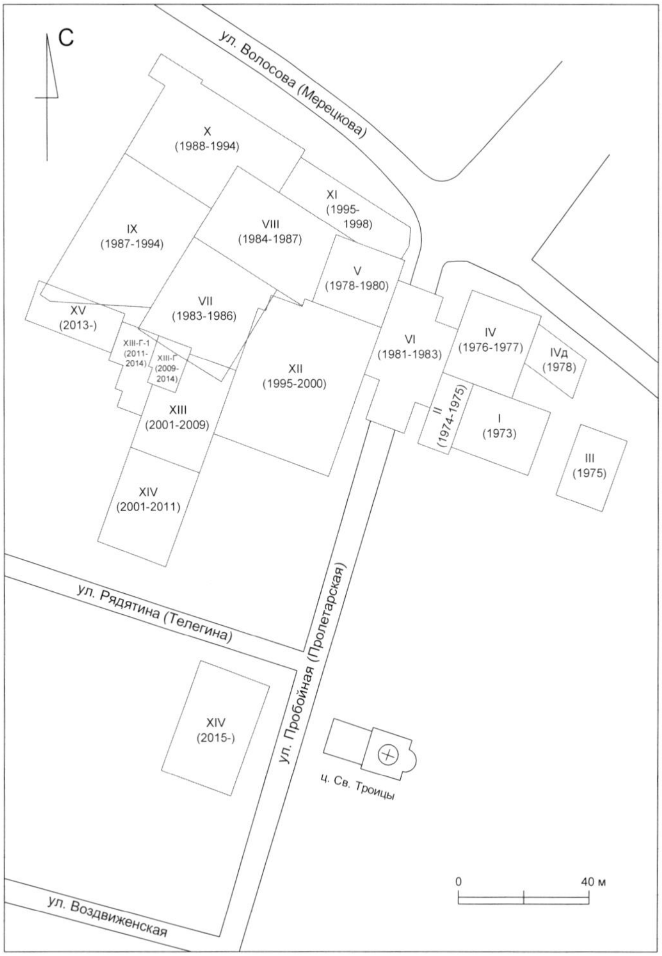

Figure 4.1: *Troitsky excavation sites showing location of Troitsky IX, X and XI. Plan supplied by Peter Gaidukov*

the extensive mandibular tooth ageing data (including samples of almost 800 cattle, over 700 pig and nearly 300 sheep/goat mandibles). Additional ageing information will be derived from epiphyseal fusion data (from nearly 4,000 cattle, and over 900 pig, 500 sheep/goat, 300 horse, 400 cat and 130 dog bones). These results, in conjunction with sexing analysis, provide information about animal husbandry strategies and the relative importance of meat, dairy produce, wool and power in the exploitation of the various domestic species. Analysis of element representation will explore whether there were biases in the abundance of different parts of the body and whether any biases were due mainly to retrieval and taphonomic factors and/or carcass processing and disposal practices. By means of an integrated analysis of the ageing and element representation evidence, we will explore whether most of the animals were brought to the town from elsewhere or whether substantial numbers were kept in the town. The deep stratigraphy of the Troitsky XI site enabled broad diachronic comparisons to be made. Assemblages from different properties from Troitsky XI have also been compared to investigate whether there were any significant variations between them. Discussions of the butchery evidence and metrical data will be discussed in Chapters 5 and 6 respectively.

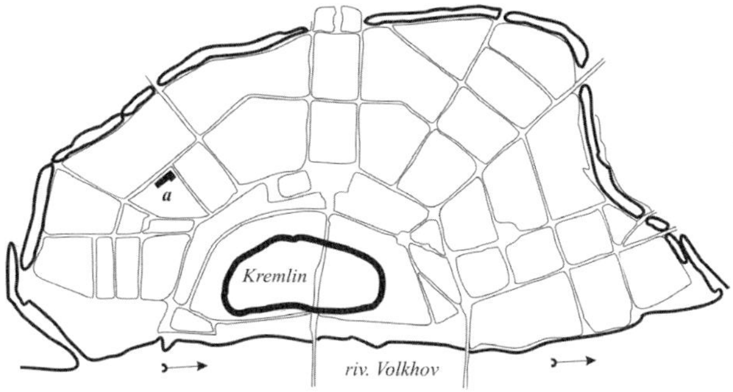

Figure 4.2: *Location of Desyatinny-1 site, Novgorod. Plan supplied by Andrei Zinoviev*

Subdivision of the samples

As discussed in Chapter 2, the Troitsky sites are located in the south-western part of the medieval town in the area known as the Lyudin End (Figure 1.3). Excavations produced evidence for a long sequence of urban properties and associated streets. Animal bones were obtained from three of the sites, Troitsky IX, X and XI. Bones from these excavations were recorded by site, depth (excavated in 20 cm spits) and, where spatial information was provided, by area (*e.g.* Troitsky X) and/or property (*e.g.* Troitsky XI – properties G, I and R). In addition, bones from several early features cut into the natural of Troitsky XI were recorded separately. Altogether, material recovered by hand excavation was subdivided into 69 separate sub-samples. Material obtained from sieving was treated separately.

In order to provide broader chronological and spatial divisions, the 69 sub-samples were often amalgamated into five main assemblage groups for analysis. Three of these consist of material from the lowest levels in each of Troitsky IX, X and XI. These are comprised of bones selected from four of the spits towards the bottom of Troitsky IX (13, 14, 16 and 18), 13 sub-samples from spits 12 to 18 from Troitsky X, and 30 sub-samples from spits 14 to 22 plus features dug into the natural on Troitsky XI. All of these date roughly to between the 10th and the mid-12th centuries. Spits 8–13 from Troitsky XI were dated to the mid-12th to early 13th centuries. Eleven sub-samples from various properties were provided. Finally, the upper fills situated between spits 1 to 7 of Troitsky XI provided 11 further sub-samples from two properties, which date approximately from the mid-13th to the early 15th centuries. For some analyses of large samples, some of these broad assemblages groups were split into smaller units. (NB: On site, spits were given negative numbers below an arbitrary datum. Therefore 'spit 20', for example, is actually 'spit-20'. These minus signs were, however, commonly omitted in the reporting of the excavations and that convention is followed here. No bones from above the datum were analysed.) Comparisons have also been made between assemblages recovered from different sites, properties and areas. Counts of domestic mammal bones recovered from the very limited sieving programme are considered separately.

Species representation at Troitsky – domestic mammals

As discussed in Chapter 2, bones from the excavations were transported from the site to the excavation headquarters where they were identified, scanned and recorded. Records of individual bones were not made, as the assessment was focused on providing general information about each sample by recording key information from each assemblage.

For each of the 69 sub-samples the following information was recorded:

a) number of individual specimens (NISP) of each element type identified to species
b) counts of gnawed, eroded, burnt and butchered specimens for each element type
c) summaries of epiphyseal fusion data by element
d) mandibular and maxillary tooth ageing data
e) metrical data

All limb bones, skull and jaw fragments, and loose teeth were identified to species where possible. Apart from a limited number of sub-samples from the lower layers of Troitsky XI, ribs and vertebrae were not identified to species apart from the atlas, axis and sacrum. As at Gorodishche, unidentified fragments were assigned where possible to large mammal (cattle, elk, and horse) or medium-sized mammal (pig, sheep/goat, and dog) categories.

No details of fragmentation or the zones of the elements present were recorded. It was therefore not possible to calculate minimum numbers of elements or individuals accurately, nor calculate whole bone equivalents. Tooth eruption and tooth wear data from mandibles of cattle, pig and sheep/goat were recorded following Grant (1982).

Overall domestic mammal species counts from Troitsky IX–XI

A total of 33,981 domestic mammal bone fragments (NISP) were identified from the hand-collected samples taken from the three sites. A further 295 fragments were identified in the sieved samples (Table 4.1). Cattle dominated the hand-collected NISP counts in all assemblage groups forming between 62% and 70% of the identified domestic mammal fragments (Table 4.1; Figure 4.3). In the lower spits, cattle were best represented in Troitsky X, followed by Troitsky IX and then Troitsky XI. There was a modest increase in the percentage of cattle in the middle (Spits 13 to 8) and upper spits (Spits 7 to 1) of Troitsky XI. However, there was not much variability between the hand-collected assemblage groups. This would indicate that beef dominated the meat supply throughout the period under consideration. This indeed was probably

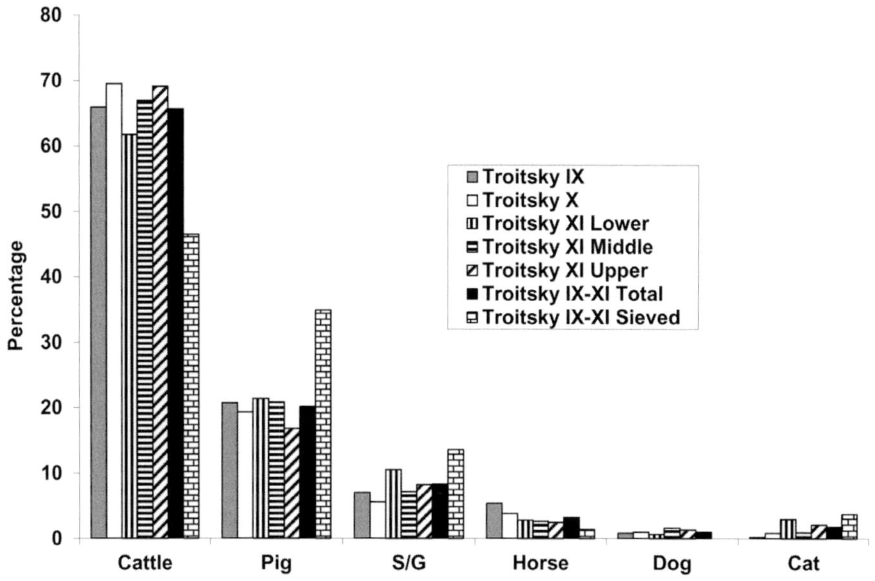

Figure 4.3: NISP percentages of domestic mammals from Troitsky IX–XI

the case but it should be noted that the sieved samples produced a significantly lower percentage of cattle (46%), indicating that the larger bones of cattle had a better chance of recovery in hand-collected excavation.

Pigs provided the second highest NISP counts in all the assemblage groups, with percentages ranging consistently between 17% and 21% in the hand-collected material (Table 4.1; Figure 4.3). The lowest percentage was obtained from the upper layers of Troitsky XI, perhaps indicating a slight decrease in the later medieval period. The percentage of pig elements retrieved from sieved samples (35%) was significantly higher indicating that these had a poorer chance of recovery using normal excavation methods.

Sheep/goat elements were poorly represented in all the hand-collected assemblage groups, providing between 6% and 11% of the NISP counts. The highest percentage was obtained from the lower spits (Spits 22 to 14) of Troitsky XI, where sheep/goat tended to be slightly more common than in Troitsky IX and X. Sheep/goat were better represented in the sieved samples (14%), again probably reflecting better retrieval standards (Table 4.1; Figure 4.3).

Horse elements provided only 3% of the domestic mammal bones from hand-collected assemblages with percentages ranging between 2% and 5%. They were even less well represented in the sieved samples (1%). Horsemeat was not an important component of the human diet in this area of Novgorod.

Dog was poorly represented contributing only 1% of the domestic mammal NISP counts in the hand-collected assemblages and was not identified in the sieved samples (Table 4.1; Figure 4.3). The presence of dogs was, however, also indicated by the high frequency of gnawed bones in the assemblage. Gnawing damage was observed on 31% of the identified mammal bones, the vast majority of which was caused by dogs, indicating that they had easy access to processing and consumption waste discarded in the properties.

Cats were slightly better represented than dogs. They were not identified in the Troitsky IX assemblage but formed between 1% and 3% of the domestic mammal bones in the other groups. They were slightly better represented (4%) in the sieved samples (Table 4.1; Figure 4.3).

Intra-site analysis of domestic mammal species counts from Troitsky X–XI

Sub-samples from Spits 16 to 18 in Troitsky X were subdivided by area. These areas represent different sectors of the excavation trench. There were minor variations in species percentages from the different areas. For example, pig and sheep/goat were slightly better represented – and cattle less so – in Areas G and A than elsewhere. Amongst the less common domestic species, horse percentages ranged between 2% and 5%, dog between 1% and 2%, and cat between 0% and 1% (Table 4.2; Figure 4.4).

It was possible to carry out a basic analysis of species variability by spit and property in Troitsky XI. Although species ranking remained largely the same throughout, with cattle bones dominating all assemblages, followed by pig and sheep/goat, there were some variations in species abundance. In the very lowest deposits (Spits 22 to 18 and features cut into the natural), the percentages of cattle elements were comparatively low, ranging between 53% and 57%, whereas pig (22%–29%) and sheep/goat (9%–13%) were both better represented than usual (Table 4.3: Figure 4.5). This phenomenon was noted in all of the properties investigated. This may indicate that there was less reliance on the consumption of beef in the period immediately following the

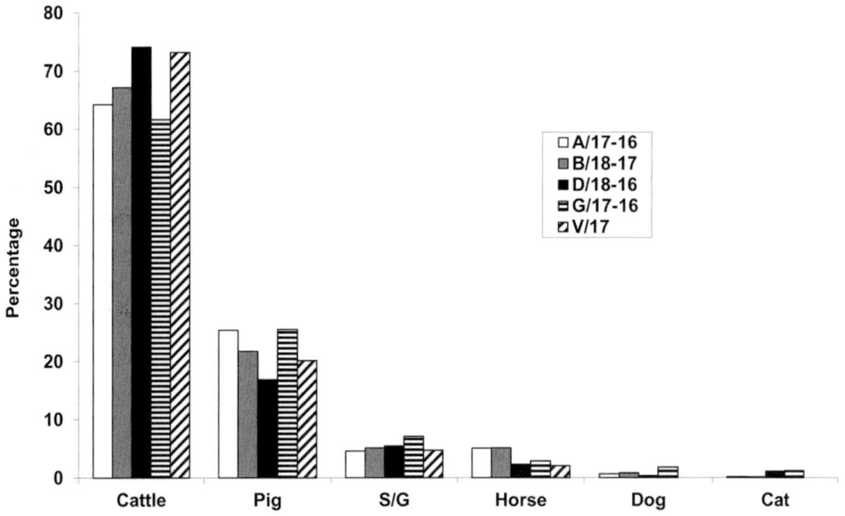

Figure 4.4: *NISP percentages of domestic mammals from Troitsky X by area and spit*

establishment of the town. This may indicate that there was more woodland favourable for pigs in the locality in the 10th and 11th century. Early clearances may also have created more small fields suitable for keeping sheep (Aleksandrovskaya *et al.* 2001). However, it is also possible that more care was taken in the recovery of bones from the lowest layers and features, allowing for the recovery of a greater proportion of bones of these medium-sized mammals.

In the subsequent spits (S17–S14), species percentages were more typical of the rest of the assemblages with cattle (70%–71%) becoming more dominant with corresponding decreases in pig (14%–15%) and sheep/goat (8%–9%). Species representation from properties G and I was very similar. The same pattern continued throughout Spits 13 to 11, with cattle slightly more dominant (69%–75%), followed by pig (16%–19%) with sheep/goat decreasing slightly (6%–7%). Again, there was little variation in species representation between the three properties involved. In the subsequent spits (S10–S8), cattle elements were still dominant but their percentages decreased a little on both Properties G and I (60% and 65% respectively), with a significant increase in pig (24%–25%) and a slight increase in sheep/goat (8%) (Table 4.4; Figure 4.6).

In the upper layers of Troitsky XI, cattle elements continued to dominate. Assemblages from Spits 5 to 7 produced between 60% and 68% cattle, with higher percentages coming from Property G. Pig (20%–25%) and sheep/goat elements (6%–8%) were found in similar frequencies to those encountered in Spits 8 to 10. The assemblages from the highest spits examined saw a slight increase in cattle (70%–71%) and a sharp drop in the frequency of pig elements (11%–16%), whereas sheep/goat increased slightly to 9%–10% (Table 4.5; Figure 4.7).

Throughout the deposits from Troitsky XI, the percentage of horse ranged between 0% and 6%, with the highest percentage being found in the early features cut into the

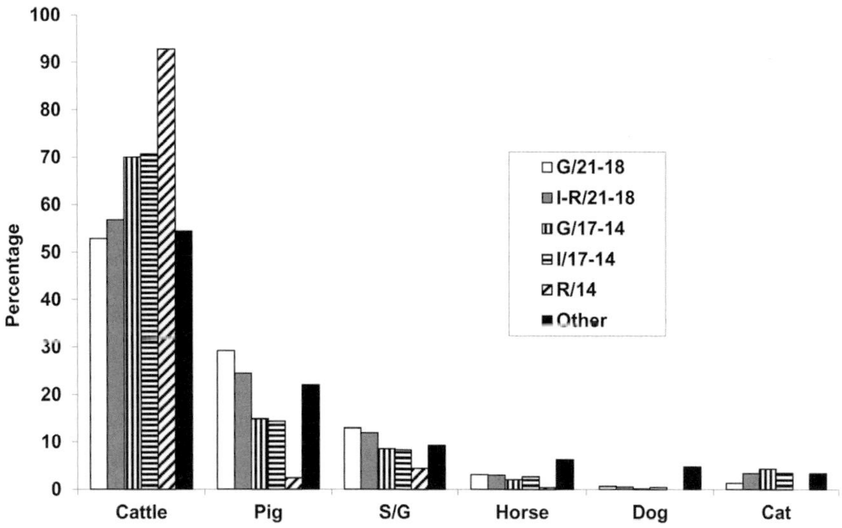

Figure 4.5: NISP percentages of domestic mammals from Troitsky XI lower spits by property

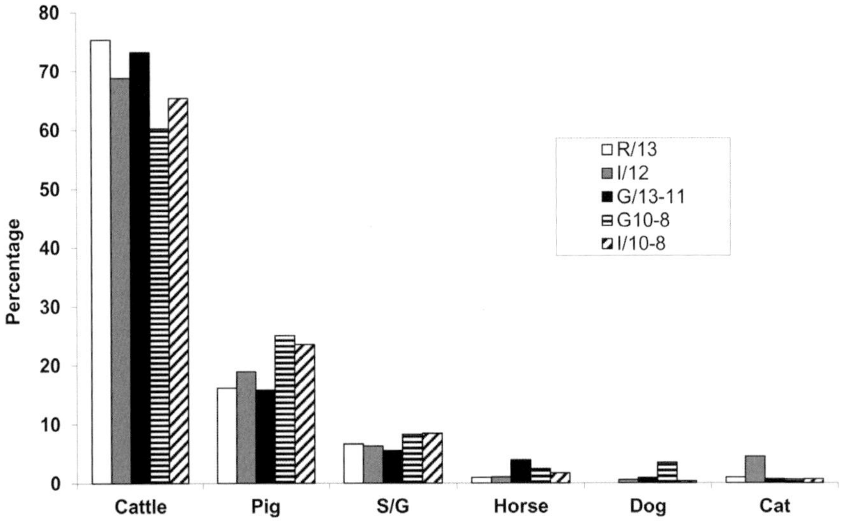

Figure 4.6: *NISP percentages of domestic mammals from Troitsky XI middle spits by property*

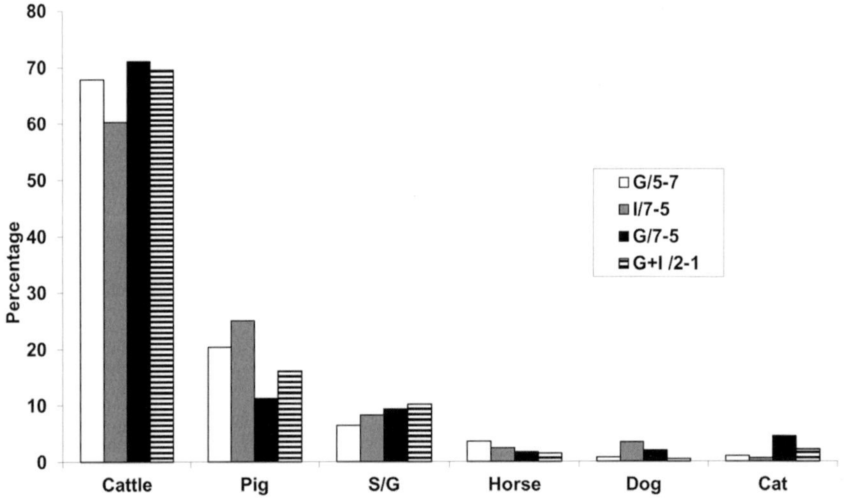

Figure 4.7: *NISP percentages of domestic mammals from Troitsky XI upper spits by property*

natural (Figure 4.5). The highest percentage of dog (5%) also came from these features, indicating perhaps that these sub-surface features were used more often for the disposal of non-food animals. In other spits, percentages of dog usually contributed 1% or less of the identified domestic mammal assemblage, occasionally rising to 3%. Cat bones were encountered slightly more frequently with percentages ranging between 0% and 5% (Tables 4.3–4.5; Figures 4.5–4.7).

Intra-site analysis therefore showed a great deal of consistency in the relative abundance of the major species both vertically and horizontally. There are some variations in the relative abundance of cattle, sheep/goat and pig, which may indicate some fluctuations in the meat diet through time for the inhabitants of these properties. However, there was no consistent pattern in the variability in species representation between properties. Indeed, the variations between them were surprisingly small. As discussed in Chapter 2, it was not possible to ascertain from which parts of the properties the bones had been found. This prevented detailed contextual analysis of the deposition of bones within the properties. The results, however, do suggest that there was not a great deal of difference in the species represented in the properties investigated. Vertical changes in species representation between spits were usually encountered across all the properties.

Intra-site analysis of cattle, pig and sheep/goat NISP counts from Troitsky IX–XI

Therefore the domestic mammal assemblages were dominated by cattle, pig and sheep/goat. These are often analysed separately from other species (*e.g.* Hambleton 1999; King 1999). Accordingly, NISP counts of these species in all of the spits are compared in Table 4.6. As it has been established that there was little variation between assemblages from different properties, the counts from each spit have been amalgamated. The results from Troitsky IX showed that cattle gradually increased from 62% to 78% between Spits 18 and 13. Pig percentages started off at around 30% but fell to 13% in Spit 13. Sheep/goat provided no more than 9% of the NISP counts in any of the four spits investigated (Figure 4.8).

In the Troitsky X assemblage, cattle were slightly more abundant, forming over 70% of the NISP counts in all five spits analysed, again showing a gradual increase to 88%

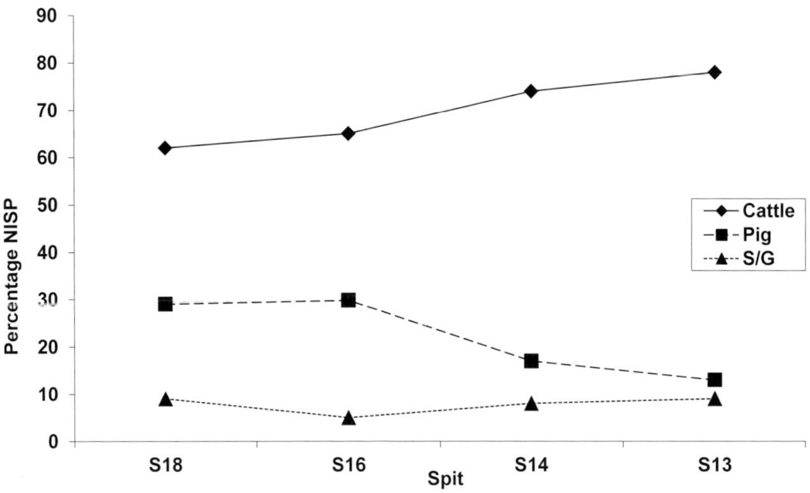

Figure 4.8: *NISP percentages of cattle, pig and sheep/goat from Troitsky IX by spit*

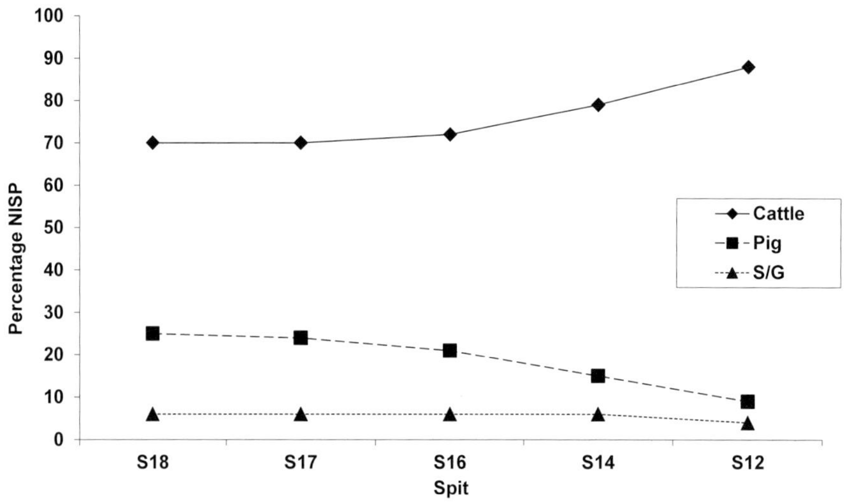

Figure 4.9: *NISP percentages of cattle, pig and sheep/goat from Troitsky X by spit*

in Spit 12 (Table 4.6; Figure 4.9). Pig percentages fell consistently from 25% in Spit 18 to only 9% in Spit 12. Sheep/goat elements remained poorly represented throughout, never rising above 6%.

Contemporary 10th- to early 12th-century deposits from Troitsky XI (Spits 22–14 plus features cut into the natural) showed similar trends to those observed in Troitsky IX and X (Figure 4.10). Cattle provided over 60% of the fragments in the very lowest spits and features. After a marked dip to 48% in Spit 20, cattle percentages rose consistently until they reached 82% in Spit 14. Pig elements were better represented in the earliest deposits, rising to 39% in Spit 20. Thereafter their percentages declined quite sharply to around 13% in Spits 16–14. Sheep/goat also gradually decreased in the same spits, falling from 15% in Spit 19 to 6% in Spit 14.

Percentages of cattle decreased through Spits 13 to 10 from 81% to 64% with corresponding increases in pig (from 13% to 24%) and, more erratically, sheep/goat (up to 12% in Spit 10). Thereafter, cattle percentages gradually rose again through Spits 9 to 3 to 79%. Pig percentages also initially rose to 27% in Spits 8 and 9 before declining gradually to 12% in Spit 3. Sheep/goat dropped to 8% in Spit 9 and thereafter percentages remained fairly constant up to Spit 5. They then increased to over 10% in most of the top layers, to levels only marginally below those of pig in Spits 4 to 2. Cattle percentages remained above 70% in most of the top spits (Figure 4.10).

Hand-collected assemblages were heavily biased towards cattle, as indicated by the higher percentages of pig and sheep/goat in the sieved samples (cattle 49%; pig 37%; sheep/goat 14%; see Table 4.6). In these circumstances it is better to compare the counts of sheep/goat and pig only, to investigate their relative abundance without the inclusion of cattle (Maltby 2010; O'Connor 2010). The results of this procedure are shown for all sites in Table 4.6, and Figure 4.11 illustrates the Troitsky XI results.

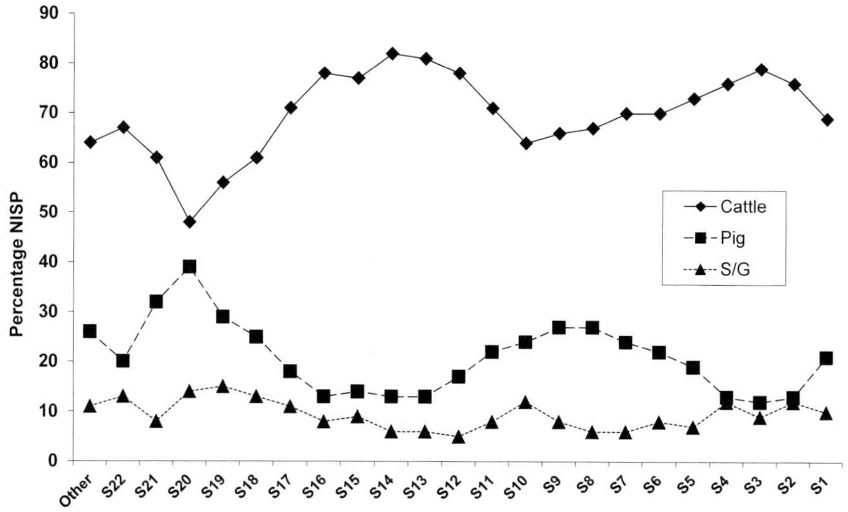

Figure 4.10: NISP percentages of cattle, pig and sheep/goat from Troitsky XI by spit

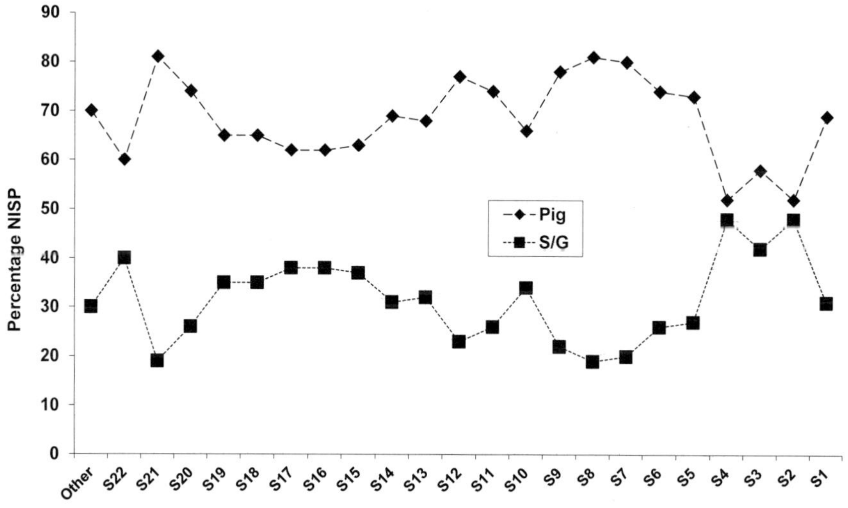

Figure 4.11: NISP percentages of pig and sheep/goat from Troitsky XI by spit

These largely confirmed the fluctuations discussed in relation to Figure 4.10. Pig usually formed over 60% of the pig and sheep/goat NISP counts but there were some spits where sheep/goat were significantly better represented, particularly in Spits 4–2, where they increased to levels of over 40%. This suggests that during the 14th century, the contribution of pork in the diet may have declined, at least in these properties. Prior to that there was generally a gradual rise in the percentage of pig elements between Spits 16 and 8.

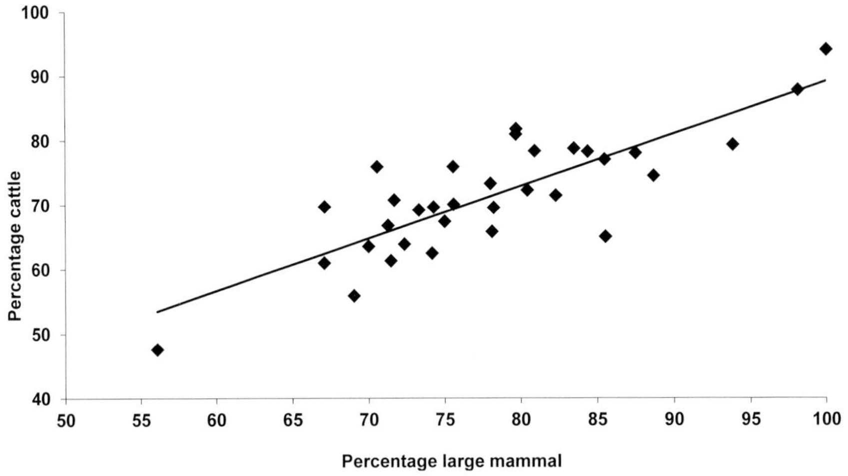

Figure 4.12: *Percentages of large mammal in the unidentified assemblages compared to the percentage of cattle in the identified mammal assemblage from Troitsky IX–XI*

Unidentified fragments from Troitsky IX–XI

A substantial proportion of the bones from the excavations were not identified. These were, however, assigned where possible to large mammal (cattle, horse, and elk) and medium-sized mammal (sheep/goat, pig) categories. Over 24,000 fragments were assigned to these categories, of which 77% were assigned to large mammal (Table 4.6). This supports the evidence from the identified material that cattle elements were by far the most commonly recovered. There indeed is quite a strong positive correlation ($r = 0.82$) between the percentage of cattle and the percentage of unidentified large mammal fragments recovered (Figure 4.12). Rather surprisingly given the higher percentages of pig and sheep/goat identified, unidentified large mammal fragments also formed 78% of the large and medium-sized mammal fragments in sieved samples (Table 4.6).

The cattle assemblage from Troitsky IX–XI

Cattle anatomical representation

The NISP counts of each anatomical element for cattle were recorded for each of the 69 sub-samples. These have been summarised by site and spit in Tables 4.7–4.10. Counts of cattle elements from sieved samples are included in Table 4.11.

The cattle anatomies were then divided into the following eight areas of the body:

1) head (horn cores; maxilla; mandible; hyoid)
2) loose teeth
3) upper limb bones (humerus; radius; ulna; femur; tibia; patella)
4) shoulder and girdle bones (scapula; pelvis) – henceforth referred to as girdle

5) carpals and tarsals
6) metapodials
7) phalanges and sesamoids
8) vertebrae (atlas; axis; sacrum only)

These counts excluded other skull fragments, as there were inconsistencies in recording cranial elements to species level in different seasons. Other vertebrae and ribs, which were included in the counts of identified elements from a few spits, were also excluded from this analysis. Totals of the excluded elements can be found within Tables 4.7 to 4.11. Counts of the areas of the body represented are provided in Tables 4.12 to 4.15.

The relative abundance of body parts was fairly consistent throughout the deposits. In Troitsky IX, upper limb bones provided 20%–29% of the NISP counts and cranial elements 20%–26%. The former outnumbered the latter in three of the four spits, the exception being the latest in the sequence, Spit 13. Phalanges were also slightly more common in that spit. Loose teeth were more common in Spit 18 than in later deposits. Percentages of other body areas varied very little between spits (Table 4.12; Figure 4.13).

In Troitsky X, the ranking of upper limb bones and cranial fragments was generally reversed, with the latter being the most common body area represented in all but the lowest spit (S18). Conversely upper limb bone fragments were more common in Spit 18 than in higher spits. They were outnumbered by metapodials in several of the later layers. The very high percentage of metapodials from Spit 13 should be treated with caution because it was derived from a small sample. However, metapodials were also quite well represented in adjacent spits. Conversely, fragments

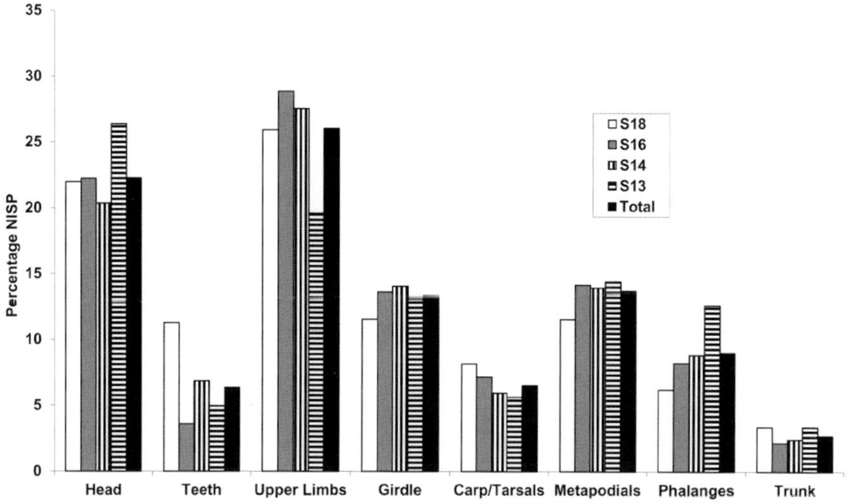

Figure 4.13: *NISP percentages of cattle body areas from Troitsky IX by spit*

of scapulae and pelvis were better represented in lower spits than in higher ones (Table 4.12; Figure 4.14).

Cattle body part representation in the lowest levels of Troitsky XI was quite similar to Troitsky IX, with upper limb bone fragments generally outnumbering cranial elements. Upper limb bones were particularly common in the two lowest spits (S21–S22) and in features cut into the natural (Table 4.13; Figure 4.15).

Percentages of upper limb bones continued to be consistently high in Spits 18 to 14 on Troitsky XI. They slightly outnumbered cranial elements throughout these

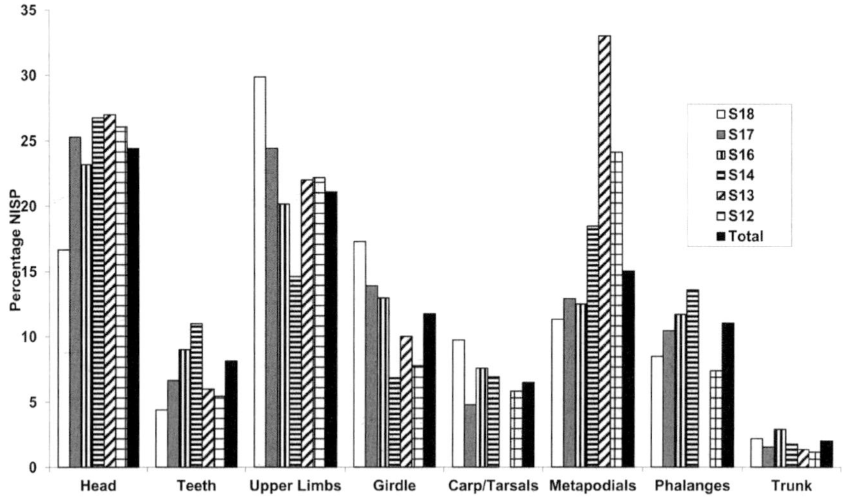

Figure 4.14: *NISP percentages of cattle body areas from Troitsky X by spit*

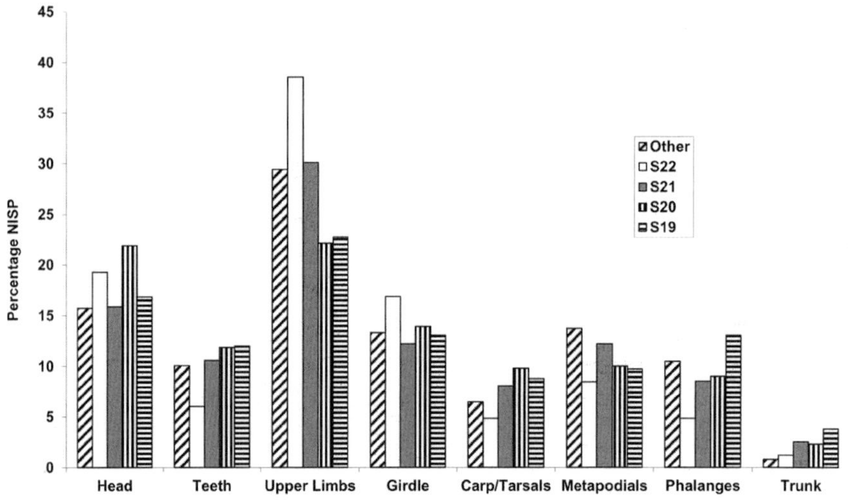

Figure 4.15: *NISP percentages of cattle body areas from Troitsky XI lowest spits*

spits. Phalanges were slightly better represented in these levels, being found about as frequently as metapodials (Table 4.13; Figure 4.16). This pattern remained largely the same through Spits 13 to 8 (Table 4.14; Figure 4.17).

This same pattern was found in Spits 7 to 5. Thereafter, there was more variability between the assemblages. Cranial fragments increased in Spits 4 to 2. Spit 2 included a substantial number of calf mandibles and loose teeth (Table 4.10). Upper limb bone fragments declined in these spits and were particularly low in Spit 2. Spit 3 included

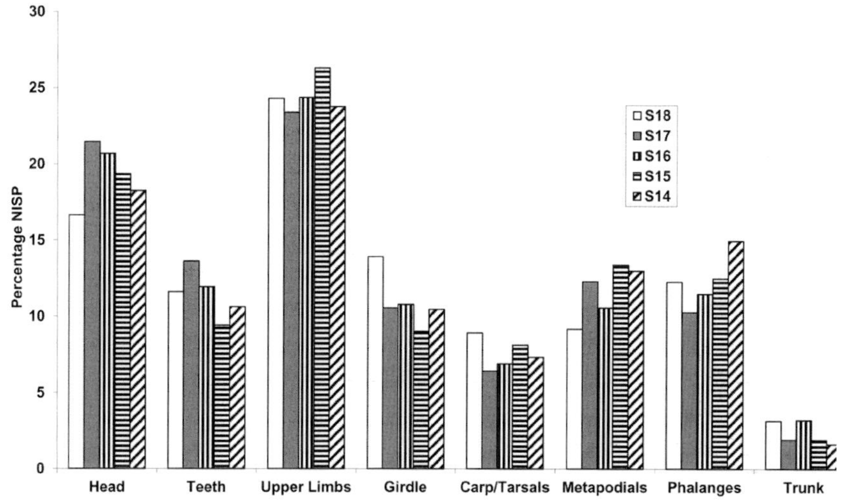

Figure 4.16: NISP percentages of cattle body areas from Troitsky XI spits 18 to 14

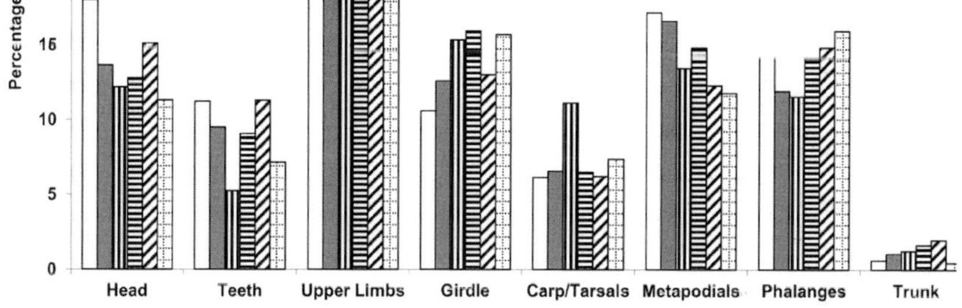

Figure 4.17: NISP percentages of cattle body areas from Troitsky XI middle spits

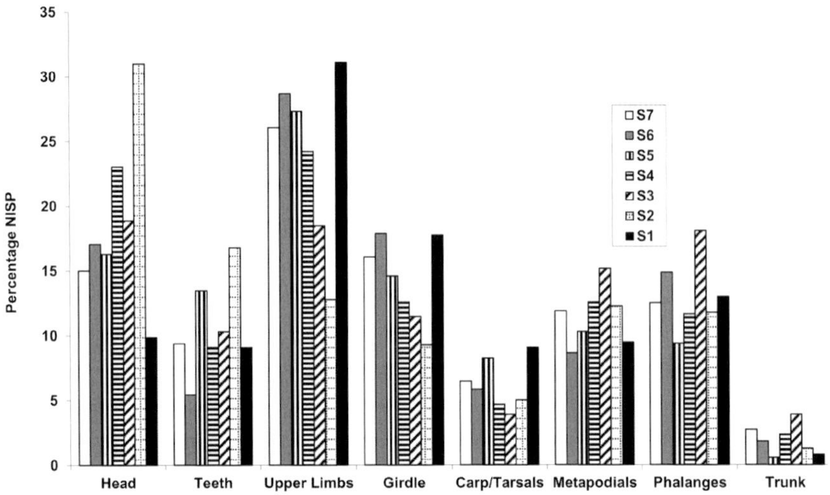

Figure 4.18: *NISP percentages of cattle body areas from Troitsky XI upper spits*

slightly higher percentages of phalanges and metapodials than usual. The cattle assemblage from Spit 1 was quite different with higher percentages of upper limb and girdle bones and a correspondingly low percentage of cranial elements (Table 4.15; Figure 4.18).

Comparisons of the relative abundance of cattle body parts from different areas of Troitsky X revealed no significant variations between the largest assemblages. There were a higher percentage of cranial fragments from Area V, but this was derived from a small sample (Table 4.16). Figures 4.19 and 4.20 and Tables 4.17 to 4.19 compare the cattle assemblages from the properties on Troitsky XI. The analysis revealed only minor variations in the relative abundance of body areas represented. The high percentage of cranial fragments from Property G, Spits 4 to 2 reflects the concentration of calf mandibles in Spit 2 discussed above (Figure 4.19).

Counts of cattle body parts from sieved samples are provided within Table 4.15. Although based on a small sample of 125 fragments, they did show significant differences from the composition of the hand-collected assemblages. Loose teeth formed a much higher proportion of the sieved assemblage, comfortably outnumbering both upper limb and cranial elements. Girdle fragments were poorly represented, being outnumbered by phalanges, metapodials and carpals/tarsals.

The analysis demonstrated that there was relatively little variation in cattle body part representation in Troitsky IX–XI. The abundance of cranial elements, metapodials and phalanges throughout the deposits indicated that much of the processing of cattle carcasses took place within the properties and it is feasible that some cattle were slaughtered within them. However, there is little evidence for significant concentrations of primary butchery waste, contrary to what was encountered at Gorodishche (see Chapter 3). Such large-scale processing was probably carried out elsewhere in the town,

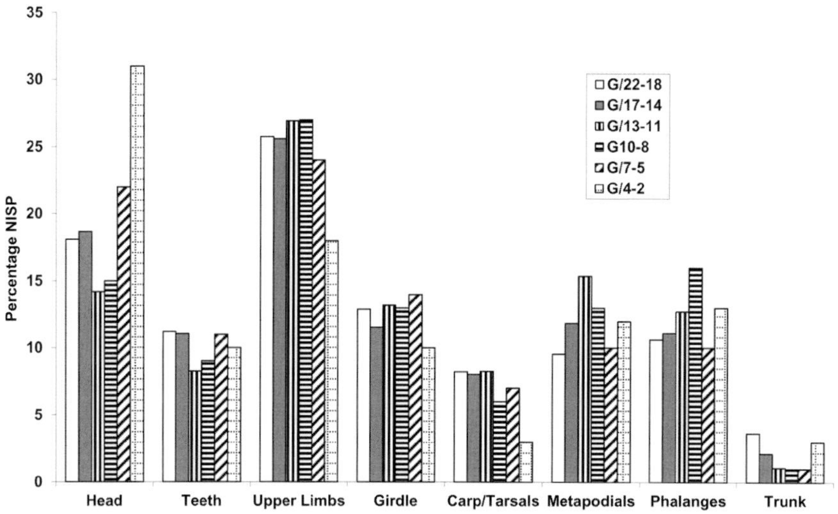

Figure 4.19: *NISP percentages of cattle body areas from Troitsky XI Property G*

but the waste from such enterprises is not likely to have been deposited within residential properties such as these. Here, primary butchery and skinning waste from cattle accumulated alongside cooking and table waste on all the properties investigated. This will be discussed further in Chapter 5. Unfortunately, the sampling strategy precluded more detailed investigation of the deposition of cattle bones. The sieving experiments also indicated that there were some biases against the retrieval of smaller elements, such as loose teeth. However, this bias was not as severe as for other species.

Cattle ageing and sexing evidence

Tooth eruption and tooth wear data from mandibles were recorded in the same way as for the hinterland sites (Chapter 3) following Grant (1982). The detailed records of 782 mandibles are provided in Table 4.20. These were then assigned to the following seven stages of eruption and wear:

Stage 1 = 4th deciduous premolars (dp4) not in wear
Stage 2 = dp4 in wear; 1st molar (M1) not in wear
Stage 3 = M1 in wear; 2nd molar (M2) not in wear
Stage 4 = M2 in wear; 3rd molar (M3) and permanent premolars not in wear
Stage 5 = M3 in wear; 4th permanent premolar (P4) not in wear
Stage 6 = P4 in wear; Grant mandible wear stage (MWS) <46
Stage 7 = MWS >45

In most cases it was possible to assign mandibles to a specific age stage. However, in some fragmentary specimens, it was only possible to assign them to broader age ranges. The results are summarised in Table 4.21 and illustrated in Figures 4.21 and 4.22.

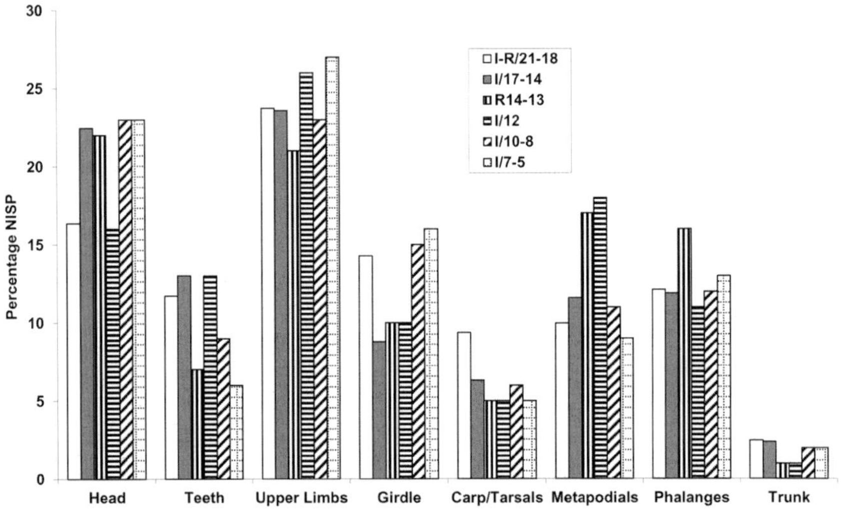

Figure 4.20: *NISP percentages of cattle body areas from Troitsky XI other properties*

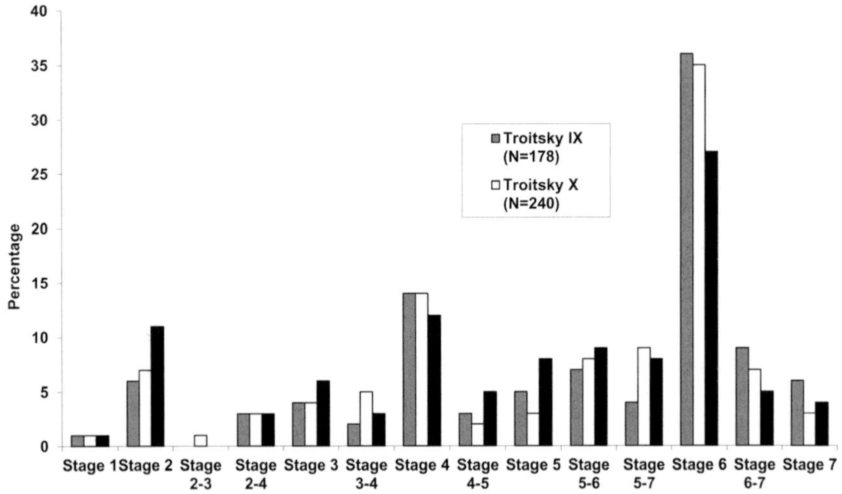

Figure 4.21: *Cattle mandible age stages from Troitsky IX–XI*

The results from the three sites were remarkably similar (Figure 4.21). All had marked peaks at Stage 6 ranging from 27% from Troitsky XI to 36% from Troitsky IX. These belonged to adult animals killed in their prime rather than at a very old age. This suggests that the most common age of slaughter of cattle may have been between four and seven or eight years of age (Jones and Sadler 2012) and that beef production was the paramount consideration in their husbandry. However, many of these cattle were old enough to have produced calves and they could have also been exploited

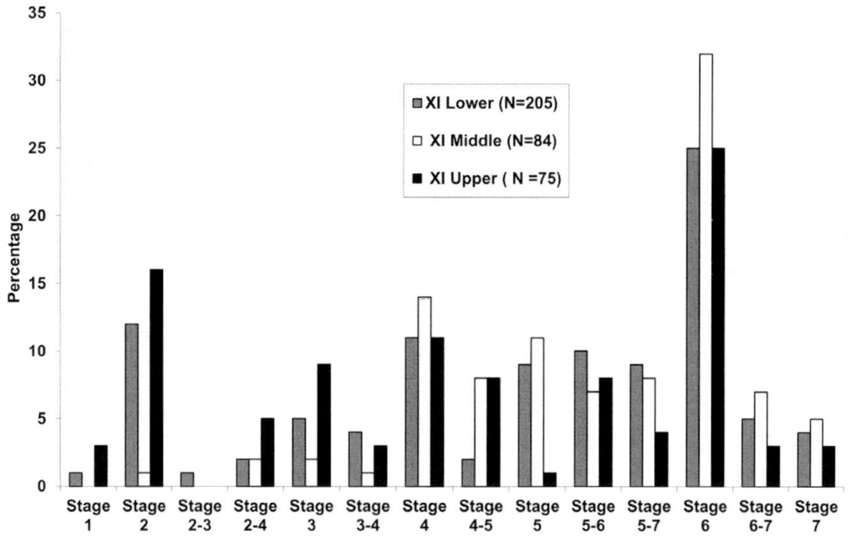

Figure 4.22: *Cattle mandible age stages from Troitsky XI*

for milk or as beasts of burden prior to slaughter. Mandibles assigned to the oldest age category (Stage 7) were much rarer (3%–6%), although some elderly animals may be included in the Stage 6–7 category. Between 12% and 14% of the cattle mandibles were at Stage 4 of the tooth eruption sequence, indicating a significant cull for the meat of cattle aged between 18 and 30 months old.

Mandibles from cattle under 18 months of age (Stages 1 to 3) also formed a significant proportion of the sample from all three sites. However, of those that could be more closely aged, only 4%–6% of the mandibles were assigned to Stage 3 (6–18 months). Significantly more (6%–11%) were at Stage 2, and belonged to calves aged between around two and six months old. In addition to the provision of veal, evidence for the slaughter of substantial numbers of young calves implies that dairy production was an important consideration in cattle husbandry.

Higher percentages of Stage 2 mandibles were recovered from Troitsky XI than from the other two sites (Figure 4.21). There were some fluctuations in the abundance of Stage 1 and 2 mandibles in different parts of the Troitsky XI sequence (Table 4.14; Figure 4.22). They were relatively common in assemblages from both the lower and highest levels but almost absent from the middle spits. This could indicate a significant increase in veal production during the later medieval period, although samples from other parts of the town are required to establish whether this was a widespread trend. Comparisons between the properties revealed few differences in cattle mortality profiles (Table 4.21).

The fusion data for each limb bone epiphysis are shown in Tables 4.23–4.24. These were divided into three broad fusion age groups (early, later and latest). The fusion evidence broadly supported the results from the mandibles apart from variations in

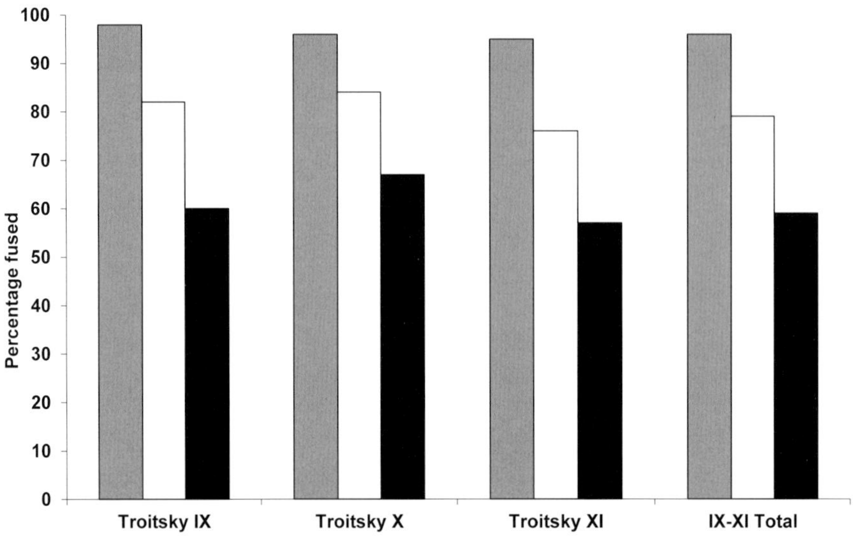

Figure 4.23: *Cattle epiphyseal fusion data from Troitsky IX–XI. Grey = early-fusing epiphyses; White = later-fusing epiphyses. Black = late-fusing epiphyses*

the abundance of calf bones. Very few (4%) of the bones in the youngest-fusing category (<12 months) were unfused. Slightly higher figures were obtained from Troitsky XI than the other two sites (Figure 4.23). These figures were lower than the percentages for Stages 1 and 2 (9%) of the mandibles. A similar discrepancy was noted in the Gorodishche assemblage (Chapter 3), perhaps indicating that calf mandibles had a better chance of survival than their limb bone epiphyses. In addition, the fluctuations in the percentages of calf mandibles recorded on Troitsky XI (Figure 4.21) were not reflected in the percentages of unfused early-fusing epiphyses, which remained low throughout the sequence (Figure 4.24). As noted above, calf mandibles were particularly abundant in Spits 2 and 3 and were also relatively more common in Spits 21 to 18 than in other spits (Table 4.20). It is possible that groups of calf mandibles were sometimes dumped together. They may have been brought to the properties attached to skins for leather-processing. However, without more detailed information about the provenance of the mandibles, this interpretation must remain speculative.

Overall, 79% of the later-fusing epiphyses were fused (Table 4.22; Figure 4.23), indicating that around a fifth of the surviving bones were from animals under 18–30 months of age (Silver 1969). Slightly higher percentages of unfused specimens were found on Troitsky XI than from Troitsky IX and X. Again, the fusion evidence does not suggest as high a level of immature slaughter as indicated by the mandibles (Stages 1–4 = 34%; see Table 4.21). Differential preservation and deposition may have again been factors.

Late-fusing epiphyses do not survive well in assemblages subjected to carnivore scavenging. This is reflected in the substantially lower numbers recorded compared

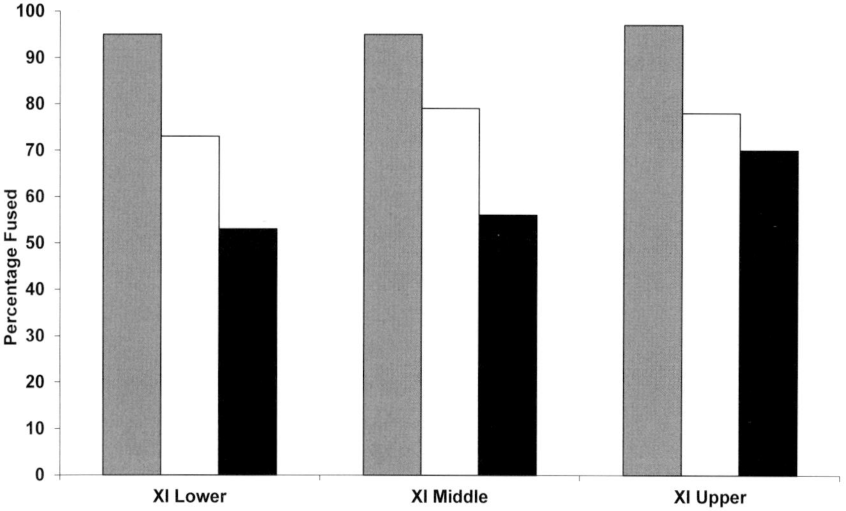

Figure 4.24: *Cattle epiphyseal fusion data from Troitsky XI. Grey = early-fusing epiphyses; White = later-fusing epiphyses. Black = late-fusing epiphyses*

with early-fusing epiphyses (Table 4.22). Of those that were preserved, 59% were fused and belonged to adult animals (Figure 4.23), a similar figure to the percentage of mandibles recorded at Stage 5 and above.

Regarding the data from Troitsky XI alone (Table 4.23; Figure 4.24), there was little difference in the percentages of fused early- and later-fusing epiphyses in the three assemblage groups. However, the upper spits included relatively more fused late-fusing epiphyses than the lower spits. This contradicts the mandible evidence which saw a decrease in the percentage of adult cattle in the latest deposits (Figure 4.22). This could again suggest that some of the calf mandibles were imported without their associated limbs, possibly attached to skins.

Sexing evidence was provided by metacarpal measurements. Greatest lengths of 68 complete metacarpals were compared with their greatest distal breadths (Table 4.24; Figure 4.25). These fell into two main groups. Those with Bd/GL indices of 0.30 or less probably belonged to cows, and those with relatively broader distal epiphyses to males. Most (72%) fell into the more slender category. Given that 80% of the distal metacarpals from Troitsky IX–XI were fused (Table 4.22) and over 24–30 months of age (Grigson 1982a), this suggests that most of the cattle slaughtered before that age were males, assuming that the metacarpals are a representative sample of the cattle population kept in and around Novgorod. Male calves not required for breeding and traction were more likely to have been slaughtered, whereas adult cows could provide both calves and milk. There were also very few specimens in the 'male' category that were exceptionally long, which suggests that few of the metacarpals were from large oxen (Figure 4.25).

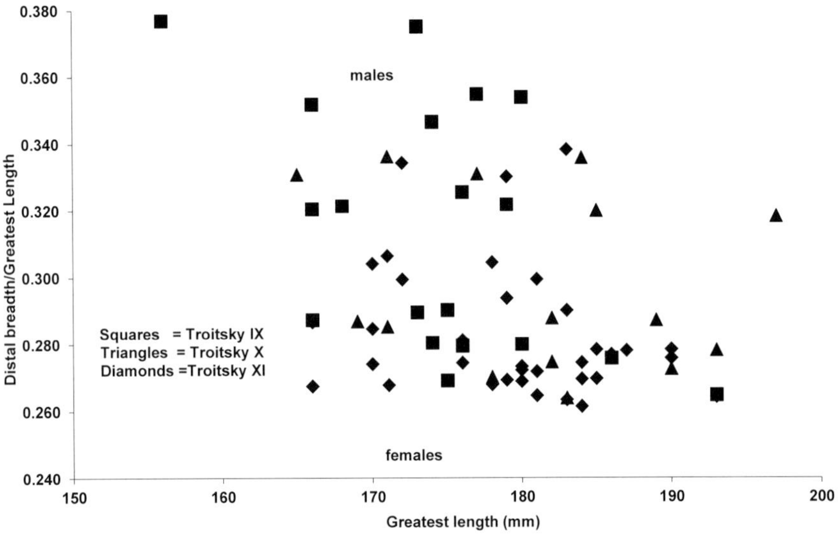

Figure 4.25: *Cattle metacarpal lengths and distal breadth indices Troitsky IX–XI*

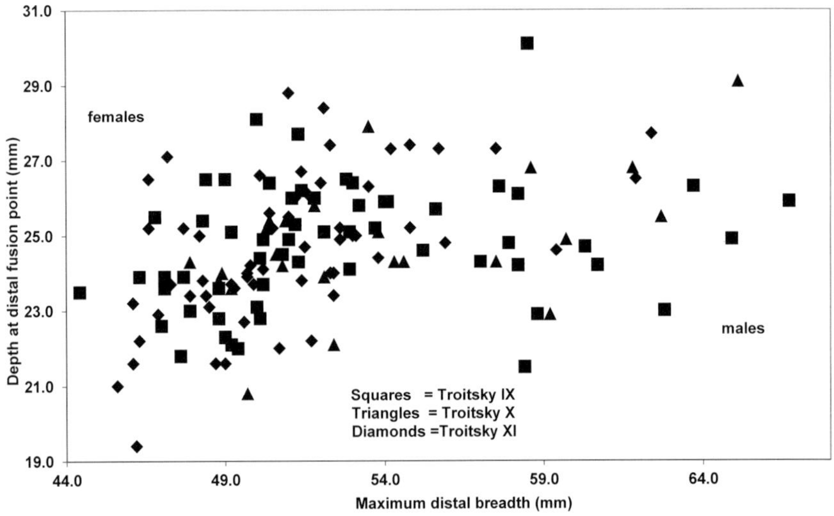

Figure 4.26: *Cattle distal metacarpal measurements from Troitsky IX–XI*

However, there are some surprising variations in the relative abundance of males and females from different sites. Nearly all the metacarpals with high distal breadth indices (male) were from Troitsky IX and X, whereas all but three of the 35 specimens from Troitsky XI have low indices (female). One explanation for this is that there was a diachronic change with very few males being represented in the later medieval deposits. Certainly, all 24 specimens from the middle and upper spits of Troitsky

XI fell into the female category (Table 4.22). However, only three (25%) out of the 12 metacarpals from the lowest spits of Troitsky XI fell into the male category, compared with 47% and 40% in Troitsky IX and X respectively.

Analysis of distal breadth and depth measurements increased the size of the sample by including broken specimens (Table 4.25; Figure 4.26). This analysis confirmed the presence of large numbers of metacarpals in the smaller size range, the majority of which would have belonged to cows. Again, nearly all the larger (male) specimens were recovered from Troitsky IX and X. From Troitsky XI, all but one (96%) of the 25 specimens from the top spits fell into the female range. Similarly, 15 (88%) of the 17 metacarpals from the middle spits (S13 to S8) fell within that group. This percentage decreased to 75% amongst the 20 specimens from the lower spits including two within the small area of overlap between male and female specimens (Table 4.24; Figure 4.26). Therefore it is possible that adult males were more commonly represented in the earlier period. However, these virtually disappear in the later periods. Further analysis of possible sexual dimorphism in cattle measurements will be considered in Chapter 6.

The pig assemblage from Troitsky IX–XI

Pig anatomical representation

The NISP counts of all the pig anatomical elements identified are presented by site and spit in Tables 4.26 to 4.29. Counts of elements recovered from sieved samples can be found in Table 4.11. Most of the pig anatomies were then divided into eight areas of the body so that broad comparisons could be made by site, spit (Tables 4.30–4.33; Figures 4.27–4.32) and, in the case of Troitsky XI, by property (Table 4.34).

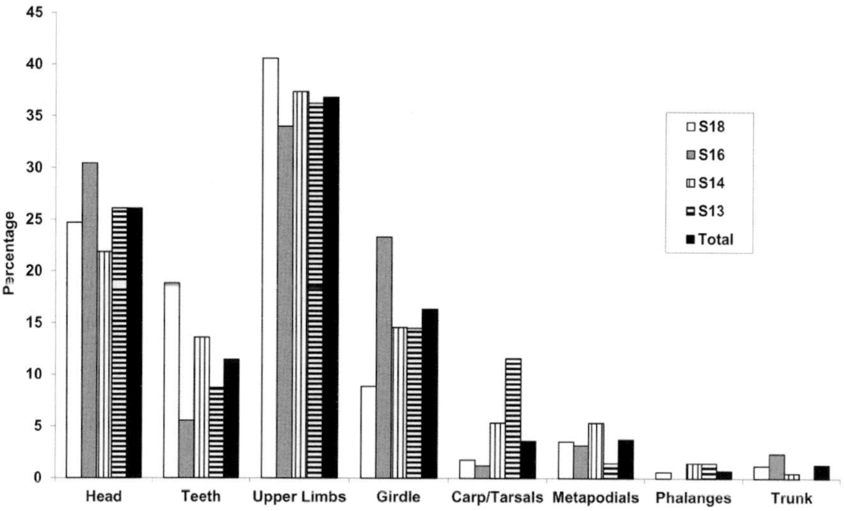

Figure 4.27: *NISP percentages of pig body areas from Troitsky IX by spit*

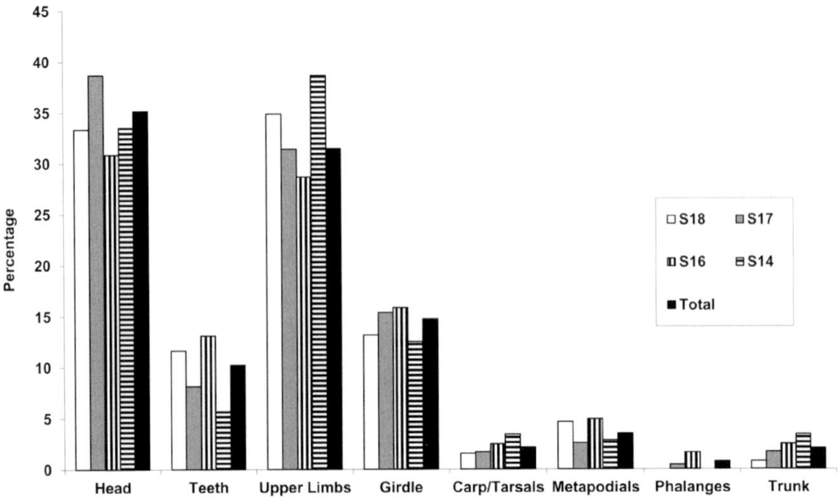

Figure 4.28: *NISP percentages of pig body areas from Troitsky X by spit*

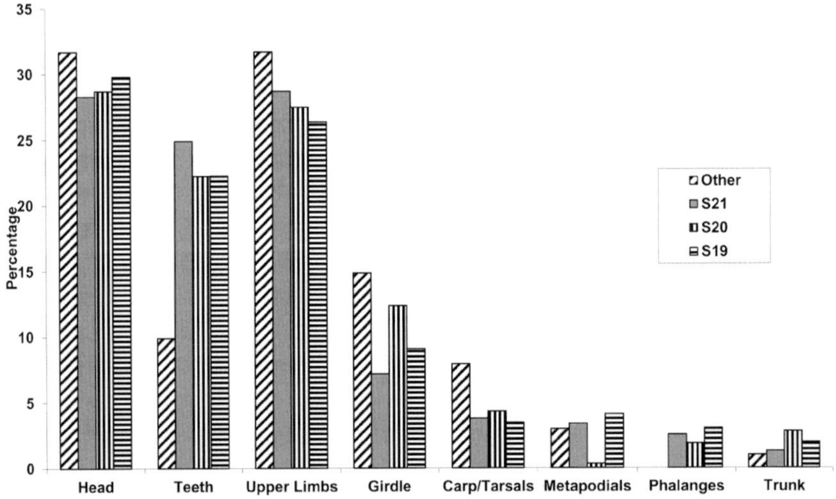

Figure 4.29: *NISP percentages of pig body areas from Troitsky XI lowest spits*

All the pig assemblages were dominated by head and upper limb bones. In Trench IX, upper limb bone counts consistently outnumbered those of the head, whereas they were generally more evenly represented in the Trench X assemblage (Figures 4.27–4.28; Table 4.30).

In Trench XI, assemblages from the lowest spits tended to have quite even numbers of head and upper limb bone fragments (Figure 4.29; Table 4.31). Thereafter, upper limb bone fragments were recovered in substantially larger numbers than

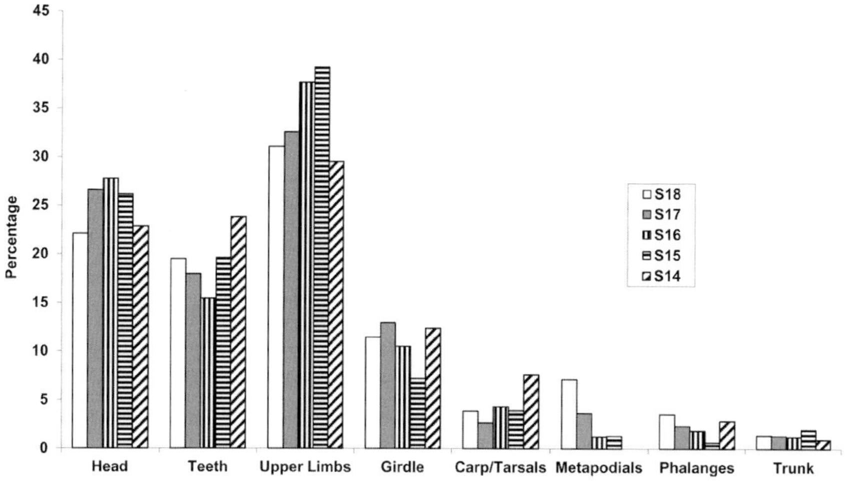

Figure 4.30: *NISP percentages of pig body areas from Troitsky XI spits 18 to 14*

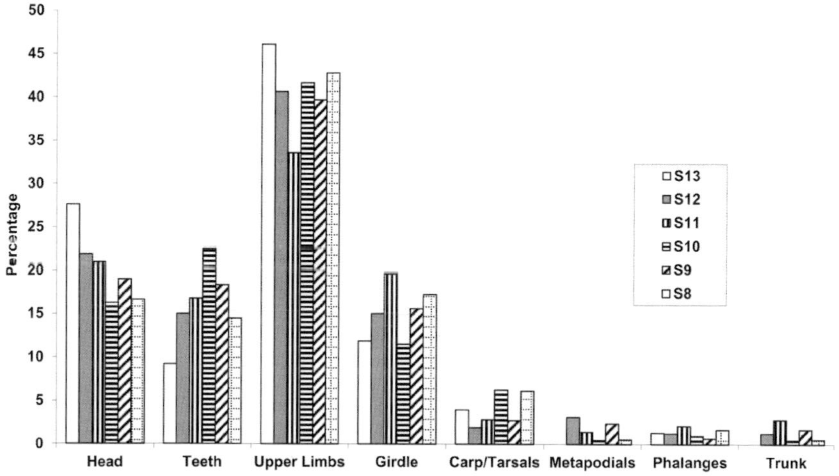

Figure 4.31: *NISP percentages of pig body areas from Troitsky XI middle spits*

cranial elements in most spits (Figures 4.30–4.32; Tables 4.31–4.33). This increase is highlighted in Figure 4.33, which compares the percentages from counts of head and upper limb bones only. Upper limb bones began to become more frequent in Spit 17 and became increasingly dominant thereafter (with only one relatively small sample from Spits 4 and 3 bucking the trend).

Loose teeth were also recovered more frequently in the lowest spits of Troitsky XI. This corresponds with the higher percentage of cranial elements. However, the abundance of loose teeth can also fluctuate markedly dependent upon retrieval rates. Significantly, loose teeth provided 38% of the pig elements retrieved from the sieve

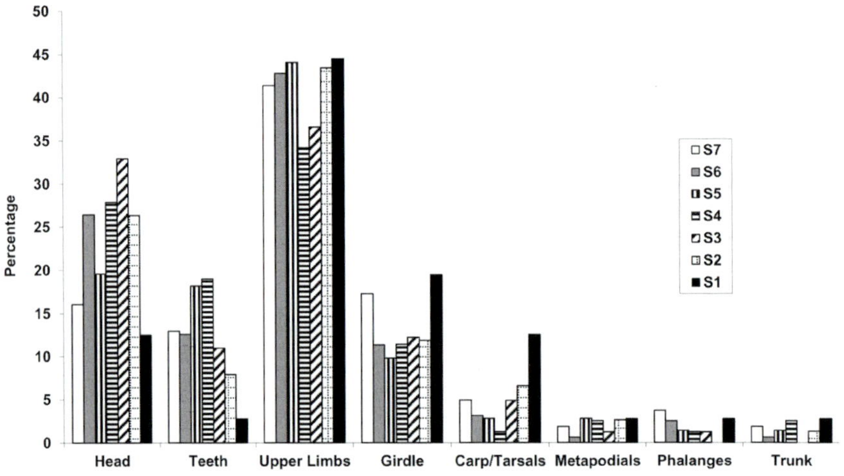

Figure 4.32: *NISP percentages of pig body areas from Troitsky XI upper spits*

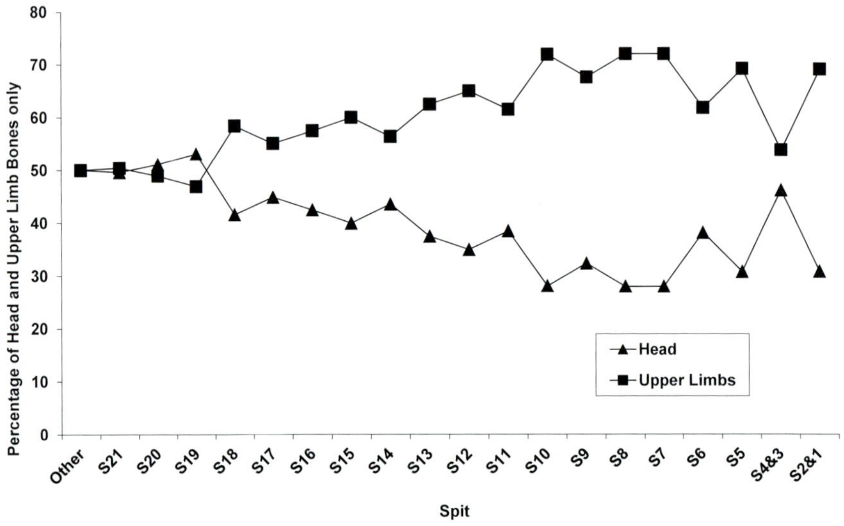

Figure 4.33: *Relative percentages of pig head and upper limb bones from Troitsky XI by spit*

(Table 4.33), indicating they were commonly overlooked during normal excavation. This could also explain their slightly lower abundance in both Troitsky IX and X, where there was less vigilance in bone retrieval.

Pig scapula and pelvis fragments were recovered in fairly consistent numbers throughout the deposits. In Troitsky XI, they tended to became marginally more frequent in the middle and upper spits (Figures 4.31–4.32), mirroring the increase in upper limb bones.

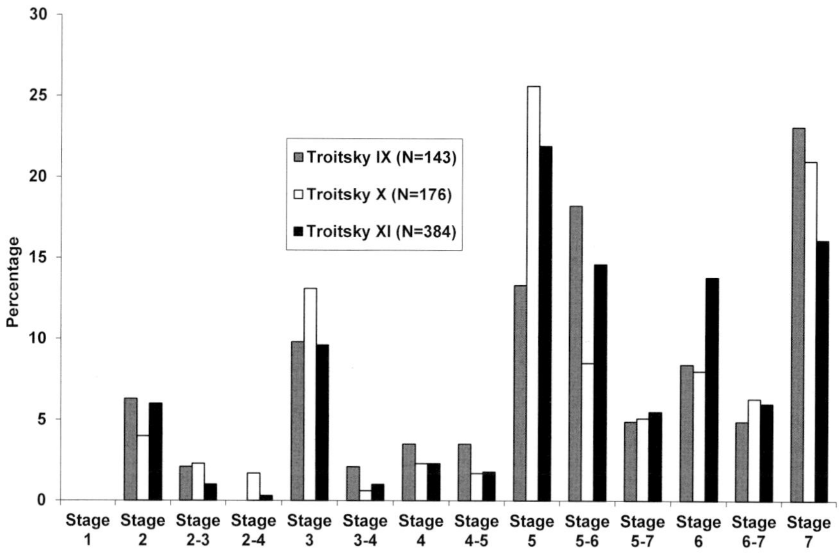

Figure 4.34: *Pig mandible age stages from Troitsky IX–XI*

Other pig elements were poorly represented. Metapodials never contributed more than 7% to any of the assemblages. Carpals and tarsals were represented almost entirely by astragali and calcanei, the largest of these bones. They were found in slightly greater numbers than metapodials in most assemblages, which is perhaps significant given that there are substantially fewer astragali and calcanei (4) than metapodials (16) in a pig skeleton.

Phalanges were even more poorly represented never providing more than 4% in any assemblage. In contrast, phalanges provided 13% of the pig elements in the sieved samples, clearly demonstrating how normal excavation methods militated against the retrieval of small bones such as these.

Despite these anomalies, the presence of all parts of their carcasses indicated that many pigs were butchered within the properties. Many of the anomalies in body part representation can be explained by the excavation methods that limited the recovery of small bones such as loose teeth, phalanges, carpals, and the smaller tarsals and metapodials. However, in the Troitsky XI sequence there were some variations that cannot be explained by retrieval bias alone. The gradual increase in the percentage of upper limb bones throughout that sequence may reflect changes in how pork, ham and bacon were acquired and processed. One possibility is that more dressed carcasses or joints of meat were introduced in the later period. The low numbers of metapodials from all the sites may also indicate that they were often not brought to the properties, although further sieving is required to establish whether this was not simply a factor of retrieval bias.

There were no significant variations between the relative abundance of pig elements found in different properties in Troitsky XI (Table 4.34).

Pig ageing evidence
Pig tooth ageing data were obtained from 703 mandibles. The detailed records can be found in Table 4.35. These mandibles were assigned to the following seven stages of eruption and wear:

Stage 1 = 4th deciduous premolars (dp4) not in wear
Stage 2 = dp4 in wear; 1st molar (M1) not in wear
Stage 3 = M1 in wear; 2nd molar (M2) not in wear
Stage 4 = M2 in wear; 3rd molar (M3) and permanent premolars not in wear
Stage 5 = P4 in wear; M3 not in wear
Stage 6 = M3 at Grant (1982) wear stages a–b
Stage 7 = M3 at Grant (1982) wear stages c–h

Where possible, mandibles were assigned to a specific age stage. In other cases they were placed in broader age ranges. The results are summarised in Table 4.36 and Figures 4.35–4.36.

No mandibles of neonatal pigs (Stage 1) were recovered from any of the sites (Figure 4.35). Given the sampling problems, this may simply be a factor of poor retrieval rates, although it may also indicate that pigs were not commonly kept within the properties. Around 6% of the mandibles were at Stage 2 and belonged to piglets under six months old (Bull and Payne 1982). Substantial numbers (10%–13%) of Stage 3 mandibles were present on all three sites. These are likely to have come mainly from pigs culled in the autumn and winter of their first year. If the mandibles assigned to Stage 2–3 are added, around 18% of the mandibles belonged to pigs under a year old (Table 4.36). In Troitsky XI juvenile pig mandibles were particularly abundant in the lowest levels (Figure 4.35; Table 4.36). Around 35% of the mandibles from Spits 22 to 19 were at Stages 2 or 3, with 23% at Stage 3, representing a marked peak of culling. Pig populations can tolerate high levels of immature slaughter. However, the proportion of mandibles from juvenile pigs decreased markedly in the later deposits of Troitsky XI. Percentages of Stage 2–3 mandibles declined to 17% in Spits 18–14 and became very rare in the assemblages from the middle and upper spits, where they formed only 4–5% of the specimens. This would appear to mark a significant change in pig exploitation in the later medieval period.

The low percentages of Stage 4 mandibles reflects the short period of time between the eruption of the second molar and permanent premolars in pigs (Stage 5). Specimens at Stages 4 and 5 mainly represent pigs culled in their second year. They provided at least 30% of the mandibles from Troitsky X, 26% from Troitsky XI and 20% from Troitsky IX (Figure 4.34), showing substantial amounts of culling of immature animals. The peak at Stage 5 may again be related to autumn and winter culls. Stage 5 mandibles were found in consistent numbers throughout the Troitsky XI sequence (Figure 4.35).

Slightly older pigs represented by mandibles with third molars in the earliest stages of wear (Stage 6) were also found frequently on all sites, particularly Troitsky XI, where they formed 14% of the mandibles. All sites also produced a substantial number of mandibles assigned to Stage 5–6 (Figure 4.34).

A substantial proportion of mandibles (19%) were at Stage 7, showing that many pigs were not killed until they were in their third year and perhaps older (Bull and Payne 1982). However, very few of the third molars had reached beyond Grant wear stage d (Table 4.37), suggesting that the majority were from pigs probably aged less than four years.

The percentage of Stage 7 mandibles increased from 5% in the lowest layers of Troitsky XI to 33% in the top spits (Figure 4.35), increasing at the expense of Stage 3 specimens. Therefore substantially more pigs aged between two and four years old were represented in the later medieval period, suggesting a less intensive cull of juvenile animals. This could suggest that there was a greater reliance on cured meat obtained from older pigs in the later period.

Interpretations of changes in pig husbandry practices must be tempered by the possibility that the pig mandibles recovered from the Troitsky properties do not necessarily accurately represent the overall mortality pattern of pigs in the Novgorod region. As has been noted above, it seems likely that relatively fewer jaws were deposited on the site in the later period when more dressed pig carcasses and joints of pork may have been imported. Conceivably these joints may have been from pigs killed at different ages.

Epiphyseal fusion analysis of the limb bones, however, did generally support the trends observed in the tooth ageing data (Figures 4.36–4.37; Tables 4.38–4.39). Overall, 18% of the early-fusing epiphyses were unfused and belonged to pigs mainly under a year old (Bull and Payne 1982). Results from the later-fusing group showed that 60% of the bones were unfused and were mainly from pigs under two years of age (Figure 3.36; Table 4.38). This percentage, however, varied markedly between

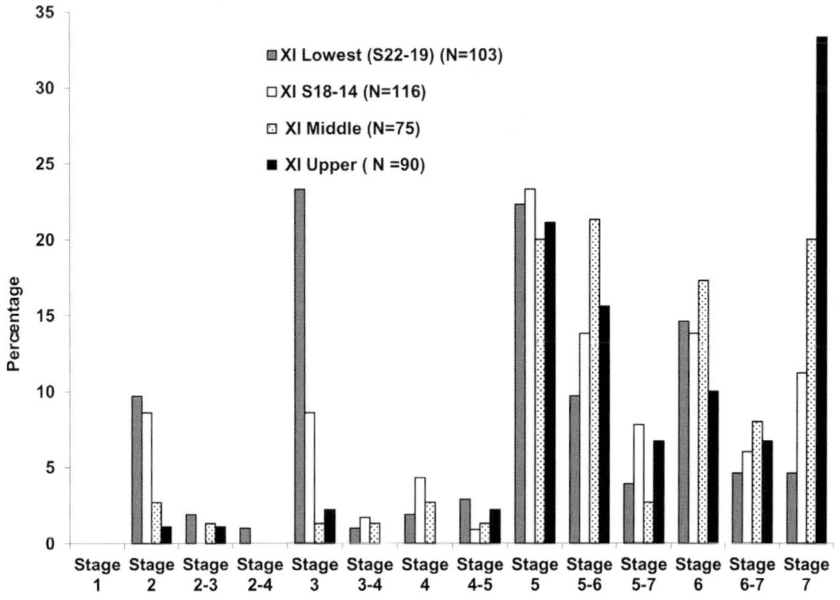

Figure 4.35: *Pig mandible age stages from Troitsky XI*

different sites, with much higher percentages of fused specimens being retrieved from Troitsky IX.

There were very few fused epiphyses (11%) in the latest-fusing group on any of the sites, which supports the mandible ageing results that indicated that very few pigs represented were older than three years of age.

Comparisons between spits from Troitsky XI revealed some variations in the percentages of unfused epiphyses in the early- and later-fusing groups (Figure 4.37;

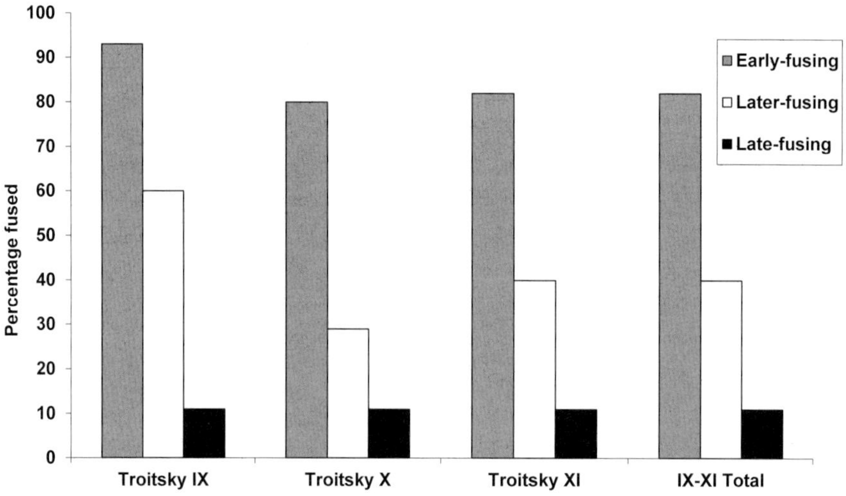

Figure 4.36: *Pig epiphyseal fusion data from Troitsky IX–XI*

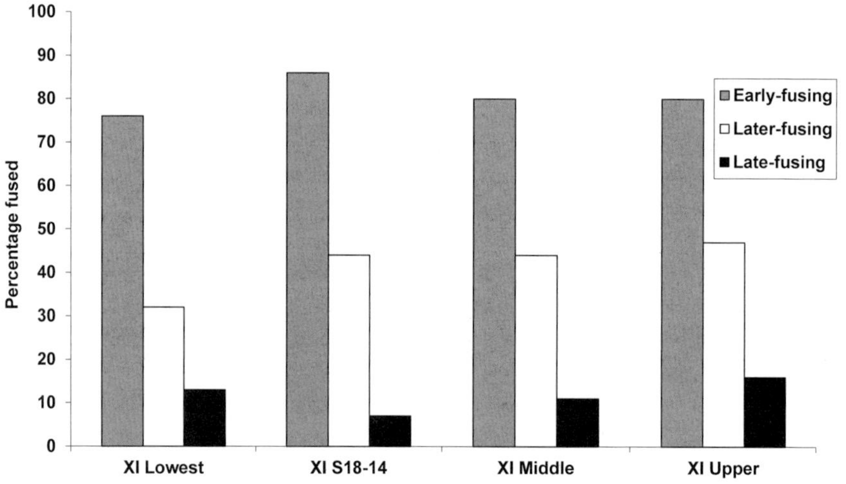

Figure 4.37: *Pig epiphyseal fusion data from Troitsky XI*

Table 4.39). In both cases the lowest spits produced the highest percentage of unfused specimens (24% and 68% respectively). This corresponds quite well with the higher percentages of Stage 2–3 mandibles in the same layers. Unfused early-fusing epiphyses decreased to 20% and below in subsequent deposits. Percentages of unfused later-fusing specimens also fell to between 53% and 56% (Figure 4.37). Therefore evidence that a greater proportion of pigs were killed in their second and third years in the later period was evident in the fusion as well as in the mandible data.

Unfortunately, no consistent records were taken for the presence of large and small canines retained within the jaws and it has not been possible to compare the mortality rates of male and female pigs.

The sheep and goat assemblage from Troitsky IX–XI

Relative abundance of sheep and goat

Although a number of bones were specifically identified as goats or sheep during the scan, there was a lack of consistency in the recording of these observations in the different seasons and a comprehensive review of the relative abundance of the two species must await further analysis.

Complete metapodials did provide an indication of the relative abundance of the two species, at least in relation to sub-adult and adult animals. Metapodials of goats are broader in relation to their lengths than those of sheep. Of the 15 complete metacarpals from Troitsky IX–XI, nine were assigned to sheep and five to goat (Table 4.40). The lengths of 14 of these are compared with their proximal breadths in Figure 4.38 and it can be seen that there is a clear separation between the two species. Seven of

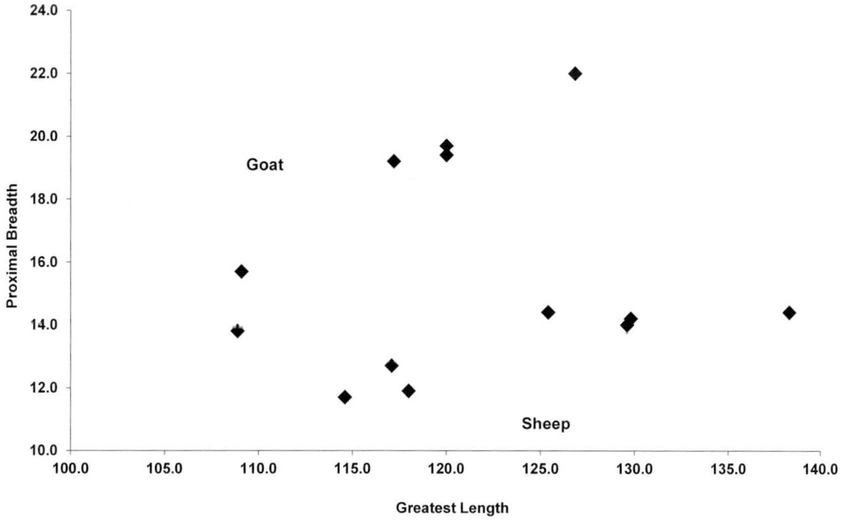

Figure 4.38: *Sheep and goat metacarpal lengths and proximal breadth measurements Troitsky IX–XI*

the ten complete metatarsals were identified as sheep compared with three of goat (Figure 4.39; Table 4.41).

Therefore 64% of the complete metapodials belonged to sheep. Measurements of the proximal ends of sheep/goat metacarpals increased the sample size to 36 by the inclusion of incomplete bones (Figure 4.40). These included 21 specimens identified

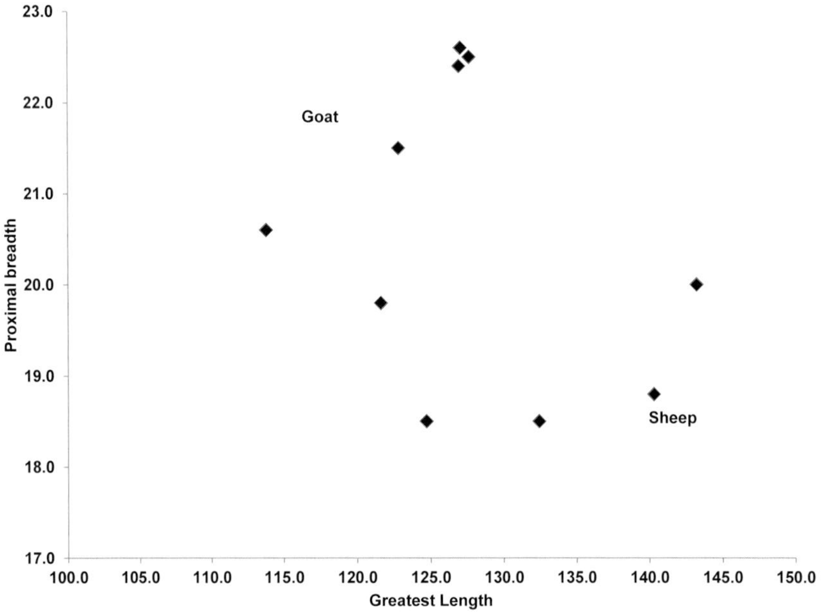

Figure 4.39: *Sheep and goat metatarsal lengths and proximal breadth measurements Troitsky IX–XI*

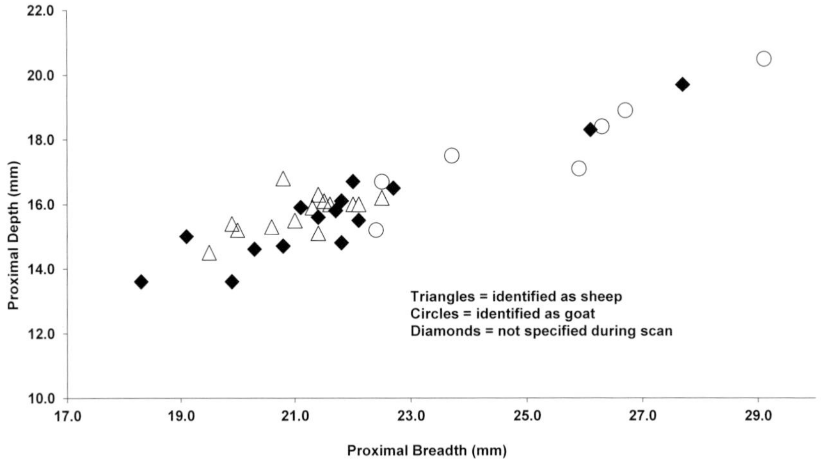

Figure 4.40: *Sheep and goat metacarpal proximal breadth and depth measurements Troitsky IX–XI*

during the scan on morphological criteria (Boessneck 1969) as sheep (14) and goat (7). There was only a slight overlap in the size, with nearly all the proximal breadths of goat metacarpals being larger than sheep. Including specimens not specifically identified during recording, around 11 belonged to goat and the remainder to sheep (69%). The impression that sheep outnumbered goat by a ratio of around 2:1 was one that was formed in observations of other bones. One exception was horn cores, where roughly equal numbers of sheep and goat were recorded. However, it is likely that many of the cores, particularly the larger ones of male goats, were from horns brought to the properties for working. There will be further consideration of the relative abundance of sheep and goat in the metrical analysis in Chapter 6.

There was no clear evidence for significant chronological changes in the relative abundance of sheep and goat in the Troitsky XI assemblage (Table 4.42).

Sheep/goat anatomical representation

NISP counts of sheep/goat anatomical elements are presented by site and spits in Tables 4.43–4.46. Element counts in sieved samples are given in Table 4.11. The sheep/goat elements were again divided into the eight areas of the body (Tables 4.47–4.50) so that broad comparisons could be made by site (Figure 4.41) and, in the case of Troitsky XI, by groups of spits (Figure 4.42).

Comparisons of the sheep/goat assemblages between Troitsky IX–XI showed relatively little variation overall (Figure 4.41). Upper limb bone elements dominated all three assemblages (35%–39%). Jaws and horn cores provided between 20% and 29% of the three assemblages. Loose teeth only provided between 4% and 6% of the assemblages. These would have been commonly overlooked during normal excavation and,

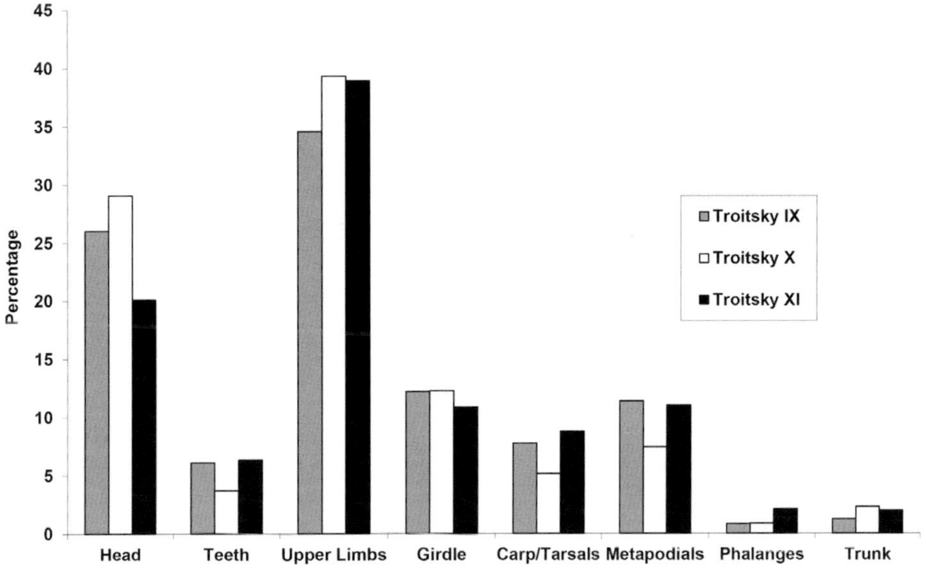

Figure 4.41: *NISP percentages of sheep/goat body areas from Troitsky IX–XI by site*

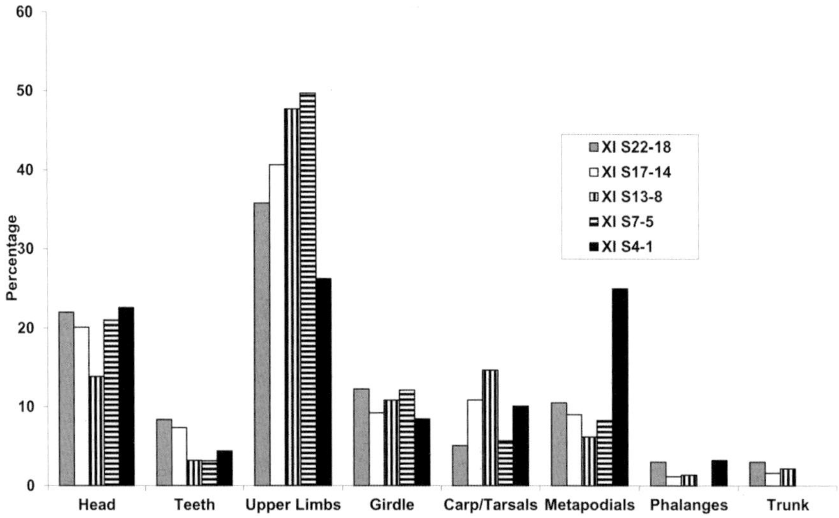

Figure 4.42: *NISP percentages of sheep/goat body areas from Troitsky XI by spits*

significantly, they were the most common elements identified (36%) in the sieved sample, albeit based on a very small sample (Table 4.11).

Scapula and pelvis fragments were found in consistent numbers throughout the trenches, providing 11%–12% of the site assemblages. Foot bones generally were not well represented. Phalanges provided only 1% to 2% of the bones. This was largely the consequence of poor retrieval of these elements. Hardly any of the smaller second and third phalanges were found (Tables 4.47–4.50), although it should be noted that phalanges were not recovered in the sieved samples either (Table 4.11). Metapodials were better represented, providing between 7% and 11% of the sheep/goat assemblages (Figure 4.41) but in most spits these were outnumbered by upper limb bones. Carpals and tarsals provided between 5% and 9% of the sheep/goat assemblages. Carpals were only found in the sieved samples (Table 4.11). The tarsals were dominated by astragali (Tables 4.47–4.50). Possible explanations for this will be discussed in Chapter 5.

However, this impression of uniformity is somewhat misleading judging by a more detailed examination of sheep/goat elements within the Troitsky XI sequence (Figure 4.42). Upper limb bones still dominated but their contribution to the overall assemblages gradually increased from 36% in the earliest deposits to 50% in Spits 7 to 5. This increase was similar to the one observed in the pig assemblage and could imply that relatively more joints of meat were imported.

Comparisons between the numbers of upper limb bones and metapodials revealed particularly interesting trends. The upper limb category is made up of more bones than the metapodial category. Excluding the rare finds of patella, each sheep skeleton contains ten upper limb bones compared with four metapodials. Given equal

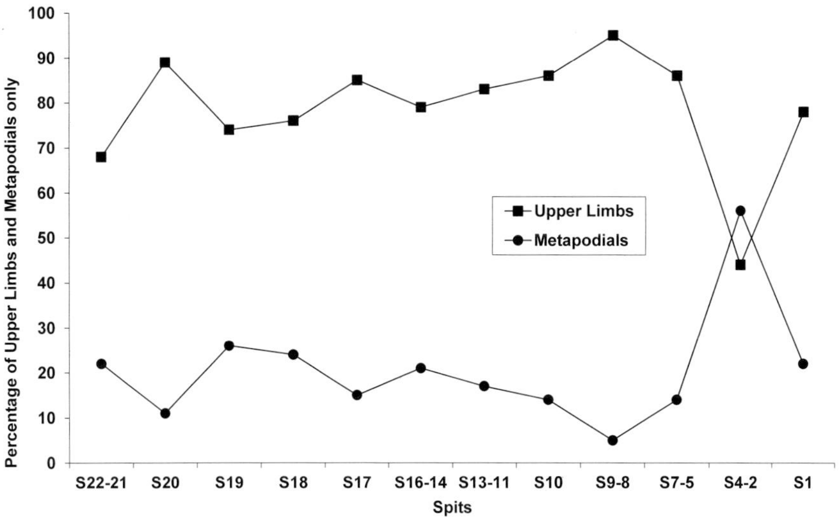

Figure 4.43: *Relative percentages of sheep/goat upper limb bones and metapodials from Troitsky XI*

deposition, fragmentation and recovery, upper limb bones should therefore provide around 71% of the total major limb bones.

In most of the spits from Troitsky XI, upper limb bones provided over 70% of the total upper limb and metapodial fragments. In most cases this figure exceeded 80% and tended to increase slightly in the later layers (Figure 4.43). This would support the contention that some of the upper limbs arrived as joints or as dressed carcasses. However, in Spits 4 to 2, there was a significant increase in the percentage of metapodials, which outnumbered upper limb bones in those layers. This implies that more bones of low food value were introduced and deposited within the properties during that period. It is possible that this reflects the importation of more skins for processing and this has parallels with the increase in the number of calf mandibles in the same spits, which may also have been brought in with the hides.

Sheep/goat ageing evidence

A total of 288 sheep/goat mandibles provided tooth ageing data (Table 4.51). The mandibles were assigned to the following seven stages of eruption and wear:

Stage 1 = 4th deciduous premolars (dp4) not in wear
Stage 2 = dp4 in wear; 1st molar (M1) not in wear
Stage 3 = M1 in wear; 2nd molar (M2) not in wear
Stage 4 = M2 in wear; 3rd molar (M3) and permanent premolars not in wear
Stage 5 = M3 in wear; M1 at Grant (1982) wear stage g
Stage 6 = M1 at wear stages h–m; M2 at stage g
Stage 7 = M1 and M2 at wear stages h–m

It is possible to differentiate between sheep and goat mandibles by observing variations of the deciduous premolars (Payne 1985) and between various morphological variations in the adult jaws (Halstead *et al.* 2002). Unfortunately such distinctions were not recorded rigorously during the scan. This is a significant handicap regarding the interpretation of the mortality data, since it is feasible that sheep and goats were exploited differently resulting in very different mortality patterns. Specimens that were explicitly identified to species during recording are indicated in Table 4.51. These identifications were made principally on immature mandibles. A total of 39 sheep and 24 goats were recorded. The percentage of sheep (62%) was similar to that recorded for the metapodials.

Mandibles of neonatal lambs and kids (Stage 1) and juveniles under three months old (Stage 2) were recovered fairly rarely from the Troitsky sites (Figure 4.44; Table 4.52), providing only around 6% of the sample overall. Sheep and goats were equally represented amongst this group but the low number of neonatal mortalities would suggest that neither of these species were kept in significant numbers within the properties. This conclusion assumes, however, that there are no major biases in the retrieval of mandibles of very young sheep and goat.

However, Stage 2 mandibles did not occur evenly throughout the deposits. None were recovered from Troitsky X, whereas they formed 11% of the sample from Troitsky IX (Figure 4.44). In Troitsky XI, they were particularly common in Spits 13 to 8 (Figure 4.45; Table 4.53), albeit in a relatively small sample. All seven Stage 2 mandibles from Troitsky XI were found between Spits 16 and 8. They were absent from both the lowest and highest spits. These fluctuations require further analysis. They may reflect changes in the intensity of dairy production in different periods.

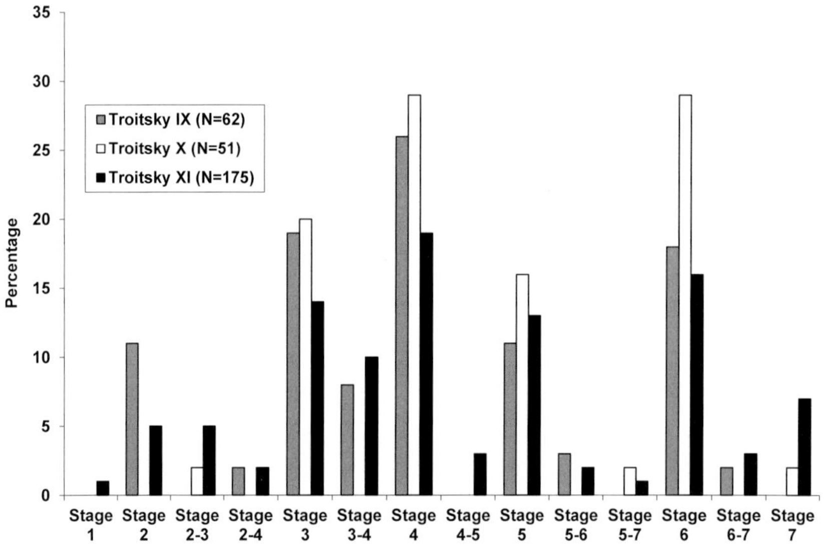

Figure 4.44: *Sheep/goat mandible age stages from Troitsky IX–XI*

Alternatively they could mainly have been associated with the skins of lambs and kids imported for processing.

Stage 3 mandibles were recovered much more consistently from all the sites, forming between 14% and 20% of the specimens (Figure 4.44). Most of these would have been from lambs and kids aged between 6 and 12 months, which, assuming Spring births, were mainly killed during the autumn and winter. Twenty-three of the 47 mandibles assigned to this stage were identified to species. Sheep/goat (15) outnumbered goat (8) but overall, there was no evidence that the slaughter pattern of the two species differed significantly. Mandibles of this age were particularly abundant in the earliest deposits of Troitsky XI, where they formed 29% of the sample (Figure 4.45). Thereafter, their abundance decreased markedly and they provided only 7% of the mandibles from the middle and upper spits (S13 to S1).

Stage 4 mandibles came mainly from animals killed for meat during their second year when they were approaching full size. These outnumbered Stage 3 mandibles in all three sites providing between 19% and 29% of the specimens (Figure 4.44). This indicates that meat production was probably the most important aspect of sheep and goat exploitation, at least in this area of Novgorod. Of the mandibles assigned to species, eight belonged to sheep and three to goat (Table 4.51). Stage 4 mandibles were relatively more common in the lowest layers of Troitsky XI (24%), decreasing to 14% in the sample from Spits 13 to 1.

There was therefore a substantial kill-off of immature and sub-adult sheep and goats aged between six and 24 months. Overall, 46% of the mandibles were assigned to Stages 3, 3–4 and 4, indicating an intensive focus on meat production (Table 4.52).

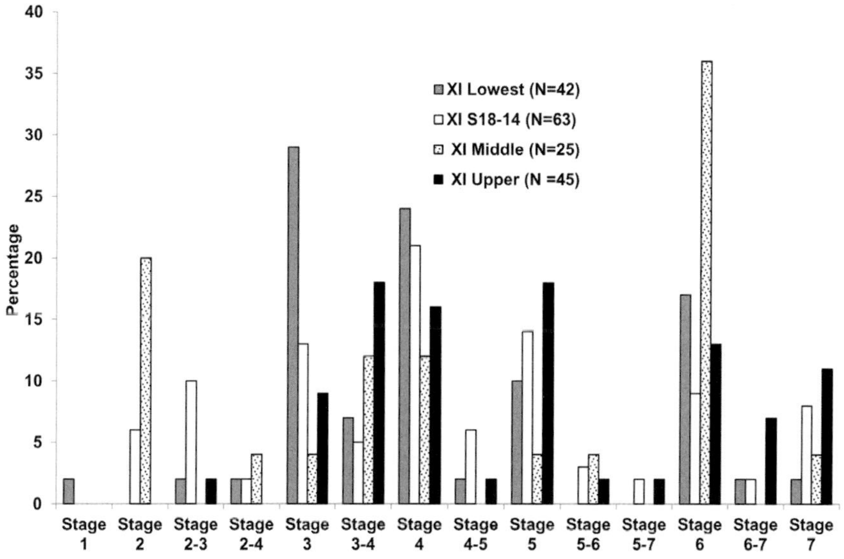

Figure 4.45: *Sheep/goat mandible age stages from Troitsky XI*

Within the Troitsky XI sequence, there was a decline from a very high percentage (59%) in Spits 22 to 19 to levels around 40% in later levels. The small assemblage from the middle spits (S13 to S8) was biased by the presence of several Stage 2 mandibles, which contributed to the lower percentage (28%) of Stages 3, 3–4 and 4 specimens.

Young adult sheep and goats (Stage 5) probably mainly aged between two and four years old were also an important component of the assemblage. Animals of this age could have provided some wool and progeny but were still killed for meat at a relatively young age. They provided between 11% and 16% of the mandibles on the three sites (Figure 4.44). Their relative abundance varied on Troitsky XI between 4% and 18%, with the highest figure being obtained from the highest spits (Figure 4.45).

Sheep and goats aged four years and above will have been kept for breeding and milk production before eventually being slaughtered. Mandibles at Stage 6 and 7 provided about a quarter of the ageable specimens from the Troitsky sites (Figure 4.44; Table 4.52). They were recorded more frequently in the Troitsky X sample (31%) than from Troitsky IX (19%). In the Troitsky XI sequence, the percentages of Stages 6, 6–7 and 7 mandibles doubled from around 20% in the lower spits (*i.e.* S22 to S14) to about 40% in the later layers (Figure 4.45).

Mandibles allocated to Stage 6 comfortably outnumbered those at Stage 7 on all the sites (Figure 4.44), indicating that meat from old individuals was not commonly consumed. It is perhaps significant, however, that the percentage of Stage 7 mandibles increased to 11% in Spits 7 to 1 of the Troitsky site, perhaps reflecting an increase in the importance of wool production in the later medieval period.

Epiphyseal fusion data broadly supported these findings. Preservation and retrieval bias probably favoured the survival of fused specimens in all fusion groups. Therefore the percentage of fused specimens recovered should be regarded as a maximum figure. Overall, about 38% of the latest-fusing epiphyses were fused and belonged to sheep and goats over three or four years old (Figure 4.46; Table 4.53). As in the case of the mandibles, relatively more bones from adult animals were recovered from Troitsky X than Troitsky IX and XI. Similarly, more fused specimens were found in Spits 7 to 1 of Troitsky XI than in earlier deposits (Figure 4.47). The anomalously low percentage of fused specimens from Spits 13 to 8 is based on only 22 specimens (Table 4.54) and the result should be treated with caution.

The percentages of fused distal tibiae and metapodials (and the few surviving proximal phalanges) were fairly consistent across the sites (60% to 68%), indicating that only around a third of these bones belonged to sheep and goats that were under 18 to 24 months old. This figure was lower than that obtained from the mandibles, which suggested that over half the sheep and goats had been slaughtered by that age (Stages 1–4 in Table 4.52). The discrepancy may relate to the poorer survival of unfused epiphyses. The percentage of fused distal tibiae (72%) was markedly higher than for the slightly later-fusing distal metapodials (48%–58%), perhaps suggesting a significant cull of animals within this age group, which could correspond to the peak

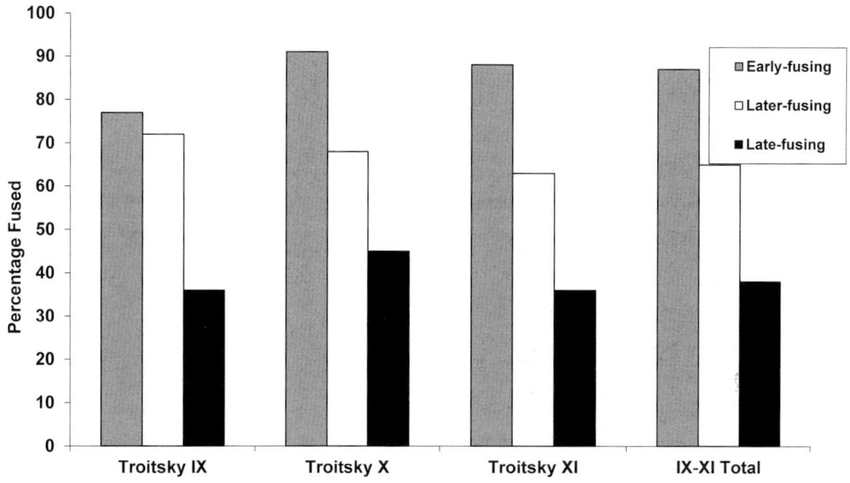

Figure 4.46: *Sheep/goat epiphyseal fusion data from Troitsky IX–XI*

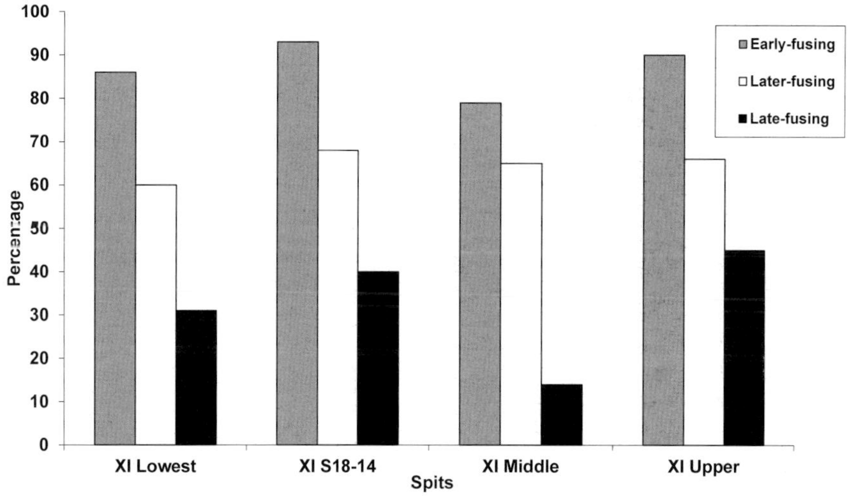

Figure 4.47: *Sheep/goat epiphyseal fusion data from Troitsky IX–XI*

of mortalities represented by Stage 4 mandibles (Figure 4.44). The percentage of fused specimens changed little within the Troitsky XI sequence (Figure 4.47).

Overall, 87% of the earliest fusing epiphyses were fused. Somewhat surprisingly, given the lack of control over retrieval, Troitsky IX provided the highest percentage of unfused specimens (23%). Elsewhere this figure was around 10% (Figure 4.46; Table 4.53). Again this was a lower percentage than that suggested by the mandible data and probably reflects retrieval and preservation biases.

The horse assemblage from Troitsky IX–XI

Relative abundance of horse and cattle

The overall abundance of horse in relation to other domestic mammals has already been briefly discussed. Generally their bones were poorly represented, nearly always providing less than 5% of the NISP counts within the broad assemblage groups (Figures 4.3–4.7; Tables 4.1–4.5). Comparisons between horse and cattle only,

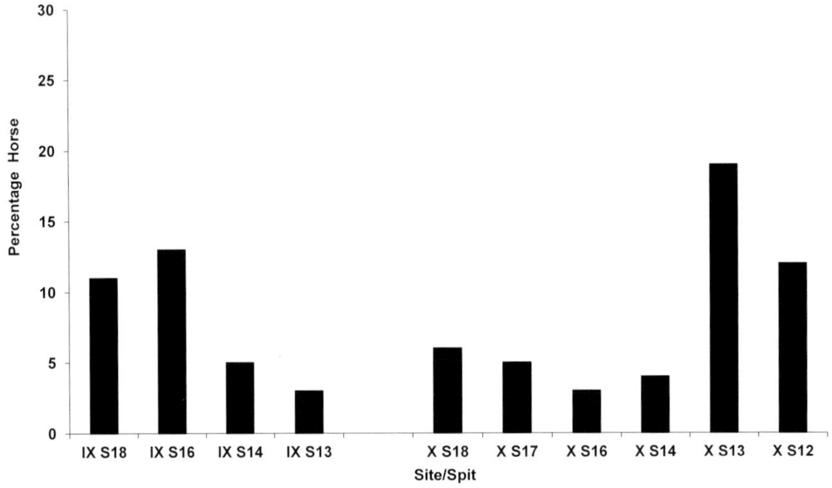

Figure 4.48: *NISP percentages of horse of total cattle and horse Troitsky IX and X by spit*

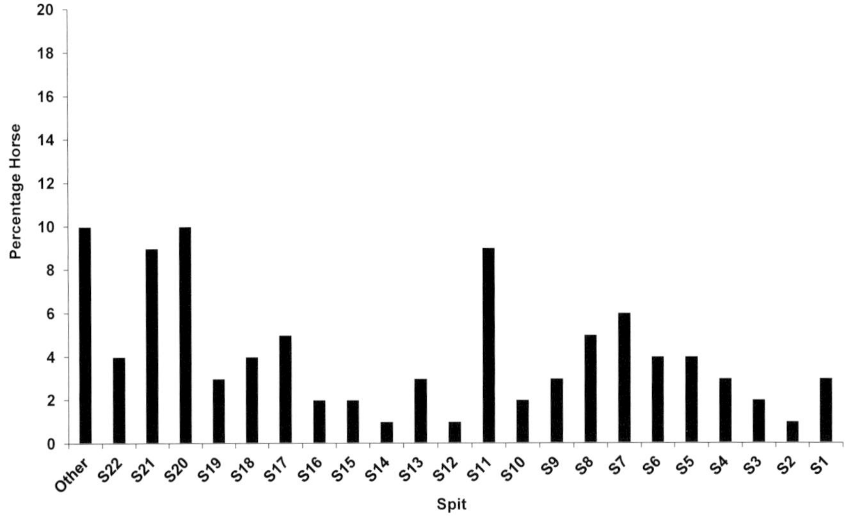

Figure 4.49: *NISP percentages of horse of total cattle and horse Troitsky XI by spit*

however, provided evidence for some minor fluctuations in their relative abundance (Table 4.55). Overall, horse contributed only 5% of the NISP counts of these two species but in some spits their abundance was significantly greater. They provided over 10% of the total in Spits 16 and 18 of Troitsky IX and in Spits 12 and 13 of Troitsky X (Figure 4.48). In Troitsky XI, horses were unusually well represented in the lowest spits and also in Spit 11 (Figure 4.49).

Horse anatomical representation

The anatomical representation of horses was completely different to those of the other domestic mammals. As was the case with the previous species, upper limb bones formed the largest category. However, cranial elements were very poorly represented (Tables 4.56–4.60; Figure 4.50). On all sites, carpals and tarsals, metapodials and phalanges were well represented. There was therefore a greater bias towards bones of the limb extremities than encountered in the other domestic mammals. To some extent this can be attributed to the larger sizes of these horse bones than their equivalents in sheep/goat and pig, making them easier to recover. However, they were also relatively more abundant than those of cattle, particularly with regard to the carpals and tarsals. Within Troitsky XI, carpals and tarsals sometimes outnumbered upper limb bones in the assemblage groups (Figure 4.51).

In addition, carpals provided a significantly higher proportion of the carpals and tarsals category (52%) than in the cattle assemblage (27%). Cattle carpals are slightly smaller than those of horse but this cannot account for all the discrepancy. Horse carpals were unusually well represented in several spits (Tables 4.56–4.59) and this clustering suggests that they were probably sometimes deposited as associated groups, although this cannot be verified.

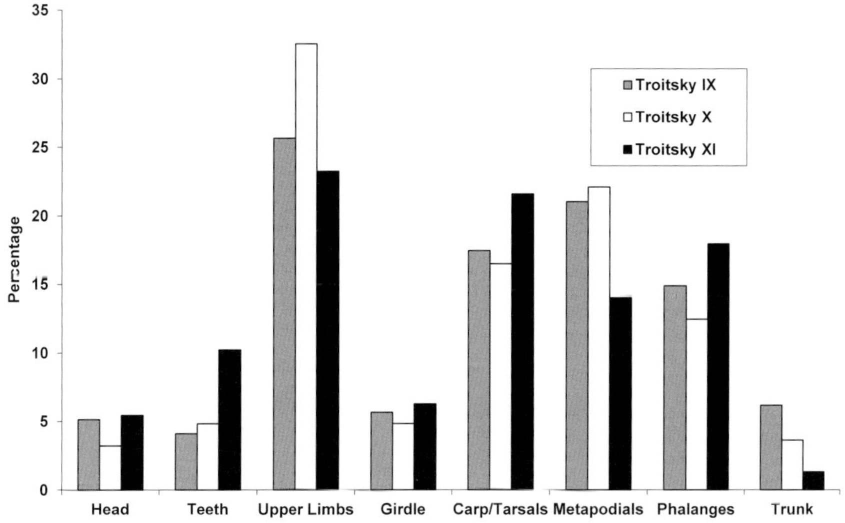

Figure 4.50: *NISP percentages of horse body areas from Troitsky IX–XI by site*

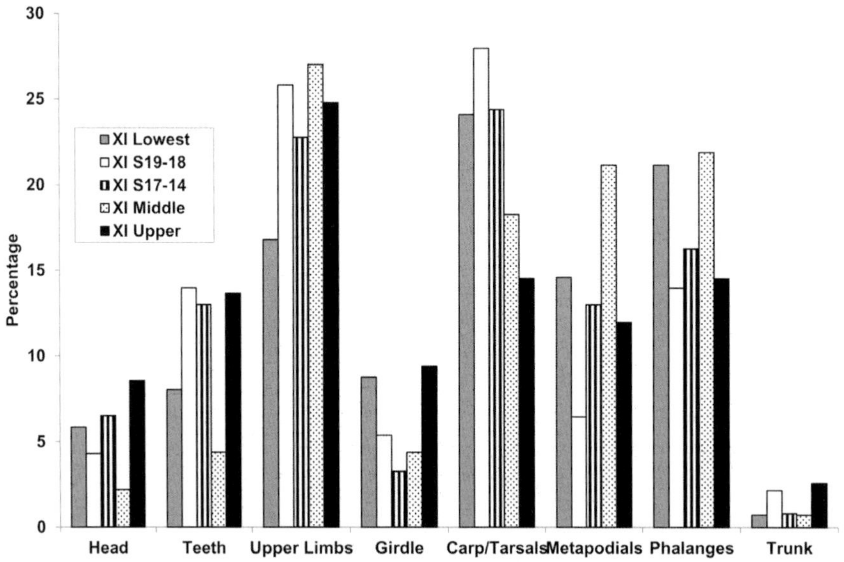

Figure 4.51: *NISP percentages of horse body areas from Troitsky XI by spits*

The low representation of head bones strongly suggests that complete horse carcasses were rarely processed within the properties. However, bones from the limb extremities were more commonly processed and deposited. These may have been associated with hides brought for processing and with the manufacture of artefacts made from metapodials and radii. This will be discussed further in Chapter 5.

Horse ageing evidence

There were too few mandibles and maxillae to justify tooth ageing analysis. Epiphyseal fusion analysis revealed that very few immature horses were represented (Table 4.61). Only 2% of the earliest-fusing epiphyses were unfused and this figure only rose to 4% in the later- and latest fusing groups. Although horsemeat was occasionally consumed, it is clear that the vast majority of horses could expect to reach maturity and be employed as beasts of burden and for riding.

The dog assemblage from Troitsky IX–XI

Dog anatomical representation

The small number of dog bones recovered has already been noted (Table 4.1). Despite clearly being active within the properties as indicated by the abundance of gnawed bones, dog carcasses were clearly not normally regarded as a source of food. Details of the number of different elements represented are provided in Tables 4.62–4.65. Although recovery methods precluded the retention of the skeletal integrity of associated bone groups, there were indications that at least some of the dog bones were from partially

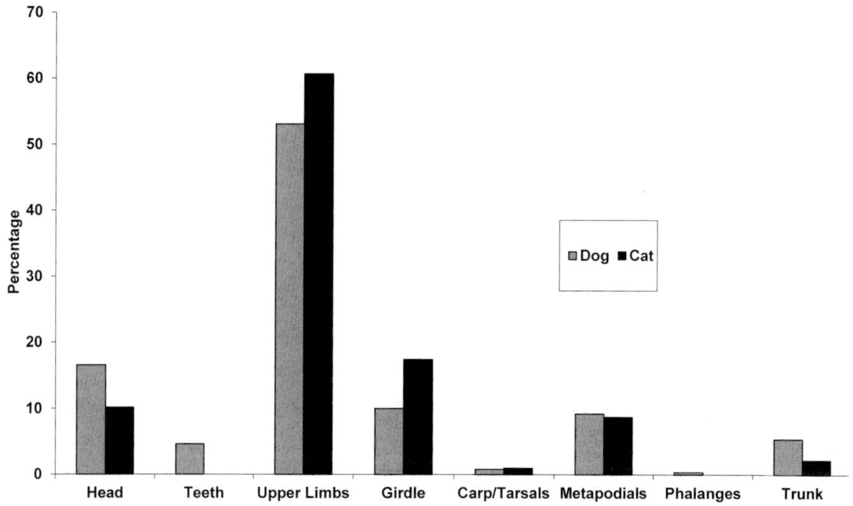

Figure 4.52: *NISP percentages of dog and cat body areas from Troitsky IX–XI*

articulated skeletons. Several sub-samples included groups of dog bones. Troitsky X, Spit 17 Area G produced 24 dog bones, including the pelvis, six vertebrae and three ribs probably from the same adult animal. Some of the other bones from this area may have belonged to the same individual. Five vertebrae from Troitsky X Spit 14 probably belonged to one animal. Property G Spits 11 to 9 produced 74 bones of dog including 42 from Spit 9. These almost certainly belonged to the same skeleton (Table 4.62).

Analysis of body parts showed the predictable bias towards larger bones with small elements such as loose teeth, carpals and tarsals, metapodials and phalanges being under-represented (Table 4.66; Figure 4.52). No dog bones were retrieved from sieved samples.

Dog ageing evidence

Fusion data were recorded from 155 limb bones (Table 4.67). Of these, 32% of the epiphyses were unfused, although this percentage varied considerably between different bones. This indicates, however, that substantial numbers of dogs died before skeletal maturity despite the fact that they were not exploited for meat. This supports the view that dogs were commonly kept in towns and indicates that mortalities of immature dogs were quite high, particularly as neonatal mortalities were not found. Some of the dogs may have been deliberately killed to control the size of the population.

The cat assemblage from Troitsky IX–XI

Cat anatomical representation

Bones of cat formed a small but consistent component of the Troitsky IX–XI assemblage, providing 2% of the identified mammal bones in the hand-collected assemblage and 4% of these in the sieved assemblage (Table 4.1). The numbers of different cat elements represented in each spit are shown in Tables 4.68–4.71. As in the case of dog, some of these bones formed associated groups but it was not possible to determine the details of each of these groups because of the limitations of the retrieval strategy. Several partial skeletons of cats were definitely present in Troitsky XI, in Spit 21, Property I–R; Spit 16, Property I; Spit 15, Property G; and Spit 4, Property G. This suggests that most cat carcasses were not utilised for their skins or meat and their bodies have sometimes survived as partial skeletons.

Analysis of body parts showed similarities with the dog assemblage with an even greater bias towards larger elements. No loose teeth or phalanges were recovered and only small numbers of metapodials, carpals and tarsals were retrieved (Table 4.72; Figure 4.52). The assemblage was dominated by upper limb bones, and scapula and pelvis fragments were also well represented. Jaws are less common although several fairly complete skulls were found, which are not shown in Figure 4.52. Most of the metapodials were found in the associated groups and elements of the trunk exclude a number of associated vertebrae. Sieving produced a small number of cat bones, again mainly from the upper limbs (Table 4.11).

Cat ageing evidence

Fusion data were recorded on 451 cat limb bones (Table 4.73). Almost half of these were unfused. There were variations in these figures between different epiphyses with small bones such as the metapodials including a higher percentage of fused specimens, probably as a result of retrieval bias. Unfused specimens consistently outnumbered fused bones in the latest-fusing epiphyses of the larger limb bones (*e.g.* proximal humerus, femur and tibia, distal radius, ulna, tibia and femur), indicating that most cats did not survive into adulthood. Given the likelihood that many bones of juvenile mortalities are likely to have been overlooked, this suggests that cats suffered a high immature mortality rate. Again the cat population may have had to be controlled by deliberate culling of stray animals and unwanted litters.

A SUMMARY OF THE ZOOARCHAEOLOGICAL EVIDENCE FROM THE DESYATINNY-1 SITE (LYUDIN END): THE DOMESTIC MAMMAL BONES (Andrei V. Zinoviev, edited by Mark Maltby)

Introduction

In 2008, excavations were carried out by the Institute of Archaeology of the Russian Academy of Science in an area to the north-west of the Kremlin, located near the Desyatinny Monastery in the Lyudin End (Figure 4.2) These uncovered a variety of medieval remains including substantial numbers of bones of domestic and wild animals. In this paper the general structure of Maltby's (2012) report has been followed. This facilitates direct comparisons between data from the Troitsky sites. The osteological material derives from the Desyatinny-1 site from layers 70–240 cm, spanning the period from the 10th to the 16th centuries. The approximate dates for the excavated spits were as follows:

Depth (cm)	Date
200–240	10th Century
190–200	Early 11th Century
170–190	Mid- to Late 11th Century
150–170	Early to Mid-12th Century
120–150	Mid-12th to Mid-13th Century
90–120	Mid-13th to Late 14th Century
70–90	15th to 16th Century

A sample of exclusively hand-collected bones included over 7,700 mammal bones, of which more than 6,300 were identified to species (Table 4.74). Although the assemblage of domestic mammal bones from the Desyatinny-1 is less than a fifth of the size of the Troitsky sample, the material nevertheless forms a valuable addition to the assemblage of Troitsky sites.

This section provides a summary of the results of studies of animal bone from this site. The detailed data and more comprehensive discussions on particular topics are provided elsewhere (*e.g.* Zinoviev 2012).

The exploitation of domestic mammals

The exploitation of cattle

As in the Troitsky sites, cattle elements formed the highest proportion of the assemblage of Desyatinny-1, providing 60%–74% of the identified domestic mammal fragments. Although, as was discussed by Maltby (2012 and above), the high percentage of cattle bones might have been influenced by retrieval standards, beef was still the most important type of meat consumed, as has been shown for other medieval Russian cities (Tsalkin 1956; 1971; Zinoviev 2008). Unlike the sequence from Troitsky XI, cattle elements in Desyatinny-1 tended to decrease slightly in relative abundance in more recent levels (Figure 4.53). However, animal bones from late medieval layers were scarce (Table 4.74) and cannot be used to trace changes in relative abundance in detail. The presence of large amounts of both high and low quality meat bones, along with parts of skulls and horn cores suggested that, as in the case of the Troitsky sites, much of the carcass processing took place within household properties.

Ageing analysis of 284 cattle mandibles brought surprisingly similar results to those from the Troitsky sites (Table 4.75; Figure 4.54). One third of the mandibles possessed fully erupted cheek teeth (Stages 6–7), thus belonging to adult cattle over four years of age (Higham 1967). As in the Troitsky assemblage, the majority of cattle were slaughtered between four and seven years of age. This indicates that many cattle were old enough to have been exploited for other uses such as dairy production, traction and transport and that meat production was not particularly intensive. However, the most common age category was Stage 6, which indicates that many animals were slaughtered prior to very old age. In addition, over half the cattle represented did not reach full dental maturity. As at Troitsky, around 14% of the cattle mandibles were at Stage 4 of the tooth eruption sequence, indicating a significant cull of third year animals. These animals would have been too young for working and therefore represent cattle raised and slaughtered primarily for beef.

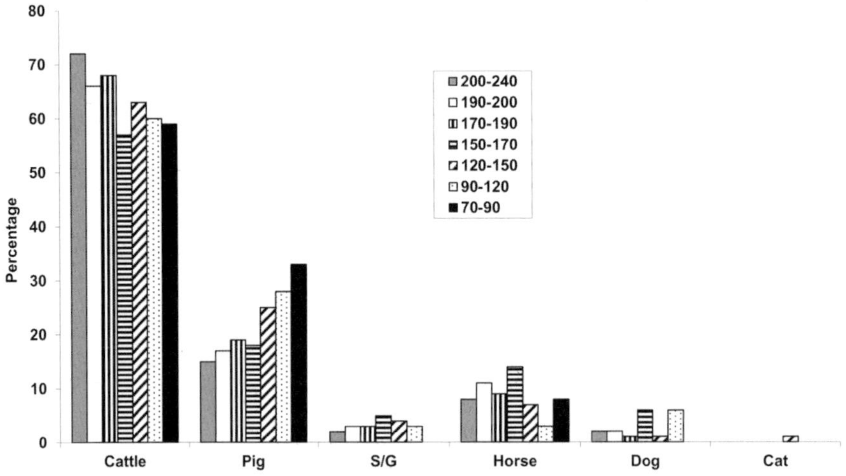

Figure 4.53: *NISP percentages of domestic mammals from Desyatinny-1 by depth*

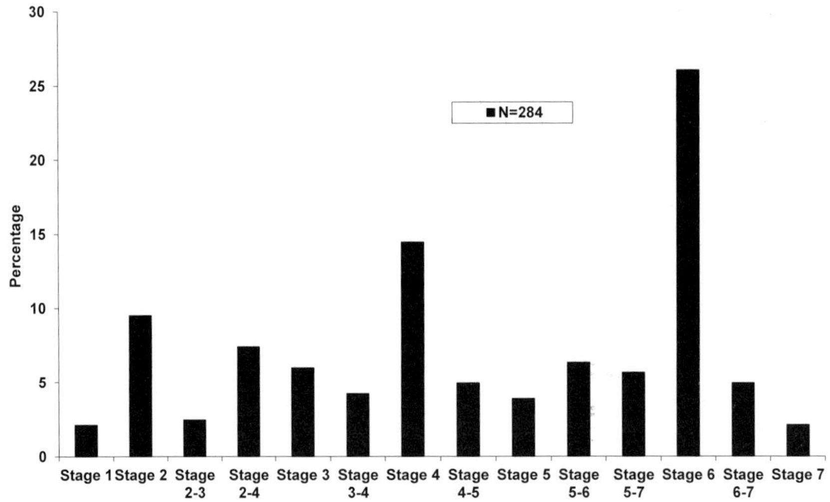

Figure 4.54: *Cattle mandible age stages from Desyatinny-1*

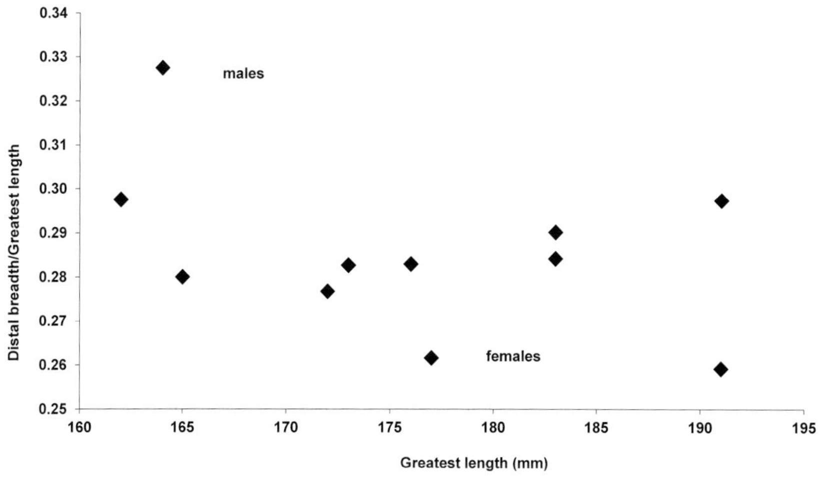

Figure 4.55: *Cattle metacarpal lengths and distal breadth indices Desyatinny-1*

Utilisation of veal has also been recorded at Desyatinny-1 by the presence of mandibles at Stage 1–3, which form at least 20% of the sample, a slightly higher percentage than at Troitsky (Table 4.21). Similarly, both samples show the dominance of animals around 2 to 6 months old (Stage 2 and early Stage 3) (Figures 4.21 and 4.53). Fluctuations in percentages within the Desyatinny-1 sequence were not as marked as in Troitsky XI. The results showed that veal production may have been more intensive in the 11th century and, after a decline in the early 12th century, from the second half of 12th century until the end of 14th century. However, the small area of excavation and the relatively small sample sizes might be responsible for such variations.

As observed in the Troitsky assemblage, analysis of the Desytinny-1 assemblage showed that veal production was probably closely connected with dairying. Metrical analysis of the metacarpals revealed that nearly all the fused specimens (over 24 to 30 months of age) from Desytinny-1 had dimensions sufficiently slender (Bd/GL<0.30) to be considered as females and unfused specimens probably mainly belonged to immature bulls and steers that were slaughtered for meat (Table 4.76; Figure 4.55). Adult cows would have been an importance source of milk.

The exploitation of pigs

The Desyatinny-1 excavations produced a considerable number of pig bones, comprising 18% of the identified domestic mammal bones in the assemblage (Table 4.74; Figure 4.53), compared with 20% from the Troitsky sites (Table 4.1). The percentage of pig, however, tended to increase in the upper layers in contrast to the Troitsky sites.

Numerous parts of the skulls, including fragments of jaws, allowed a precise determination of age of a number of pigs (Table 4.75; Figure 4.56). The presence of a substantial number of juveniles (Stages 2 and 3) as well as the dominance of animals

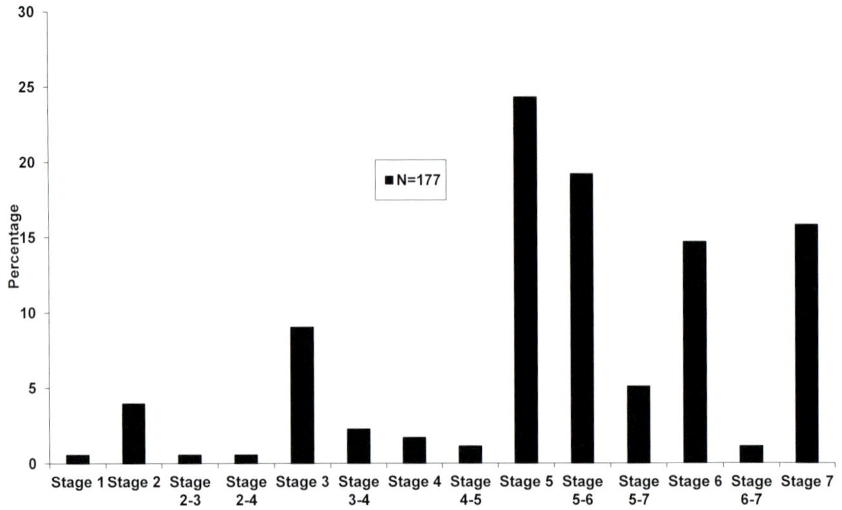

Figure 4.56: *Pig mandible age stages from Desyatinny-1*

culled at the end of their second year (Stage 5 and 6), is a common feature for kitchen remains of pigs (Zinoviev 2008). Thereafter pigs do not effectively gain sufficient weight to justify feeding costs (Volkopyalov 1968). One of the peaks of pig culling was at Stage 3 (Figure 4.56), approximately in the autumn and winter of their first year of life (Bull and Payne 1982). This is also a feature of Troitsky sites (Figure 4.34).

The exploitation of sheep and goats

Positive identifications were based not only on features of the skull (Payne 1985; Halstead *et al.* 2002), but also on numerous characteristics of postcranial elements (Boessneck *et al.* 1964; Boessneck 1969; Gabler 1985; Gromova 1953; Kratochvil 1969; Prummel and Frisch 1986). Whereas sheep were more dominant in the Troitsky assemblage, goats dominated the Desyatinny-1 assemblage. In this study both species were united within one category to make the result comparable with those from the Troitsky sites, from where bones of sheep/goat comprised less than 10% of the domestic mammal elements (Table 4.1). The same was true for Desyatinny-1, where these animals provided less than 4% of the assemblage (Table 4.74).

Lamb, mutton and goat-meat have never been dominant in the diet of people from medieval towns of the forested European part of Russia (Tsalkin 1956; 1971; Zinoviev 2008). The reason, as was noted by Maltby (2012, 362), is that 'swamps and wet meadowlands around the area would have been more suitable for cattle grazing and the substantial forests in the area would not have favoured keeping sheep in large numbers'. The percentage of sheep/goat at Desyatinny-1 did not show any tendency to increase in relation to pigs in later deposits, contrary to what was observed in Troitsky XI. This might be connected to the fact that most of the sheep/goat remains at Desyatinny-1 belonged to goats, which have often been kept in towns in larger numbers.

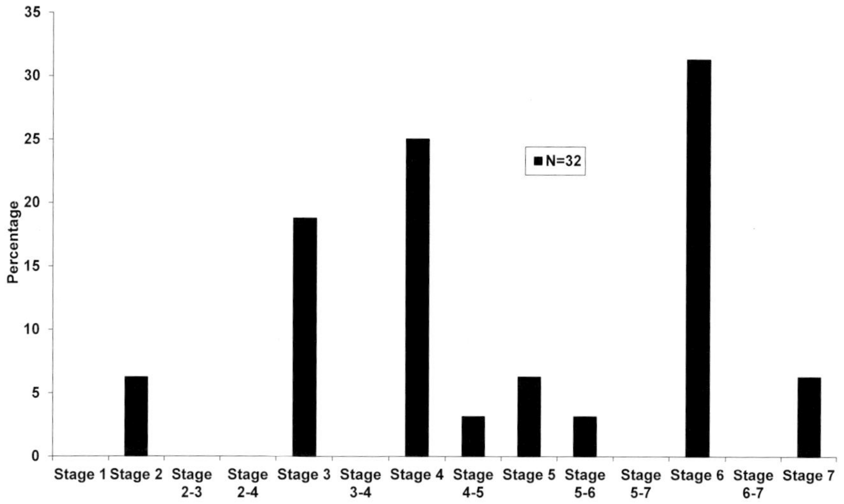

Figure 4.57: *Sheep/goat mandible age stages from Desyatinny-1*

Lambs and kids were both represented on the Desyatinny-1 Site (Table 4.75). Separation of mandibles of both species gave a higher percentage of immature lambs. This might indicate that sheep were predominantly exploited for meat production, whereas goats could have been used more in dairy production. The presence of young kids, slaughtered prior to weaning (Stages 1–2), raises the possibility that some goats were dairy producers and supports the contention that goats were often reared in the town itself, whereas lambs were more often culled at later ages. The highest numbers of sheep/goat were culled at Stages 4 and 6 (Figure 4.57).

Other domestic mammals

Horse bones were surprisingly numerous on the Desytinny-1 Site, outnumbering those of sheep and goats (Table 4.75). Although some of their remains bear butchery marks, most of the bones were found intact. This fact agrees with Christian taboos on consuming horsemeat in many parts of medieval Europe, as has been discussed by Maltby (2017a). Intact bones originated from adult animals with fused epiphyses and were mostly represented by girdles, limb bones and vertebrae. As at Troitsky IX–XI, cranial fragments were rarely found. Although some of the horses might have been exploited for meat, the majority of them lived relatively long lives to serve other purposes. Remnants of at least two horses show pathology related to horseback riding practice, when several lumbar vertebrae were fused to each other and to the first sacral vertebra.

Bones of dogs and cats were also found relatively rarely on the Desytinny-1 site (Table 4.75). Whilst the less numerous remains of cats were mainly represented by separate limb bones of different individuals, those of dogs generally belonged to animals buried in their entirety. No butchery marks were observed on bones of either species. Most of the kitchen bones from Desyatinny-1 bear traces of dog gnawing.

In contrast to the Troitsky assemblage (Table 4.67), most of the dog bones found on Desyatinny-1 belonged to adult animals, which were buried or dumped after death. The majority of the dogs found were over six years old, suggesting relatively long service to their owners. Bones of older dogs often showed traumatic and arthropathic changes. The cat assemblage did include a number of unfused bones, showing that more cats than dogs failed to attain old age.

DOMESTIC MAMMAL BONES FROM THE KREMLIN EXCAVATIONS 1996 (Mark Maltby)

A small excavation within the Kremlin took place in 1996. Bones from some of the later medieval layers (Spits 4 to 7) were sampled. These produced an assemblage of 255 animal bone fragments including 89 identified to domestic mammals (Table 4.77). Percentages of cattle (63%), pig (27%), sheep/goat (6%) and horse (3%) were similar to those recorded on the Troitsky and Desyatinny-1 sites. The cattle assemblage included quite a broad range of elements including 14 scapula and pelvis fragments and eight phalanges. The types of bone of other species were biased towards larger elements. However, the sample was too small to merit further analysis.

-5-

CARCASS PROCESSING OF DOMESTIC MAMMALS FROM NOVGOROD AND SITES IN ITS IMMEDIATE HINTERLAND

Mark Maltby

INTRODUCTION

Analysis of butchery and fragmentation patterns can provide information about the intensity and nature of carcass processing for the different domestic species. Carcass processing encompasses skinning, segmentation, filleting, meat preservation, marrow extraction, and bone-, horn- and antler-working. Used in conjunction with anatomical representation data (Chapters 3–4), analysis of butchery data can be used to investigate whether and where specialist butchers, tanners, horners and bone-workers were operating within a settlement. Examination of the types of butchery mark can provide information of what implements (knives; cleavers; large blades; fine blades; saws) were being utilised in various aspects of processing.

Any observed knife cuts, chop, blade and saw marks were recorded on all the bones recorded in detail from the hinterland sites discussed in Chapter 3. Fragmentation data were also recorded from these sites. These observations were recorded by Mark Maltby and Sheila Hamilton-Dyer for the Gorodishche samples collected before 1998. Butchery marks observed in the 1998 sample were recorded by Ellie Hambleton in 2003. Butchery from other hinterland sites was recorded by Mark Maltby.

The majority of the bones from Troitsky IX–XI and the Kremlin in Novgorod were not recorded in detail. However, detailed records were made for selected assemblages from lower levels of Troitsky XI by Mark Maltby and these can be directly compared with the results from Gorodishche and the other hinterland sites. Butchery evidence from the Desyatinny-1 site in Novgorod, the 2000–2004 assemblage from Gorodishche and the Minino assemblage has not been considered in this chapter, as methods of recording were not directly comparable.

Where detailed records were made, the location and nature of the marks were noted. The types of marks can be broadly divided into the following categories:

Severing: this is where a bone has been completely broken by a cleaver or a heavy blade. No distinction was made about which type of tool was used. Many other bones were deliberately broken to obtain marrow and for pot-boiling but only those specimens that displayed a clear processing mark were recorded as butchered.

Longitudinal splitting: limb bones that bore evidence of splitting along the longitudinal axis, principally to gain access to the marrow cavity, were assigned to these categories.

Superficial chops or deep blade marks: these were marks that had not completely severed the bones. Again, no distinction was made about which type of tool was used in each case but most of these marks were made with a heavy blade rather than chopping with a cleaver.

Incisions: these were typically fine incisions made with a metal knife, although some deeper cuts also fell into this category. No distinction was made about whether the point or the blade of the knife had been used.

Saw marks: these were rare and generally only observed on bones, antlers or horn cores that had been used in the manufacture of artefacts.

Classifications, which combine location and nature of the marks, are provided by species and element in Tables 5.12–5.58. These are broad classifications and they do not account for all of the variations regarding the exact locations of the marks and the nature of the incision, chop or blade mark. It should also be emphasised that these classifications merely record the marks on the bone and not the purpose of them. Interpretations of the marks are discussed in the text.

GORODISHCHE

Fragmentation of major domestic mammal bones

Fragmentation of bones can be the result of butchery practices or other post-mortem alterations to the bone, including damage resulting from scavenging, erosion, trampling and modern breakage during or after excavation. In most archaeological assemblages the degree of fragmentation will be the result of the combination of these factors and it is often difficult to determine which was paramount.

The fragmentation rates of the major bones of larger mammals at Gorodishche are given in Tables 5.2–5.5, where the whole bone equivalents (WBE) of the bones have been divided by the total number of their fragments (NISP). There were significant differences in the fragmentation patterns of cattle bones from different parts of the body (Table 5.2). The majority of mandibles, scapulae, humeri, radii and femora were heavily fragmented and rarely was half or more of the bone recovered. Apart from the assemblage from the 1979 excavations, the mean WBE of all these bones was never higher than 0.23 of a complete bone and in many cases the average lay below 0.20 (Figure 5.1). The mean sizes tended to be higher in the 1979 sample, reflecting the less rigorous sampling procedures, which favoured the recovery and retention of larger elements. Cattle metapodials tended to survive in a more complete state in all of the assemblages. Although many were broken open for marrow, metapodials were less likely to be broken into smaller pieces during subsequent butchery and cooking.

Similarly, the small metapodials of pigs were generally much less fragmented than those of the upper limb bones. The mandibles and scapulae were also heavily fragmented. The bones collected in 1979 were generally less fragmented than those in the sieved samples from later excavation seasons (Figure 5.2).

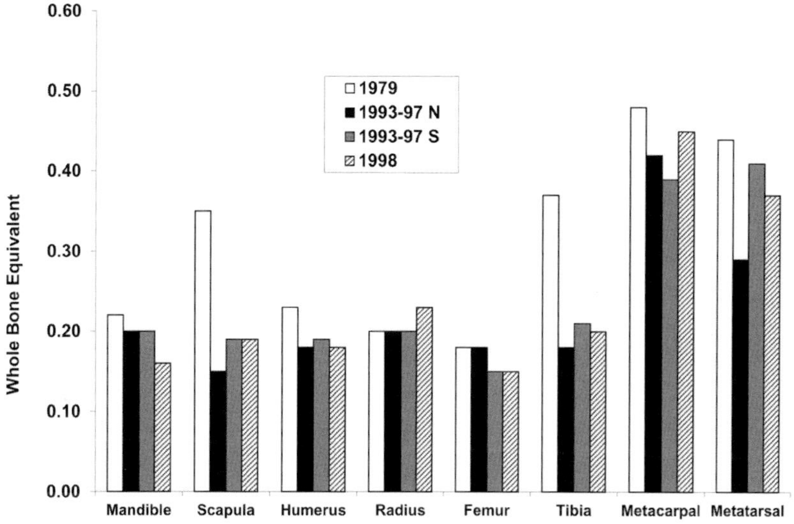

Figure 5.1: *Gorodishche 1979–1998: average fragment size of cattle elements expressed as whole bone equivalents*

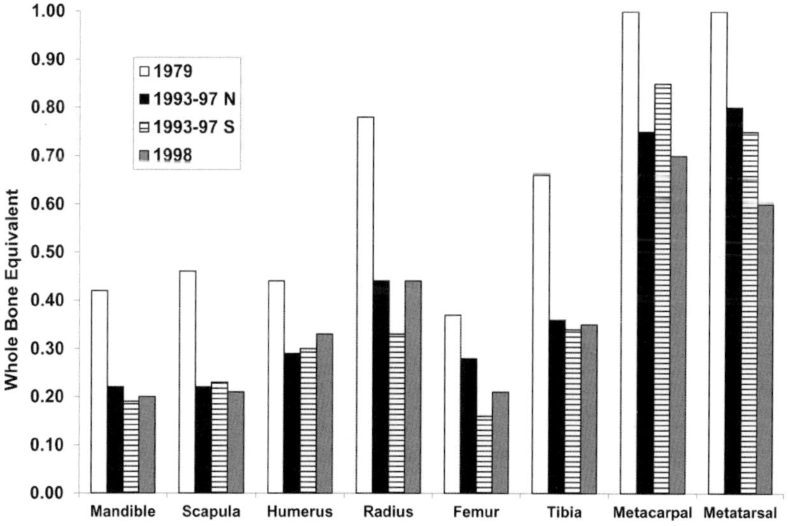

Figure 5.2: *Gorodishche 1979–1998: average fragment size of pig elements expressed as whole bone equivalents*

Overall, counting all the eight elements analysed in this exercise, cattle were the most fragmented (mean WBE 0.25), followed by pig (0.35), horse (0.40), sheep/goat (0.42) and dog (0.57). The variations mirror the differences in butchery frequencies of these species. The contrast between cattle and horse is most marked. Cattle bones were much more heavily fragmented. This cannot simply be a factor of differential preservation. Horse bones were less likely to be fragmented because their meat and marrow were not as intensively exploited. Sheep/goat elements were less heavily fragmented than pig, reflecting their smaller size. Unsurprisingly, dog bones generally survived in a more complete state. These counts include bones found in associated bone groups and, as described below, there is only occasional evidence for processing of their carcasses.

There were variations in fragmentation levels between assemblage groups. The 1979 assemblage was the least fragmented for all species as a result of the less rigorous retrieval strategy. The cattle and pig samples were more fragmented in layers than in the pits, particularly in the case of cattle in the Black layer (Tables 5.2–5.3). These differences reflect better preservation conditions in the pits where the burial of the bones restricted the damage caused by erosion and trampling.

Cattle processing marks

Full details of butchery and fragmentation data are recorded in the archive. Table 5.1 summarises the number of observations of butchery marks made on domestic mammal bones. Altogether, 413 records were made, representing 11% of the domestic mammal assemblage excluding loose teeth and unidentified fragments. They were most commonly found on cattle bones (15%), reflecting the high intensity of butchery on their carcasses. Classifications of the butchery marks from cattle can be found in Tables 5.12–5.35.

Ten (29%) of the cattle horn cores bore chop or deep blade marks. These were mainly located at the junction with the frontal bone (Table 5.12). These were made when the horn was detached from the skull as a prelude to horn-working. Three of the butchered horn cores were found in close proximity in the South bank brown layer, suggesting that sometimes a number of horns were gathered together for processing.

Butchery marks were found infrequently on other parts of cattle skulls (Table 5.12). Four occipital condyles had been chopped through centrally, indicating that the skulls involved had been split open to remove the brain. An incision above the teeth of a maxilla was probably made during the removal of cheek meat. Chop and/or heavy blade marks associated with the separation of the skull from the vertebrae and mandibles respectively were found on a basisphenoid and temporal, whereas fine incisions were observed on a zygomatic (underneath the eye socket), an occipital and a petrous, showing that different tools were used to achieve the same goals.

Twenty-one (13%) cattle mandible fragments were recorded as butchered, including seven from Pit 1 from the South bank. Four of a large group of cattle jaws from that pit bore fine incisions, two of which were located in a position indicative of the removal of the tongue. Incisions were only observed on three other mandibles from elsewhere in the excavations. All these were located near the condylar process of the ramus and made during separation from the skull. Chop and deep blade marks, however, were more prevalent in this area. In six cases these were located on or near the condylar process. In eight other cases, the ramus had been severed obliquely or vertically nearer to the teeth. This would also have been made when detaching the skull from the mandibles but would also give access to the tongue at the same time. Two mandibles bore chop marks on the diastema between the incisors and premolars made when the two sides of the mandibles were separated (Table 5.13).

The incidence of butchery marks tended to be highest on those bones which bear the greatest meat weight. Twenty-one (25%) cattle scapulae were recorded as butchered. Only chop or deep blade marks were observed (Table 5.14). Most of these were found on the blade of the scapula, which was commonly severed, often on or just above the neck of the bone where the shoulder was separated from the humerus. Several scapulae bore chop or blade marks on or around the glenoid, also made during disarticulation.

Nineteen (27%) of the cattle humeri were recorded as butchered. Seventeen of these consisted of chop or heavy blade marks (Table 5.15). These were found mainly around the distal end and in most cases the bone had been completely severed where the carcass had been segmented at the joint between the humerus and radius/ulna. The proximal ends of two humeri bore deep blade marks made during separation from the scapula. The discrepancy in the frequencies of butchery observations between the proximal and distal ends is mainly a factor of differential preservation. Many more distal ends survived and more observations of butchery were duly recorded. Incisions were observed on only two humeri. One specimen bore a cut near the proximal end of the shaft, indicative of filleting. A knife cut was also observed on the distal joint surface of a calf humerus.

Butchery marks were less frequently inflicted on lower forelimbs. Observations were made on only 12 (16%) radii and ten (22%) ulnae (Tables 5.16–5.17). Five radii and four ulnae bore chop or deep blade marks on or near the proximal end made during disarticulation from the humerus. Only one example of a chop through the distal end of a radius was observed. A deep blade mark running across the shaft was found on one radius. Another specimen bore a fine knife cut in the same location. An incision was recorded on one unfused distal radius. Incisions were also found on four ulnae: two of these, which may have been preparatory marks made prior to disarticulation with a heavier blade, were located near the proximal end; the other two were found on the shaft.

Butchery marks were found on 23 (27%) pelvis fragments, all but three of which consisted of chop or deep blade marks (Table 5.18). Twelve were found on various facets of the acetabulum, which was commonly severed when the hindlimb was detached. These marks correspond to those observed on the proximal ends of some of the femora. Six pelves bore chop or deep blade marks on the pubis. These were also associated with dismemberment. Occasionally butchered ilium and ischium shafts were also encountered. In one case a serrated blade was used. One acetabulum bore a fine knife cut and incisions were also found on a pubis and ischium of another specimen.

Butchery marks also appeared frequently on cattle femora. They were recorded on 26 (24%) specimens. These mainly consisted of chop or heavy blade marks (Table 5.19). Five were located near the proximal end. These were made during separation from the pelvis. Twelve femora bore similar marks on and around the distal end. Most of these had been severed vertically between the distal condyles. The only two specimens bearing fine incisions near the proximal end were again from calves. However, several femora of other immature cattle bore deeper blade marks.

Butchery marks were found much less frequently on the lower hindlimbs. Only eight (9%) cattle tibiae were recorded as butchered. Five had been severed in various locations whereas an unfused proximal epiphysis and a distal end bore incisions (Table 5.20). Butchery observations on tarsals were also surprisingly rare. Only two astragali (8%), four calcanei (11%) and one centroquartal (5%) were recorded as butchered (Tables 5.21–5.23). Both of the astragali and two of the calcanei had been severed transversely when the foot was detached. However, it does seem that this was an infrequent practice. It is possible that the shaft of the tibia was a more frequent segmentation point, although butchery evidence is slight. Fine incisions on two calcaneal tubers, a centroquartal and a carpal indicated that alternative methods of disarticulation around these joints were sometimes employed.

Butchery evidence was found more commonly on cattle metapodials (Tables 5.24–5.26). Twenty-five metacarpals (28%) and 22 metatarsals (23%) bore processing marks. In addition, seven small fragments from the distal ends of indeterminate metapodials had been severed obliquely or vertically and two others bore incisions. Similar marks were observed on the distal ends of seven metacarpals and five metatarsals. In some cases these bones had been split vertically along the whole or a substantial length of the shaft, allowing access to the marrow cavity and providing lengths of cortical bone for possible bone-working. Similar splitting was observed on the proximal end or shafts of a further ten metacarpals and five metatarsals. In several other cases the action had removed only a small portion of the medial or lateral aspect of the distal end. This was the result of disarticulation from the phalanges, as the marrow cavity remained intact. Incisions were only recorded on the proximal ends of one metacarpal and one metatarsal. In both cases these had subsequently been split longitudinally.

Eight cattle first phalanges (7%) bore butchery marks. In two cases these had been split longitudinally, probably when they were separated from the metapodials. Horizontal incisions were found on five first phalanges (Table 5.27). The location of most of these was consistent with marks made when a knife was inserted preparatory to skinning.

Butchery marks were found on nine cattle second phalanges (9%). Five of these bore horizontal incisions associated with skinning preparation. Two had been split and superficial chop or deep blade marks were found on three specimens (Table 5.28). One third phalanx has been split longitudinally.

There was substantial evidence for butchery of the axial skeleton (Tables 5.29–5.35). Nine (43%) atlases were recorded as butchered. Most had been severed vertically, probably when the vertebral column was split. Three (30%) axes and nine (53%) sacral vertebrae bore chops or deep blade marks. In six cases the sacrum had been severed transversely during separation from the lumbar vertebrae. In three cases the marks were located on or near the auricular articulation. These were inflicted during separation from the pelvis.

A substantial number of unidentified large mammal ribs and vertebrae also bore butchery marks. Most of these probably belonged to cattle, although a few butchered horse vertebrae were also noted. Most butchery consisted of chops through the bone. Transverse marks across the vertebrae indicated where the spine had been segmented. Chops running along the length of the spine tended to be offset slightly from the midline (paramedial). Here, the vertebrae are less thick and easier to sever. Several cattle thoracic vertebrae had fine cuts on the lateral process made when they were disarticulated.

As expected, the frequency of butchery indicates that cattle were intensively utilised for meat. Most butchery was carried out with heavy blades and the marks were mainly made during carcass segmentation. Surprisingly few filleting marks were recorded. Surface erosion of the bones may sometimes have obliterated fine incisions. Other marks indicated that cattle marrow, skins and horns were routinely processed. Metapodials were commonly split open for marrow. Marrow was probably commonly obtained from other limb bones as well, but breakage was not made with cleavers to the same extent. The accumulation of several mandibles butchered in an identical manner in a pit from the South bank indicates that specialist butchers may have been involved in the initial processing of some of the cattle carcasses in the area that produced high percentages of cattle butchery waste (Chapter 3).

Pig processing marks

Butchery marks were observed on 117 (9%) of the pig bones (Table 5.1). The carcasses of medium-sized mammals require less segmentation than those of cattle and therefore butchery marks are usually less common. The types of butchery marks observed can be found within Tables 5.36–5.58. Only ten pig skull fragments (4%) were recorded as butchered including three cases where the skull had been split open to remove the brain. Superficial chops were found on two occipital condyles, made during separation from the spine (Table 5.36). Three of the butchery marks that were observed on eight mandibles (4%) consisted of heavy blade marks that had separated the two sides of the jaw through the diastema. One mandible had been severed obliquely through the ramus; another was severed between the molars and premolars. Fine incisions made during disarticulation from the skull were observed on the rami of three other mandibles (Table 5.37).

Most of the butchery marks were again observed on the upper limbs, scapula and pelvis. Fifteen (19%) scapulae bore butchery marks. These mainly consisted of superficial chop or deep blade marks located on the blade or the neck (Table 5.38). In several cases the scapula had been completely severed indicating that the carcass was often segmented at the shoulder, as in the case of cattle. In only one instance were disarticulation cuts observed on the glenoid, suggesting instead that the distal half of the scapula and either the whole or proximal portion of the humerus often formed one joint of meat.

Butchery marks were observed on 22 (21%) humeri. Eight consisted of chops or deep blade marks on or near the distal end made during detachment from the radius/ulna. However, nine humeri had been severed (usually obliquely) through the shaft showing that the upper forelimb was commonly separated slightly higher up the leg. Similar marks were observed near the proximal end of two other humeri. Fine transverse incisions were observed on four pig humeri. All these, including two on unfused specimens, were located near the distal end.They were made during disarticulation (Table 5.39).

Butchery marks were rarely encountered on pig radii and ulnae (Tables 5.40–5.41). One distal portion of a radius had been severed through the shaft. Separation from the humerus was indicated by chop marks on the proximal articular surfaces of two ulnae, which had been severed. A third ulna bore both superficial blade marks and fine incisions around this area. A fourth ulna had a fine incision on its shaft.

Fifteen (25%) pig pelvis fragments were recorded as butchered. In three cases, the acetabulum had been damaged by a heavy blade. Other chop or heavy blade marks were recorded on two iliac crests and the shafts of five ischia and a pubis. All these marks were associated with segmentation of the carcass in the pelvic area. Fine incisions were observed on the shafts of two ischia and on an acetabulum (Table 5.42).

Surprisingly, butchery marks were observed on only ten pig femora (11%). There was only one example where the head of the femur had been completely severed. Deep blade marks and a finer knife cut were found on the proximal end of a complete femur of a piglet. The latter was probably a preparatory cut to expose the bone prior to severing it from the pelvis. Deep blade marks were recorded on another proximal femur, whereas knife cuts associated with filleting were observed on three shafts. Three femora had been severed near the distal end and one shaft had been split open longitudinally (Table 5.43).

Seventeen (17%) pig tibiae bore butchery marks. Fourteen of these, mainly chop or heavy blade marks, were located on the shaft. These had sometimes severed the bone. Two specimens had deep blade marks at the proximal end. Two tibiae had been severed near the distal end. All these were made during segmentation. Incisions were observed on only three tibiae and were all likely to have been made during filleting (Table 5.44).

The evidence suggests that pig upper hindlimbs were often segmented from the foot by chopping through the shin (tibia) above the ankle bones (tarsals). Butchery marks on the pelvis indicated that this was where the hindlimbs and the vertebral column were regularly separated.

Butchery marks were found infrequently on pig lower limb bones (Table 5.1). Fine incisions were observed on only two calcanei, an astragalus and a cuboid. One carpal had been chopped through longitudinally and two calcanei also bore deep blade marks (Tables 5.45–5.47). No marks were observed on pig phalanges. The distal end of a fourth metatarsal bore knife cuts on its anterior surface, probably resulting from disarticulation (Table 5.49). A peripheral metapodial bore a superficial blade mark that had removed a thin slither of the shaft indicating defleshing. However, there is little evidence that the trotters were heavily utilised for meat.

Chop and blade marks through a number of vertebrae of unidentified medium-sized mammals showed how the flanks of pigs and sheep/goat were divided into roughly equal halves. Two pig atlases had been severed where they had been detached from the skull (Table 5.52).

Sheep/goat and dog processing marks

Butchery marks were observed on 15 sheep/goat bones, mainly from the 1998 assemblage. This represents only 7% of the assemblage (Table 5.1). The shafts of three tibiae had been severed indicating that the leg joint was separated from the feet in the same area as commonly observed on pig (Table 5.44). The shafts of two femora (Table 5.43), an acetabulum (Table 5.42) and glenoid (Table 5.38) had also been severed during dismemberment of the upper limbs. One humerus had been severed just above the distal end, whereas another had fine incisions in the same area (Table 5.39). Incisions were also observed near the proximal ends of two radii, indicating that knives may have been more commonly used for disarticulating the elbow joint than in the case of pig (Table 5.40). An astragalus also bore incisions across the anterior surface where the feet had been detached (Table 5.45). Other incisions were found on an axis (Table 5.53). Two goat horn cores had been severed (Table 5.36), indicating that raw material for horn-working was being acquired.

A lumbar vertebra, which had been severed, and a humerus that bore both fine incisions and superficial chop marks near the distal end were the only examples of dog bones with butchery marks. Although very rarely encountered, these observations do indicate that dog flesh was occasionally consumed.

Horse processing marks

The contrast in the treatment of horse and cattle carcasses is apparent in the differences in the frequency of butchery marks recorded. In addition to having a much higher proportion of associated bone groups (see Chapter 3), only 14 horse bones (4%) were recorded with butchery marks (Table 5.1). The ramus of one mandible had been severed vertically in a similar manner to several cattle specimens (Tables 5.13 and 5.37). The distal end of a humerus had been completely severed from the radius and ulna, again in a manner also observed on several cattle humeri (Tables 5.15 and 5.39). The blade of a scapula (Table 5.38) and a patella had also been severed during dismemberment. A radius had been split longitudinally, perhaps for marrow extraction (Table 5.40). A complete radius and tibia had both been cut with a knife at their distal ends, presumably when the upper limbs had been separated from the feet at the carpal and tarsal joints respectively (Tables 5.40 and 5.44). A corresponding incision was found on an astragalus (Table 5.45). A shaft of an ilium had been severed and this specimen also had been cut (Table 5.42). Two first phalanges bore fine cuts, one on the anterior surface near the proximal end and the other running vertically up from the distal end (Table 5.50). The location and direction of these cuts are not compatible with skinning marks. They may have been made when the toes were being separated from the metapodials and the hooves. An atlas and a sacrum had been chopped through during segmentation of the spine (Tables 5.52 and 5.57). The associated group of vertebrae and ribs from the 1998 excavations included a lumbar vertebra bearing a superficial blade mark on the cranial articulation that was made during the segmentation of the spine (Table 5.56).

The evidence suggests that at Gorodishche some horses were butchered using methods also practised on cattle carcasses. However, carcass processing of horses was carried out less frequently and intensively. Their major economic and social importance lay elsewhere.

OTHER HINTERLAND SITES

Georgii

Fragmentation of major domestic mammal bones

Fragmentation data are provided in Table 5.6. The assemblages of sheep/goat and horse are too small for analysis of individual elements and fragmentation analysis of cattle and pig is also limited by the small samples. Overall, combining all eight elements analysed, cattle bone fragmentation at Georgii (Mean WBE 0.22) is slightly greater than at Gorodishche, suggesting that many bones were deliberately broken open to extract marrow or used as pot boilers. It should also be noted, however, that substantial fragmentation was also caused by animal scavenging. The pig mean WBE from Georgii is 0.26 compared with 0.35 at Gorodishche, showing that the Georgii pig assemblage was also more highly fragmented due to a combination of butchery and preservation factors. The horse mean WBE is 0.32, is also lower than at Gorodishche, but this is based on a sample of only 15 bones.

Processing marks on bones of domestic mammals

Butchery marks were observed on 33 domestic mammal bones from Georgii, representing 10% of the total identified domestic mammal assemblage excluding loose teeth (Table 5.7). As at Gorodishche, butchery marks were most commonly observed on cattle (14%). Classifications of the butchery marks from cattle at Georgii can be found within Tables 5.12–5.35 and for the other domestic mammals within Tables 5.36–5.58.

Many of the butchery marks observed on cattle bones from Georgii were similar to those encountered in the Gorodishche assemblage. For example, one mandible bore a superficial chop mark on the caudal aspect of the ramus made during disarticulation from the skull (Table 5.13). Three first phalanges bore fine knife cuts on their shafts made during skinning (Table 5.27). Incisions associated with filleting, which were observed on the shafts of a humerus (Table 5.15), radius (Table 5.16) and femur (Table 5.19), also have parallels at Gorodishche, as do the incisions found on the olecranon of an ulna (Table 5.17), made during preparation for disarticulation from the humerus. The severed shaft of a pubis and deep blade marks on an acetabulum provide evidence for dismemberment of the hip joint (Table 5.18). The proximal ball joint of a femur had also been severed during the same process (Table 5.19). However, superficial oblique blade marks near the distal end of a tibia (Table 5.20) and on the anterior of a centroquartal (Table 5.23) have no parallels in the Gorodishche assemblage. Both were probably associated with the dismemberment of the hock. A fine incision found on the anterior of the proximal end of a metacarpal (Table 5.24) was made during the removal of the foot from the forelimb. A cervical vertebra had been split down its midline and had also been cleaved transversely where the neck was segmented (Table 5.31).

Butchery marks were observed on only nine (6%) pig bones from Georgii (Table 5.7). Two associated lumbar vertebrae had been split (Table 5.56) and an atlas had been severed transversely during beheading (Table 5.52). A scapula bore deep blade marks near the glenoid inflicted when it was separated from the humerus. It also had similar marks running across the blade indicating further segmentation (Table 5.38). A humerus shaft had also been severed during dismemberment (Table 5.39). Deep blade marks were also found on the caudal aspect of a mandibular ramus (Table 5.37). These were made during separation from the skull. Incisions running across the anterior of an astragalus (Table 5.45) and at the distal end of a peripheral metapodial were inflicted when the feet were removed.

Several of these butchery marks have parallels at Gorodishche. Again, heavy blades were the most commonly utilised butchery tools although fine knife blades were also employed.

Butchery marks were observed on only one sheep/goat bone (5%) from Georgii. This was a humerus that had fine cuts on its distal articular surface made when it was disarticulated from the elbow joint. A similar specimen was recorded at Gorodishche (Table 5.39).

Four horse bones (9%) from Georgii produced evidence of butchery marks, a higher incidence than at Gorodishche but based on a much smaller sample (Table 5.7). Cut marks across the shaft of the first phalanx in the associated group of 12 foot bones of a foal indicate that its carcass was skinned (Table 5.50). Fine incisions were also recorded on two pelvis fragments, one of which also bore deeper blade marks near the acetabulum made during dismemberment (Table 5.42). Filleting is evidenced by the presence of fine incisions on the blade of a scapula (Table 5.38). As at Gorodishche, horse carcasses were less heavily exploited than those of cattle but there is evidence that at least some of them were skinned, and that horsemeat was consumed. No butchery marks were observed on the three dog bones recovered from Georgii.

Prost and Vasilievskoye

Only three observations of butchery were recorded in the small sample from Prost. A cattle femur shaft bore oblique incisions associated with filleting. The proximal end of a cattle metacarpal had been chopped through transversely during disarticulation. A horse first phalanx bore several incisions on various aspects of its shaft indicating skinning, similar to the specimen from Georgii. Cuts were also noted near the proximal anterior surface of this bone, suggesting that the toes were also separated from the metapodials.

The only butchered domestic mammal bone observed at Vasilievskoye was a pig tibia, whose shaft had been chopped through during dismemberment of the hind limb, in a manner commonly observed at Gorodishche.

Summary of butchery evidence from hinterland sites

There were a lot of similarities in the butchery marks recorded on all the sites investigated. Heavy blades including cleavers were the most commonly employed and the consistency of butchery of cattle and pigs at Gorodishche in particular suggests that specialist butchers were operating there. Segmentation of the limb bones tended to be carried out through the shafts of bones and the blades of scapulae rather than at the articulations and there is convincing evidence that joints of meat were being prepared consistently. Many of these would have been preserved by salting and smoking rather than for immediate consumption. Three of the sites had evidence that at least some of the horses were processed for their skins and flesh, although not as intensively as cattle. At Gorodishche, there is also evidence for the occasional butchery of dog carcasses.

NOVGOROD BUTCHERY

Introduction

As discussed in Chapter 2, only the presence of butchery marks was recording during the rapid scanning of the assemblages from the Troitsky IX–XI sites. Given the rapidity of the scanning and poor illumination in the recording rooms, these figures should be regarded as minimum figures as some marks, particularly fine incisions, could have been overlooked. Similarly, recording of fragmentation was not routinely carried out, so direct comparisons with fragmentation data from Gorodishche and Georgii cannot be made. Comparisons of fragmentation patterns would have been problematic in any case because of the differences in recovery techniques.

However, a substantial sample of butchered bones from the lower spits of Troitsky XI was recorded systematically in 2001 and 2002 and the analysis of these data will provide the bulk of the discussion in this section.

Frequency of butchery marks on domestic mammal elements

Introduction

The raw counts obtained from the scans for each bone type are provided in Table 5.8–5.11 for cattle, pig, sheep/goat and horse respectively. The results are presented by site and, in the case of Troitsky XI, by lower, middle and upper spits. A summary of the overall results for all the main domestic species is provided in Figure 5.3.

Cattle butchery frequency

A large percentage of the cattle bones bore evidence of butchery in the form of heavy blade marks, cleaver marks, fine incisions and occasionally saw mark (Figure 5.3). Overall, butchery marks were observed on 30% of the cattle elements (excluding loose teeth) from the three sites, with a very high percentage (40%) from Troitsky IX compared with Troitsky X (28%) and Troitsky XI (30%). What also emerges very clearly is that the incidence of observed butchery is significantly greater in the Troitsky sites than at Gorodishche (15%) and Georgii (14%). This could be the result of several factors. The retrieval of a higher proportion of less fragmented bones from the Troitsky sites provided more opportunities for butchery marks to be observed. The preservation of bones from those sites was exceptionally good, which made observations of butchery marks easier. Fine incisions were recorded more commonly on bones from the Troitsky sites than at Gorodishche (Tables 5.12–5.35), which may also be linked with better preservation of the bone surfaces.

The frequency of carcass processing marks varied. As expected good meat-bearing bones such as the vertebrae, ribs, humerus, scapula and pelvis included high percentages of butchered specimens (Table 5.8). However, high percentages of butchery observations on bones with much less meat value, such as the mandible and metapodials (Figure 5.4), indicated intensive exploitation of all parts of the carcass for a variety of products.

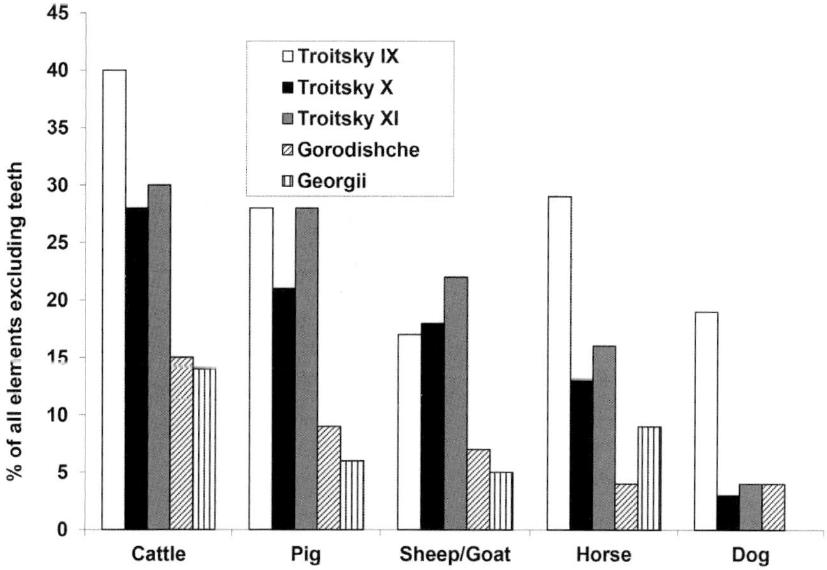

***Figure 5.3:** Novgorod, Gorodishche and Georgii: butchery frequencies on domestic mammal bones*

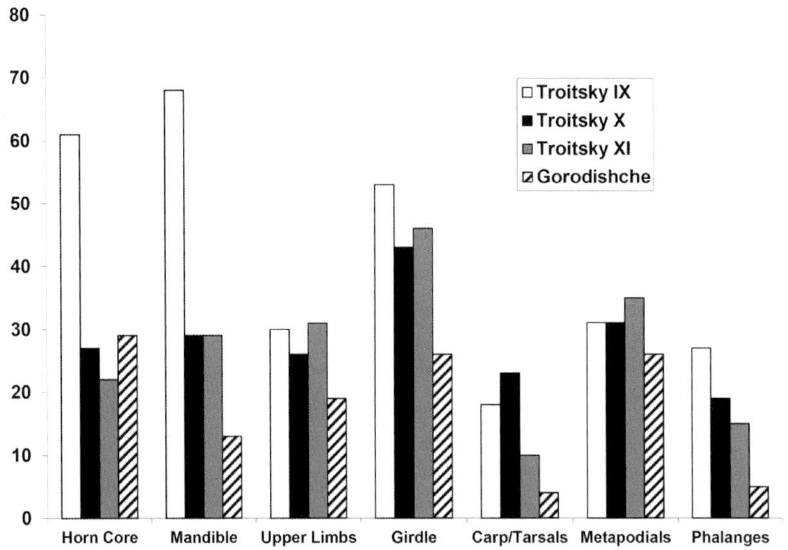

Figure 5.4: *Novgorod and Gorodishche: percentages of cattle butchered elements by body area*

Pig butchery frequency

Butchery marks were observed on 27% of the pig elements (excluding loose teeth) from the Troitsky sites (Table 5.9), reflecting high percentages on Troitsky IX and XI in particular. Again these percentages were substantially higher than encountered at Gorodishche (9%) and Georgii (6% – Figure 5.3). As with cattle, the trend is reflected across all areas of the body (Table 5.9; Figure 5.5). The highest percentages of butchery marks on pig bones came from areas with high meat content, particularly the vertebrae, ribs, girdle and upper limb bones. Relatively few marks were recorded on the metapodials and phalanges.

Sheep/goat butchery frequency

Although sheep and goats were relatively poorly represented, their carcasses were intensively exploited. Butchery marks were observed on 21% of their elements from the Troitsky sites (Table 5.10). This is a lower percentage than for cattle and pig, as the smaller carcasses of sheep and goats require less segmentation. Again, much higher percentages of sheep/goat bones were recorded as butchered than at Gorodishche (7%) and Georgii (5% – Figure 5.3).

The frequency of butchery marks on sheep/goat bones was fairly consistent across the three Troitsky sites. Again, bones associated with areas of high meat content were the most heavily butchered (Table 5.10; Figure 5.6). No butchery marks were observed on any of the phalanges and they were infrequently recorded on metapodials and mandibles. Horn cores, particularly from Troitsky X–XI, often showed marks associated with horn removal.

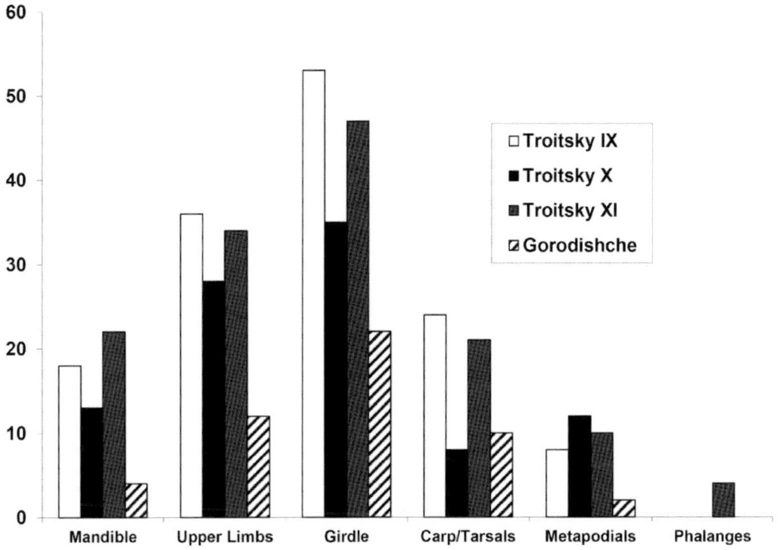

Figure 5.5: *Novgorod and Gorodishche: percentages of pig butchered elements by body area*

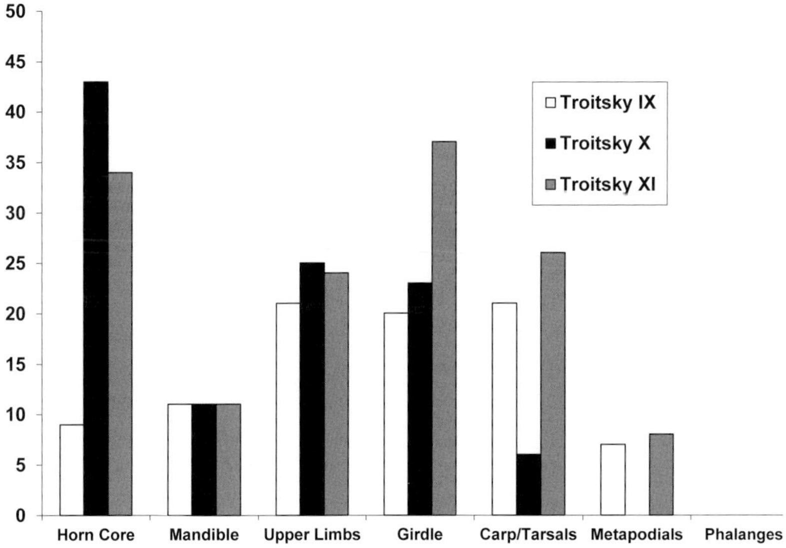

Figure 5.6: *Novgorod and Gorodishche: percentages of sheep/goat butchered elements by body area*

Horse butchery frequency

Carcass processing marks were recorded on 18% of the horse elements excluding loose teeth. They were particularly prevalent on bones from Troitsky IX but they were also commonly observed on other Troitsky sites (Table 5.11; Figure 5.3). Although their carcasses were less heavily exploited for meat and marrow than

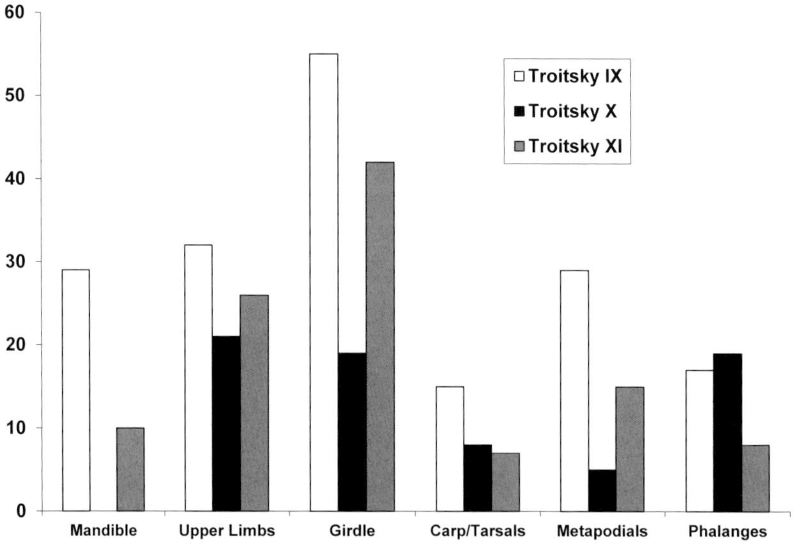

Figure 5.7: *Novgorod and Gorodishche: percentages of horse butchered elements by body area*

cattle, there is nevertheless substantial evidence that processing of horse carcasses for various products took place within Novgorod throughout the medieval period. Again, butchery marks were observed on a higher percentage of horse bones from all the Troitsky sites than in the Gorodishche (4%) and Georgii (9%) assemblages (Figure 5.3).

Fluctuations in the frequency of processing marks on horse mandibles and girdle bones on the Troitsky sites (Figure 5.7) are probably a consequence of small sample sizes plus the retention of more large bones from Troitsky IX. The high incidence of butchery marks on bones such as the scapula, pelvis, the upper limb bones and vertebrae attests to the processing of horsemeat. Some of the marks on the radius and metapodials are associated with boneworking, particularly in the creation of the burnished limb bones often interpreted as skates or runners. Skinning is evidenced on some of the marks observed on the phalanges.

Dog and cat butchery frequencies

Butchery marks were observed on 16 (5%) of the dog bones from the Troitsky sites. Again, the highest incidence was recorded on the bones retained from Troitsky IX (Figure 5.3). Only five cat bones (0.9%) were recorded with processing marks. Therefore, there appears to have been little processing of their carcasses.

Discussion of butchery frequencies

The scanning of butchery marks from the Troitsky sites revealed several consistent patterns. For most of the major domestic species, notably cattle, horse and dog, higher percentages of butchered specimens were recorded from Troitsky IX than Troitsky

X–XI (Figure 5.3) One suspects this is linked either with preferential retention of 'interesting' bones or the recovery of greater numbers of largely complete bones that had a greater chance of having butchery marks observed somewhere on the bone than in the case of more fragmentary specimens.

There was a greater incidence of butchery marks of all the main domestic species in comparison with the assemblages from the hinterland sites of Gorodishche and Georgii (Figure 5.3). Most of cattle and pig elements from Gorodishche produced lower percentages of butchered specimens than their counterparts on the Troitsky sites, irrespective of the area of the body involved (Tables 5.8–5.9; Figures 5.4–5.5). Several reasons can be postulated to explain this. The first is that the use of heavy blades and cleavers may have been more prevalent in Novgorod, creating more easily observable butchery marks. However, there is no clear evidence that such implements were more commonly utilised when the butchery marks are considered in detail. The superb surface preservation of the Novgorod bones undoubtedly made observations of butchery easier than at the hinterland sites. However, surface preservation was also good in the anaerobic deposits at Gorodishche.

It is also possible that the greater fragmentation of bones at Gorodishche may be a factor, with butchery marks being less likely to be present on small fragments. However, this theory cannot be fully tested as fragmentation data was not recorded for the Troitsky assemblage. In addition, differential fragmentation does not explain the higher incidence of butchery on small bones such as the carpals, tarsals and phalanges on the Troitsky sites, as these bones tended to be found largely complete in all assemblages. It is possible therefore that there was a greater intensity of carcass processing in Novgorod than at Gorodishche.

The preliminary survey of butchery marks from the Troitsky sites indicated that, as expected, the carcasses of cattle, pig and sheep/goat all bear abundant evidence of processing for meat and other products such as marrow, skins, bone and horn. Less expected was the high frequency of processing marks on horse bones. The implications of this will be discussed in more detail below but it clear that many horses were not simply buried after death. Their carcasses were commonly utilised for a range of products including meat. In contrast, attitudes to dogs and cats were different. Their carcasses were much less likely to be processed, although this does not necessarily mean that they were all given discrete burials and there is evidence that dogs, in particular, were occasionally skinned and butchered.

Cattle processing marks Troitsky XI

Introduction

The systematic recording of processing marks on cattle bones from selected spits from the lower layers at Troitsky XI produced a total of over 1,600 separate records, including specimens which had been subjected to multiple types of butchery. The individual records were assigned to a series of broader classification categories. These classifications follow those devised originally for the analysis of Iron Age and Romano-British butchery (Maltby 1989) with some additions. Summaries of these classifications are provided in Tables 5.12–5.35 where they can be compared with the butchery classifications from the hinterland settlements.

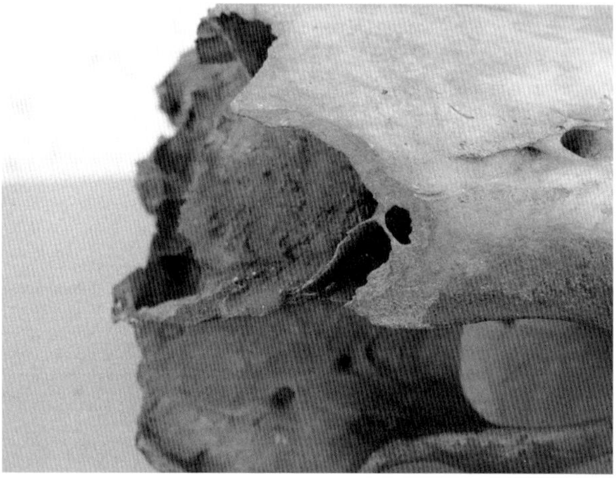

Figure 5.8: *Cattle skull from Troitsky with evidence that the horn core has been severed. Photograph by Krish Seetah*

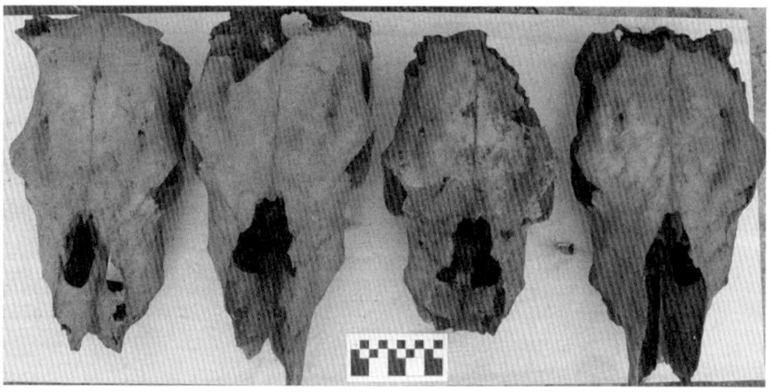

Figure 5.9: *Cattle skulls from Troitsky sites with horn core removed. Photograph by Krish Seetah*

Horn cores

Forty-seven instances of damage inflicted by heavy blades or cleavers were recorded on cattle horn cores or on the adjacent part of the skull, from which the horns had been detached. In most cases the core had been completely severed (Figure 5.8; Table 5.12). In only four cases were finer incisions noted in this area. There were no large accumulations of horn cores to suggest that they had been brought to the properties for processing but it is feasible, given the evidence for the import of other osseous materials such as elk antler (Smirnova 2001), that some cattle horns were imported. However, the impression was gained that many of the cattle carcasses were butchered within the properties and the horns routinely removed as part of this process, as indicated by the presence of many fairly complete skulls from which the horns have been removed (Figure 5.9).

Other skull fragments

Excluding horn cores, the most common butchery damage noted on cattle cranial elements were skulls that had been split open transversely across the frontal. There were at least 16 examples. Several others had superficial marks adjacent to the break in the skull. This process would have provided access to the brain. The presence of these skulls strongly suggests that the slaughter and primary butchery of cattle carcasses must have taken place regularly within the properties. This method of transverse cleaving was not recorded at Gorodishche but seems to have been practised from the earliest phases within Novgorod (Table 5.12).

Evidence of separation of skulls from the neck vertebrae using a heavy blade was witnessed on several occipital condyles, some of which were completely severed. Several other skulls had been cleaved transversely through the maxilla or premaxilla. Superficial chop marks or heavy blade marks were recorded on several other areas of the skull including the temporal/zygomatic border, where such marks were associated with separating the skull from the mandibles (Table 5.12).

Incisions with a fine blade accounted for only 39 butchery observations on cattle skulls. They were most commonly located on the premaxilla and maxilla and in the frontal/parietal areas. Many of these were probably made during skinning. Less frequently, cut marks were noted on the zygomatic, temporal and nasal bones.

Mandible

A total of 126 butchery records were made on cattle mandibles, 40% of which consisted of fine incisions whilst the remainder were heavier blade or cleaver marks (Table 5.13). Over a third of the butchery records consisted of heavy blade or superficial chop marks located on the ramus, mainly on or near the caudal or lateral aspects just beneath the condyle (Figure 5.10). Most of these marks were associated with the separation of the mandibles from the skull. Ten specimens had finer incisions in the same area, showing that knives were also used for this procedure, although there was no evidence for chronological trends in the relative frequency of these marks. At least nine specimens bore incisions on the lingual aspect of the body of the jaw indicative of the removal of the tongue (Rixson 1989) and other incisions on the medial/ventral part of the ramus may also have been associated with this process. Other incisions found quite commonly on the lateral surface of the body and the ramus are more likely to have been associated with the removal of cheek meat. In seven cases the mandible had been completely severed across the body in a similar manner to specimens from Gorodishche (Figure 5.11). Such segmentation would have released marrow from the cavities underneath the molars.

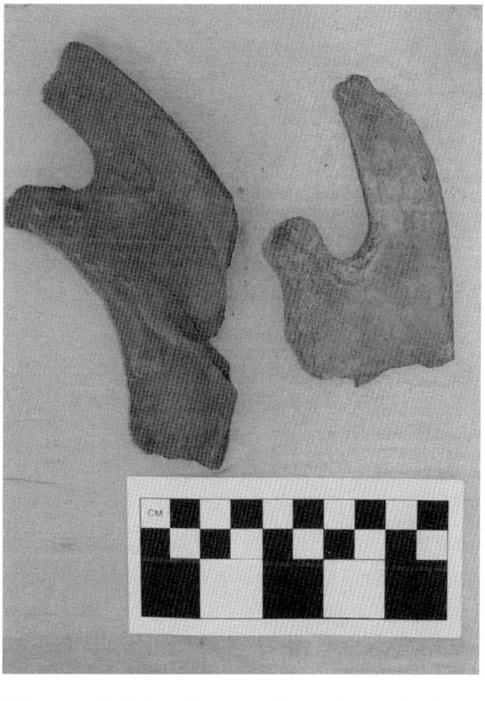

Figure 5.10: *Cattle mandibles from Troitsky with typical butchery marks on ramus. Photograph by Krish Seetah*

Figure 5.11: *Severed cattle mandible from Troitsky. Photograph by Krish Seetah*

Seven mandibles bore incisions on various parts of the diastema. Superficial chop or heavy blade marks were found in the same area of five other specimens and five had been completely severed, mirroring butchery found on some premaxillae. The presence of large numbers of butchered mandibles again indicates that many cattle carcasses were processed within the properties, as it seems unlikely that portions of previously butchered mandibles would have been imported.

Hyoid

Butchery marks were recorded in detail on 16 hyoids. Thirteen were damaged by fine incisions. Heavier blade marks were found on three other specimens. Although hyoids can be damaged when the throat is cut, most of these marks are more likely to be associated with the removal of the tongue. The presence of butchered hyoids again indicates that cattle heads were commonly processed on the properties, presumably from animals slaughtered there.

Scapula

A total of 125 separate butchery events were recorded on cattle scapulae (Table 5.14). In a number of cases several different types of mark were observed on the same specimen. The majority (79%) of the butchery records consisted of superficial chops or heavy blade marks. As at Gorodishche, by far the most common (39 examples) were marks found on scapula blades that had been completely severed transversely or obliquely. A further 20 had been severed through the neck just above the glenoid. Five scapulae had been severed transversely near the proximal (cranial) end of the blades. Ten bore superficial transverse or oblique blade marks on the scapula blade, usually on the lateral surface. Given the amount of meat attached to both the medial and lateral sides of the scapula, it is likely that segmentation would usually have been carried out with a sharp heavy blade. A cleaver may sometimes have been required to chop through the thicker neck of the bone. The evidence implies that the scapula blade may have been a joint that was separated from the glenoid and the upper forelimb. Eight specimens bore incisions on the neck of the scapula. These could have been preparatory marks for cleaving but in most cases they were found in isolation, indicating that knives were used for disarticulation from the proximal humerus in these cases. This would also account for the transverse incisions found adjacent to the glenoid cavity on three other scapulae. Longitudinal or oblique chops or deep blade marks were found on 16 glenoid cavities, usually near their edges. These also represent evidence for segmentation from the humerus.

The scapula may have subsequently been boned out, often perhaps after the meat had been preserved by salting and/or smoking. However, no clear examples of perforations in the blade were found indicative of hanging the shoulder joint. Evidence for filleting from the scapula was also quite limited. Only three specimens bore longitudinal blade marks created when slivers of bone were removed with the meat and only seven examples had fine longitudinal incisions on the medial or lateral sides of the blade. Transverse incisions found on the blades of five scapulae may also have filleting marks. Incisions associated with filleting decreased above Spit 18. They were almost absent from the Gorodishche assemblage (Table 5.14).

Humerus

Seventy-seven records of butchery were made, of which 66% were marks made with cleavers or heavy blades (Table 5.15). As at Gorodishche, the most commonly recorded type (13 examples) consisted of distal ends that had been severed transversely or obliquely when separated from the radius and ulna. Oblique chops through the medial aspect of the distal end (7 examples) may have served the same purpose. Ten humeri bore blade marks in the vicinity of the distal end providing further evidence that heavy blades were commonly used for segmentation of the forelimb. Similar marks situated on, or adjacent to, the proximal end were recorded much more infrequently (3 examples) but this mainly reflects the vulnerability of this end of the bone, which had often been destroyed by dog gnawing. Transverse chop and heavy blade marks were found on seven humerus shafts including several that had completely severed the bone. This would have allowed access to the marrow cavity and it is likely that these marks were made when the humerus was broken down to a suitable size for pot boiling. Only four humeri were recorded with longitudinal splits which had completely opened the marrow cavity. However, many others bore spiral fractures indicating breakage for marrow extraction and pot boiling.

Fine incisions on humeri mainly occurred near the distal end (12 examples). Several of these were found on bones that had been severed and were probably preparatory marks that removed the soft tissue prior to cleaving. In a few cases, however, the marks were located in positions that indicated that the ligaments had been severed, enabling the separation of the humerus from the radius and ulna without using a cleaver. Most of the fine incisions (11 examples) on shafts were associated with filleting, as were marks made by heavier blades on three further specimens. Fine incisions were encountered much less frequently in the Gorodishche sample (Table 5.15), either reflecting better survival of such marks in Novgorod or more intensive filleting. There were no clear chronological variations in butchery patterns on cattle humeri in the Troitsky XI assemblage.

Radius and ulna

A total of 68 records of butchery on cattle radii were made, of which 78% consisted of chop or heavy blade marks (Table 5.16). Such marks were found on or near the proximal end of 16 specimens including seven that had severed the bone. These correspond to those found on the distal humerus and were made during segmentation. Five radii bore fine incisions on the medial aspect of the proximal end also made during disarticulation. One radius was chopped transversely through the distal end. Chop/heavy blade marks were also commonly recorded on the shaft. Eleven of the 21 examples had completely severed the shaft and most of the superficial marks were located adjacent to the break (Figure 5.12). These could be where the upper forelimb was separated from the foot during initial butchery. Alternatively, they could also have been made during the breaking up of the bones for pot boiling. This type of butchery was not recorded on radii from Gorodishche (Table 5.16). Only seven examples of longitudinally split radii were recorded in the Troitsky sample. Filleting marks made with heavy blades were recorded on six specimens. Some of the 13 incisions found on various parts of the shaft could also have been made during filleting, although some located relatively close to the proximal end are more likely to have been made during disarticulation.

Figure 5.12: *Severed cattle radius from Troitsky. Photograph by Krish Seetah*

Only 17% of the 52 detailed butchery records of ulnae consisted of fine incisions (Table 5.17). Twenty-eight ulnae had been severed through, or just above, the proximal joint surface, corresponding with marks on the distal humerus and proximal radius, confirming that this joint was commonly severed with a cleaver. Superficial chop or heavy blade marks in the same area were also observed on three ulnae. Chop marks around the centre of the shaft were found on eight specimens, corresponding to those observed on the shafts of radii. These were also absent from the Gorodishche assemblage, indicating a change in carcass processing. Five of the nine records of fine incisions on ulnae from Troitsky XI were located on the lateral aspect of the olecranon, on or adjacent to the proximal articular surface. Two other ulnae had similar marks on the medial surface. These were also made during disarticulation.

Pelvis

Butchery marks were frequently observed on pelvis fragments, as is to be expected from this area of high meat content. Chop and heavy blade marks accounted for 87% of the 134 butchery classifications (Table 5.18). Seventy-two specimens had been completely severed, mainly, as at Gorodishche, on or around the acetabulum (35 examples). Superficial marks in this area were recorded in a further ten cases. Other areas of severance included the shaft of the ilium (11 examples); the pubis (7), the shaft of the ischium (12) and the ischial tuberosity (4). Superficial chop or heavy blade marks were also found in all these locations. Twelve of the 16 chop and heavy blade marks recorded on the sacro-iliac joint had severed the joint. Cleavers were clearly used commonly to separate the pelvis from the upper hind limb and the vertebral column. Most of the specimens with incisions were found in the lowest spits and were located mainly on the shaft of the ischium. These may have mainly been filleting marks but their locations and forms were quite variable.

Femur

A total of 76 butchery records included a higher incidence of fine incisions (41%), although these were still outnumbered by chop and heavy blade marks (Table 5.19). The marks were quite diverse with no particular types dominating. Superficial chop and heavy blade marks on various parts of the shaft were the most commonly observed (10 examples). Most of these were located near breaks in the shaft and were probably made during bone breakage for pot boiling. Seven shafts had been completely severed during the same process. Eighteen femora bore fine incisions on disparate parts of the shaft running in various directions. Most of these probably derived from filleting, as did the shallow skim marks found on six shaft fragments, which became more prevalent in later deposits. Five femur heads had been completely severed from the pelvis but superficial chop and deep blade marks around the proximal end were rarely encountered. In contrast, seven specimens were recorded with incisions on various aspects of the proximal end. A similar diversity was encountered at the distal end, although heavy blade marks were the most common in this zone. Longitudinal splitting was recorded on five specimens. Similarly split bones were also encountered at Gorodishche.

Tibia

Fifty-one butchery records were made, of which a high proportion (45%) consisted of fine incisions (Table 5.20). All bar one of these were located on various parts of the shaft or near the distal end. Some of the latter were disarticulation marks but the majority were made during meat removal. Filleting marks made with a heavier blade were found on five specimens, including some from early deposits. Twelve tibiae had been completely severed; three at the distal end, one through the proximal end and eight through the shaft. The latter may again be associated with the preparation of bones for pot boiling, as do the superficial chop marks found on three other shafts near to where they were broken. At least six tibiae had been longitudinally split for marrow extraction.

Carpals

Butchery evidence was recorded on 16 specimens, mainly on radial and intermediate carpals. In ten cases these consisted of fine transverse or occasionally oblique incisions on various aspects of the bones. Deeper transverse blade marks were found on two other carpals. All of these would have facilitated the separation of the metacarpal from the upper forelimb. In four cases longitudinal chop marks were recorded. These had severed the bone completely in two cases. It is probable that these carpals were split whilst still attached to the radius when it was split open for marrow.

Astragalus

Most (60%) of the 43 butchery classifications consisted of fine incisions (Table 5.21). Most of these were located on the anterior, either across the central area (10 examples) or towards the distal end (7 examples). They were also quite commonly located on the medial surface (7 examples) and occasionally on the posterior and lateral sides (1 example each). These were all associated with disarticulation of the upper hindlimb from the feet at the hock joint. Heavier transverse blade marks were found on five examples and one had been cleaved through transversely but it would appear that knives were the implements most commonly used to remove the feet, mirroring the evidence from the carpals. Curiously, no incisions were observed on any of the astragali from Gorodishche.

Longitudinal chops were recorded on nine astragali, in six cases splitting the bone into front and rear portions. Such marks may have been inflicted when the tibia was split open whilst the astragalus was still attached.

Calcaneus

Most (72%) of the 29 butchery classifications consisted of fine incisions (Table 5.22), mainly on the clacaneal tuber. These marks were also made when the feet were detached from the upper hindlimb. The other butchered specimens had heavier blade marks located on various parts of the bone including two that had completely severed the bone.

Centroquartal

Eight of the ten butchered specimens bore knife cuts, usually on the anterior surface (Table 5.23). These were disarticulation marks made when separating the hock joint from the metatarsal. The other two had been longitudinally split, possibly when still attached to the metatarsal.

Metacarpal

Only 19% of the 75 detailed butchery records on metacarpals consisted of fine incisions (Table 5.24). Relatively few of the butchery marks were located at the proximal end. These included four specimens with fine incisions and four others with deeper blade marks, all made during disarticulation from the upper forelimb. In contrast, five metacarpals had been severed through the distal articular surface during separation from the phalanges and there were 11 cases where superficial cleaver or heavy blade marks were located in the same area. Two other specimens had finer incisions across the distal end. Nine longitudinally split specimens were recorded. They were therefore found relatively less frequently than at Gorodishche and most of the split specimens came from the lowest spits, perhaps indicating that the practice became less common in the later periods. Metacarpals were much more commonly broken transversely, sometimes with the aid of a cleaver. Fifteen shafts had been completely severed and superficial chops were recorded on nine specimens. Seven metacarpals had knife cuts on various aspects on the shaft. These were generally not located near breaks and were therefore not preparatory cuts prior to cleaving. There is little meat adhering to these bones, so they are unlikely to be filleting marks. It is probable that many of them were made at the onset of skinning.

Metatarsal

Butchery classifications were recorded on 112 metatarsals, of which 27% consisted of fine incisions (Table 5.25). The overall pattern of butchery was very similar to the metacarpals. Marks on or near the proximal end (8 with incisions and 4 with heavier cuts) were relatively uncommon compared with the distal end, where 14 metatarsals had been completely severed and superficial chop marks were also recorded on or near the distal end of ten other specimens. Most of the 16 metatarsals with longitudinal splits were found in the lowest layers but they were not as prevalent as at Gorodishche. Transverse breakage was much more common and cleavers had completely severed the shafts of 18 metatarsals, although relatively few of these were found in the lowest spits. Superficial chops, usually located near the breaks were also recorded in 18 cases, again mainly from Spit 17 and above. Fourteen metatarsals had fine horizontal incisions on the shaft, many again probably associated with skinning.

Metapodials

Butchery marks were found on a further ten small fragments of distal metapodials that could not be further identified. The marks were similar to those observed on more complete metacarpals and metatarsals (Table 5.26).

Phalanges

Fifty-seven classifications of butchery were recorded on first phalanges. The most common of these (17 examples) were heavy blade or superficial chop marks on the posterior of the proximal end. These would have facilitated the separation of the foot from the distal metapodials. This type of mark was not encountered in the Gorodishche assemblage. Three further examples had similar marks on the peripheral aspect of the proximal end. There were only eight examples of first phalanges that had fine incisions on the shaft. These signify where initial cuts of the skin were made prior to stripping the skin from the carcass. Their rarity strengthens the case that the incisions commonly found on the shafts of metacarpals and metatarsals were associated with this procedure. Eight first phalanges bore blade marks that had removed small slivers of the shaft as the blade was run longitudinally down the bone. These marks were not observed in the Gorodishche assemblage (Table 5.27). Their purpose is probably associated with skinning. If the initial incisions made to free the skin were commonly made on the metapodials, a blade could have been used to free the skin from the phalanges either before or after they had been disarticulated from the metapodials. Twelve first phalanges bore fine incisions on various aspects adjacent to the proximal end. These could also conceivably have been skinning marks but are more likely to have been made during the separation of the phalanges from the metapodials.

Only 19 detailed butchery records were recorded on cattle second phalanges (Table 5.28). Twelve of these consisted of fine incisions, mainly located on the peripheral aspect of the proximal end (6 examples) or on the anterior aspect of the shaft (3 cases). The former are likely to have been made during disarticulation from the first phalanges; the latter were probably associated with skinning. Heavy blade marks were found on seven second phalanges including three running longitudinally, the latter probably also associated with skinning.

Butchery marks were recorded on seven third phalanges, three of which consisted of fine incisions and the remainder of heavier blade marks. The incisions were all located near the articulation with the second phalanx and indicate the separation of the two bones. The blade marks all trimmed the edges of the bone, suggesting that they were made when the skin and the hoof were removed.

Vertebrae and ribs

Twenty-seven detailed records of butchery were recorded on cattle atlases (first cervical vertebrae). Eighteen of these consisted of dorsal-ventral chops through the bone running cranio-caudally. In six cases these was fairly central; in four cases the splitting was slightly off-centre and in eight cases the blade had merely trimmed the edge of the bone (Table 5.29). These marks imply that the atlas was still attached to other cervical vertebrae when the neck was split into sides. Two specimens bore fine transverse incisions made when the animal was beheaded and two other examples had been chopped through transversely during the same process.

Transverse chops were the most common butchery marks observed on the axis, providing eight of the 19 classifications recorded (Table 5.30). All the other records consisted of specimens that had been split axially when the neck was divided into two sides. Most of these marks were located slightly off-centre. A similar pattern of segmentation was observed on other cervical vertebrae (Table 5.31). Thirty-four of the 72 butchery classifications consisted of transverse chops through the bone where the vertebral column had been segmented. A total of 25 cervical vertebrae had been split cranio-caudally, mainly off-centre. Five had been chopped through horizontally. Only one record of a fine incision was recorded but superficial chop marks or heavy blade marks were observed on various parts of seven vertebrae.

Thoracic vertebrae were butchered in the same way as cervical vertebrae. Of the 58 butchery records, 21 consisted of transverse chops through the bone and 34 consisted of cranio-caudal chops through the bone where the carcass had been split into two sides. Again, only in seven cases was this split down the midline (Table 5.32).

Only four of the 78 butchery records on lumbar vertebrae consisted of fine incisions (Table 5.33). Transverse chops through the bone (29 examples) were common and cranio-caudal chops through the bone, mainly on the edge of the body were observed on 39 specimens. Superficial chop marks were recorded on six specimens (Figure 5.13).

Cranio-caudally chops mainly located near the lateral aspect of the bone accounted for 16 of the 23 observations of butchery on cattle sacral vertebrae. Transverse chops through the bone were observed on six specimens (Table 5.34).

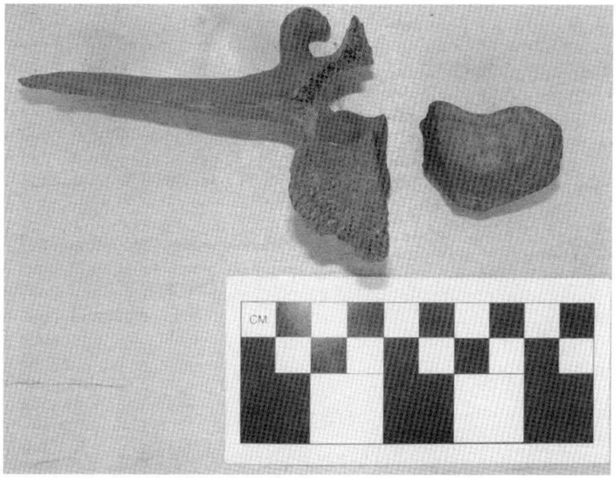

Figure 5.13: *Examples of severed cattle lumbar vertebrae from Troitsky sites showing splitting down the midline (left) and splitting off-centre (right). Photograph by Krish Seetah*

The butchery evidence on the vertebrae indicated that the trunks of cattle were often split into sides but usually not symmetrically. Chopping through the midline, where the vertebrae are at their thickest, would have taken more effort than splitting the carcass slightly off-centre. The transverse cleaving of vertebrae indicates that the backbone was routinely segmented into sections.

Thirty of the 50 butchery records on ribs consisted of chop or heavy blade marks on and around the dorsal end, in 18 cases severing all or part of the rib head (Table 5.35). These marks correspond to those found on the lateral parts of thoracic vertebrae. Breakage of the ribs through the shaft using a heavy blade was encountered on ten specimens. Three other shaft bore heavy blade marks on their medial surface. Such segmentation is likely to have been linked with pot boiling. Fine knife cuts were observed on only seven of the ribs recorded in detail. These were also mainly located on the medial aspect of the shafts.

Pig processing marks Troitsky XI

Introduction

The detailed recording of butchery marks on pig bones from Troitsky XI produced over 485 records, including specimens that had multiple records. The individual records were again assigned to broader classification categories. Summaries of these classifications are provided in Tables 5.36–5.58. Because of the smaller sample, the butchered pig bones from all the spits examined are treated together.

Skull

There were 31 butchery records. As with cattle, the most common were skulls that had been transversely chopped through across the frontal bone giving access to the brain (Table 5.36). In four further cases splitting had been achieved by chopping along the midline of the skull. Nine maxillae or premaxillae had been severed transversely to remove the front of the skull. A zygomatic and an occipital condyle had also been severed, the latter during the act of beheading. Only five records of fine incisions were made, three on the maxilla (probably associated with the removal of cheek meat) and one each on the zygomatic and temporal (associated with separation from the mandible). Deeper skims made with a blade were found on a maxilla and parietal.

Mandible

Sixty-five butchery events were recorded (Table 5.37), of which the most common consisted of mandibles that had split into two halves along the diastema (13 examples). Two other examples had been severed transversely across the diastema. Superficial chop marks were found on various aspects of three other diastemas. Nine mandibles had been severed through the body usually near the ramus. Ten further examples of superficial chop or heavy blade marks were observed on various parts of the ramus, including four located near the condyle, made when the skull was detached. Superficial chops were also occasionally found on either side of the cheek teeth.

Finer incisions accounted for 38% of the butchery classifications on pig mandibles. Most of these were located on various parts of the ramus (9 examples) and diastema (8 examples). Cuts beneath the cheek teeth were occasionally observed on the buccal (5 examples) and lingual surfaces (3 examples), signifying the removal of the cheek meat and the tongue respectively.

Scapula

A total of 56 records of butchery were made (Table 5.38). As with cattle, the most common were scapulae that had been severed across the blade (13 examples). Transverse chops through the neck (3 examples) and the proximal end (1 example) were observed more rarely. Two scapulae had longitudinal chops through the glenoid, also associated with segmentation. The scapula was therefore divided into joints. At least two specimens bore perforations in the blade where a hook had been inserted for hanging up the shoulder joint, probably for smoking. Superficial chop or deep blade marks were recorded on 13 scapulae. Nine of these were also made during segmentation of the blade; the other four (located on the neck above the glenoid) were made during separation from the humerus.

Eighteen (32%) of the butchery records on the scapula consisted of fine incisions. Eight of these were situated on the rim of the glenoid cavity or on the neck just above it. These have been interpreted as disarticulation marks. The remainder were located on various aspects of the blade, including six running longitudinally, which were made during filleting. Three specimens bore deeper skims on the lateral aspect of the blade, also indicative of meat being stripped from the bone.

Humerus

A wide range of butchery marks were also observed on pig humeri (Table 5.39). Of the 66 records, 22 consisted of transverse or oblique chops that had severed the bone either on or near the distal end (9 examples), the shaft (12) or the proximal end (1). These were all associated with segmentation of the forelimb. In two cases the distal end had been split longitudinally, which would have severed the joint and allowed access to marrow. Two specimens bore superficial chop marks near the distal end or on the shaft usually adjacent to places where the bone was severed.

Fine incisions accounted for 45% of the humerus butchery classifications, including 19 situated on various parts of the shaft and ten near the distal end. The former were mainly associated with filleting and the latter with disarticulation from the radius and ulna.

Radius and ulna

Only ten butchery records were made on pig radii (Table 5.40). Five of these consisted of shafts that had been completely severed. Three others had superficial chop marks located near to where the shaft had been broken. Only two bones had evidence of fine incisions.

Butchery marks on pig ulnae were more common (Table 5.41). Eight of the 25 butchery records were associated with the severing of the olecranon or proximal joint surface when the ulna was separated from the humerus. Three had been severed through the shaft. Superficial chop marks were occasionally recorded both at the proximal end and on the shaft. Fine incisions accounted for 40% of the butchery classifications. Six of these were located on the lateral or medial aspects of the olecranon where ligaments had been cut during disarticulation. Two were found on the proximal articulation and two on the shaft.

Pelvis

Butchery marks were frequently recorded on the pelvis. Many of the 45 records consisted of bones that had been completely severed. Locations included the sacro-iliac joint (4 examples); the ilium shaft (5); the acetabulum (8); the pubis (2); the ischium shaft (4) and the ischial tuberosity (2). Superficial chops or heavy blade marks were also occasionally found in some of these zones (Table 5.42). A third of the butchery observations consisted of fine incisions, most commonly located on the shafts of the ischium (8 examples) and ilium (5). The abundance of butchery marks on the pelvis reflects the high meat content of this area of the carcass.

Femur

Fine incisions were more commonly found on the femur, accounting for half of the 44 butchery observations (Table 5.43). Most (17) were found on the shaft and were associated with filleting. Two femoral heads had been completely severed and superficial chop marks were found on a third specimen. In contrast, seven femora had been severed around the distal end and four others had been chopped through the shaft during segmentation of the hindlimb. Several superficial chop marks were recorded in the same zones.

Tibia and fibula

Sixteen of the 40 butchery records on pig tibiae consisted of chops through the shaft and it is clear that the upper limb was commonly severed at this point. Four other tibiae bore superficial chops in the same area and a corresponding chop was found on the shaft of a fibula (Table 5.44). Two proximal and two distal ends had also been completely severed. Fine incisions accounted for 40% of the butchery records on the tibia. Ten of these were found in central parts of the shaft and six near the distal end. Most of those on the upper part of the shaft are likely to have been filleting marks. Two fibula shafts also bore fine incisions. Heavier skims on another tibia were probably also made when meat was stripped from the shaft. One tibia had been split open longitudinally to release the marrow.

Tarsals

All but three of the 18 butchery records on pig astragali consisted of fine incisions located mainly on the anterior aspect on the central (10 examples) or distal part of the bone (3). Cuts were also occasionally observed on the medial and lateral aspects of the bone (Table 5.45). All of these would have facilitated the separation of the feet from the upper limbs. In two cases heavier blade marks were encountered running across the anterior aspect. One astragalus had been longitudinally split.

Butchery marks were recorded on only five pig calcanei. Four bore knife cuts, usually on the calcaneal tuber. The fifth had a blade mark than had removed a sliver of bone from the tuber (Table 5.46). All the marks were associated with the disarticulation of the hock.

Metapodials and phalanges

Only four metapodials were recorded as butchered. All the marks consisted of fine incisions associated with disarticulation (Tables 5.48–5.49). Superficial longitudinal blade marks were recorded on a first phalanx (Table 5.50).

Vertebrae and ribs

Sixteen detailed butchery records were made on the atlas (Table 5.52). Six had been severed transversely during beheading. One had been severed longitudinally towards its lateral border. The remaining marks consisted of fine incisions. Five of these ran transversely and were also likely to have been made when the skull was detached.

An axis had been chopped through axially but not exactly along the centre of the bone (Table 5.53). Five thoracic vertebrae and a lumbar vertebra had been severed in a similar manner when the carcass was split into roughly two halves (Tables 5.55–5.56). Three lumbar vertebrae had been severed transversely during segmentation of the hindquarters. Superficial chops and a fine incision were found on other specimens.

Many pig ribs displayed butchery marks. Fifteen of the 39 records consisted of chop marks near the dorsal articulation (Table 5.58). These were inflicted during separation from the thoracic vertebrae. Six ribs had been severed through the shaft and heavy blade marks were observed on the medial aspects of five shafts, indicating further segmentation often took place once the ribcage had been separated. Fine incisions accounted for a third of the recorded butchery marks, occurring both near the dorsal end (7 examples) and on the medial and lateral aspects of the shaft.

Comparisons with pig butchery marks at Gorodishche and Georgii

Tables 5.36–5.58 allow butchery records of pig bones from Troitsky XI to be compared with those from the hinterland sites. Detailed comparisons are limited by the small sample sizes from the latter. More types of butchery marks were recorded in the larger Troitsky sample. Common butchery marks that were not found in the Gorodishche sample include maxillae and premaxillae that have been severed vertically (Table 5.36). However, most of the common types were also recorded in small numbers at Gorodishche and butchery methods were generally very similar. Knife cuts were found in relatively greater numbers in the Troitsky sample. Excluding vertebrae and ribs, incisions accounted for 41% of the butchery observations from Troitsky compared with 21% in the Gorodishche assemblage. Better surface preservation of the bones allowing clearer visibility of cuts probably partly accounts for this but it may also reflect more intensive exploitation of pigs in Novgorod (Tables 5.36–5.58).

Sheep/goat processing marks Troitsky XI

Introduction

The small sheep/goat assemblage meant that only 184 detailed records of butchery marks were made on their bones from Troitsky XI, including bones that had multiple records. The individual records were assigned to the same broad classification categories as cattle and pig (Tables 5.36–5.58).

Cranial elements

Cleaver marks were observed near the bases of three horn cores, one of which belonged to a four-horned sheep and one to a goat. Horn-working was probably commonly practised but there is no evidence for large-scale processing of sheep and goat horns within these properties. Three sheep skulls had been split open axially to remove the brain. Deep blade marks and finer incisions running transversely were observed on a sheep parietal. A maxilla also bore knife cuts (Table 5.36).

Six of the nine butchery marks recorded on sheep/goat mandibles consisted of fine incisions (Table 5.37). Three of these were located on the ramus, one on the diastema and one on the buccal (outer) aspect of the body of the jaw. Two specimens bore superficial chop marks near the mandibular condyle made during separation from the skull. A knife cut on a hyoid indicated tongue extraction.

Scapula

Fine incisions accounted for ten of the 18 butchery records on sheep/goat scapulae (Table 5.38). Three of these were located near the glenoid and were disarticulation marks. The remainder were found on various aspects of the scapula blade and were mainly made during filleting. Four blades had been severed transversely or obliquely in a similar manner to that more commonly observed on cattle and pig scapulae. Two had been severed through the neck of the bone during separation from the humerus.

Humerus, radius and ulna

The 12 records of butchery on humeri included chops through two shafts, one distal end and one proximal end (Table 5.39). Six records of fine incisions were also made on various parts of the bone. Heavier blade marks associated with filleting were observed on two specimens. Only six radii bore butchery marks of disparate types and fine incisions were observed on four ulnae (Tables 5.40–5.41).

Pelvis

Nine of the 18 butchery classifications consisted of bones that had been severed during segmentation of the hip joint (Table 5.42). Four of these were located on the iliac tuberosity, three on the pubis and two on the ischium. In contrast to cattle and pig, none of the acetabula had been severed. Superficial chop marks were found on two other ilium shafts. Only three records of fine incisions were made.

Femur and tibia

Fourteen butchery records on sheep/goat femora included nine consisting of fine incisions located on various parts of the bone (Table 5.43). Chop marks were found near two proximal ends. One distal end had been completely severed as had one shaft but such segmentation was much less common than observed on cattle and pig femora, presumably because of the smaller carcass sizes of sheep and goat.

Two-thirds of the butchery records on sheep/goat tibiae consisted of horizontal fine incisions either on the shaft or near the distal end (Table 5.44). The latter may have been mainly associated with disarticulation of the ankle joint. However, some of these cuts and most of the incisions on the distal part of the shaft could have been initial skinning marks. Only four tibiae had been severed, one through the distal end and three through the shaft. A few superficial chop marks were observed in the same areas.

Tarsals

Sheep/goat astragali were unusually well represented in the Troitsky assemblage in comparison with other small bones (Tables 4.43–4.46), perhaps reflecting their acquisition and use as artefacts. There is abundant evidence for butchery on them. Nineteen records of butchery were made, mainly consisting of incisions located on the anterior of the bone made during disarticulation from the tibia. Occasionally knife cuts were observed on other surfaces (Table 5.45). No detailed butchery marks were recorded on calcanei and the only mark observed on the rarely recovered centroquartals consisted of a transverse incision on the anterior aspect made when detaching the metatarsal (Table 5.47).

Metapodials and phalanges

Chop marks were found on the shafts of three metacarpals (Table 5.48). In two cases these bones had been subsequently worked. Two of the six metatarsals with butchery marks (Table 5.49) also showed evidence of working. Normal processing generally did not produce butchery marks on metapodials. No marks were observed on any of the rarely recovered sheep/goat phalanges.

Vertebrae and ribs

Three sheep/goat atlases were recorded with butchery, all consisting of incisions associated with beheading (Table 5.52). By contrast, four of the five butchery marks recorded on axes consisted of chop marks that had severed the bone either transversely (3 examples) or axially (Table 5.53). Transverse cleaving also accounted for eight of the 11 butchery records on other cervical vertebrae indicating segmentation of the neck into joints (Table 5.54). Three of the six butchery marks on thoracic vertebrae consisted of chops running axially through the vertebral body towards the lateral part of the bone (Table 5.55). Three lumbar vertebrae had been butchered in a similar manner (Table 5.56). As in the case of cattle and pig, splitting of the trunk was not exactly symmetrical. Five rib heads had also been chopped during this process (Table 5.58). Five other ribs bore fine incisions in the same area. Three shafts had been completely severed and knife cuts were observed on the medial aspects of four shafts during subsequent butchery of the ribcage.

Horse processing marks Troitsky XI

Introduction

As discussed previously, butchery marks were observed on an unexpectedly high percentage of horse bones from the Troitsky sites (Table 5.11). However, only 19 butchery marks were observed in the assemblage selected for detailed recording from Troitsky XI. To supplement these data, nine butchery marks from bones recorded in detail in previous seasons from Troitsky IX and X have been added. The individual records were assigned to the same broad classification categories as the other domestic mammals (Tables 5.36–5.58).

Cranial elements

Horse cranial elements were poorly represented and the only specimen on which butchery marks were observed was a mandible from Troitsky XI Spit 21 bearing three axial knife cuts on the lateral aspect of the ramus just behind the third molar.

Scapula, humerus and radius

Four horse scapulae, all from early deposits in Troitsky XI (Spits 20–22) bore butchery marks. Two had fine incisions located on the medial aspect of the blade or neck. Two other specimens bore deeper blade marks in the same areas. The marks were made during the disarticulation of the scapula from the humerus and possibly the filleting of meat (Table 5.38). A humerus from Spit 21 bore knife cuts on both the posterior and medial aspects of the shaft indicative of filleting.

A radius from Troitsky XI Spit 18 had been chopped through the distal end during segmentation of the feet. A radius from Troitsky IX Spit 13 bore both blade marks and finer knife cuts on the shaft. Although it was not completed, these marks may represent evidence for initial preparation of this bone for tool manufacture.

Pelvis, femur and tibia

An ischium fragment from Troitsky XI Spit 21 bore fine incisions on two aspects of the shaft. A femur from Spit 20 bore six incisions on the medial aspect of the shaft. Filleting is the most likely explanation for these marks.

The distal half of a tibia from Troitsky XI Spit 18 bears longitudinal blade marks that have smoothed the bone. These are reminiscent of marks on some of the radii and metapodials that have been worked. Distal tibiae are less suitable for skates but the marks could be related to burnishing of some sort, as Zinoviev argues below. Other processing marks recorded on horse tibiae are shown in Table 5.44. A specimen from Troitsky IX Spit 13 bore a superficial chop mark and had then been completely severed just above the distal end in a manner similar to that commonly encountered in cattle butchery. A fine incision was found on the shaft of a distal tibia from Troitsky X Spit 14 and from the same spit another tibia bore a deeper blade mark on the shaft.

Tarsals

Troitsky XI Spit 20 produced a small associated bone group consisting of horse tarsals and metatarsals. Both the astragalus and calcaneus in this group bore cleaver marks made when the foot was detached from the upper limb. The proximal end of the astragalus had been completely severed and superficial marks made in the same action were found on the calcaneus. Superficial chop marks were found on another astragalus in Troitsky XI Spit 15 (Table 5.45).

Phalanges

Two first phalanges bore processing marks. Fine incisions were located on the lateral aspect of the distal end of a specimen from Troitsky X Spit 16. These may have been skinning marks but they would also have facilitated the separation of the phalanges. A group of three phalanges was found in Troitsky XI Spit 15. A crude chop mark was found on the medial surface of the shaft of the first phalanx, perhaps made when the metapodial was detached. In the same spit the front of a third phalanx had been chopped off presumably when the hoof was removed. Another third phalanx from Troitsky X Spit 16 had been chopped through transversely near the proximal end.

Vertebrae

An axis from Troitsky XI Spit 18 and a thoracic vertebra from Spit 21 had both been severed transversely and axially towards the lateral side of the body in a manner identical to that commonly encountered on vertebrae of the other domestic species (Tables 5.53 and 5.55).

Dog processing marks Troitsky X–XI

Eight records of butchery were recorded on dog bones examined in detail. These again included some bones from Troitsky X. A mandible from Troitsky X, Spit 18 bore a knife cut on the buccal aspect below the teeth, perhaps indicating removal of flesh, although such marks can also be produced during skinning. A humerus from Troitsky X, Spit 16 had an oblique knife cut on the medial aspect situated just above the distal end. Such marks are usually associated with disarticulation from the radius and ulna. Ulnae from Troitsky X, Spit 16 and Troitsky XI, Spit 19 both bore fine incisions on the lateral surface of the olecranon made during disarticulation from the humerus. A humerus from Troitsky XI, Spit 11 and a radius from Spit 12 had been severed through their shafts. A fibula from Troitsky XI, Spit 18 had been trimmed along its shaft with a blade. These marks may have been associated with boneworking. Finally, an atlas from Troitsky XI, Pit 170 bore an oblique knife cut on its ventral surface, possible associated with beheading.

Cat processing marks Troitsky XI

Evidence for butchery marks on cats was very rare and it is clear that most cat carcasses were not processed. However, cuts were observed on five bones from Troitsky XI. A skull from Property G, Spit 21, bore knife cuts on the frontal above the orbits, clearly indicating skinning. A knife cut on the front of a mandible from Property G, Spit 14 may also have resulted from skinning. A radius from the same spit had a knife cut on the distal end. This could indicate the removal of the feet possibly with the skin attached. Two femora from Spit 12 had both been chopped through the shaft. These were not the result of skinning and indicate that the carcasses had been dismembered. In one case, there is evidence of polishing, suggesting that the bone may have been worked. This may also account for the severing of the second femur.

Comments on worked bones from Desyatinny-1 (Andrei Zinoviev)

Horn cores where removed from skulls of cattle (32) and goats (4) prior to butchery. Whether they were removed to get horns themselves is not clear, since the number of butchered cores was too low to establish the presence of a horn-working industry in the area covered by Desyatinny-1.

Within the assemblage were slightly worked and polished long bones of horses and astragali of cattle. Polished astragali of cattle have been interpreted as burnishers (Figure 5.14), which are suitable for holding in hand. Worked long bones (radii and metapodials) of horses have often been considered as skates or sled runners (see for example Maltby 2017a). However, there are problems with such interpretations.

Besides not having traces for fastening, the polishing does not conform to that produced by snow or ice. The character of polished surface suggests that they could also have been burnishers, used for more extensive surfaces (Figure 5.15).

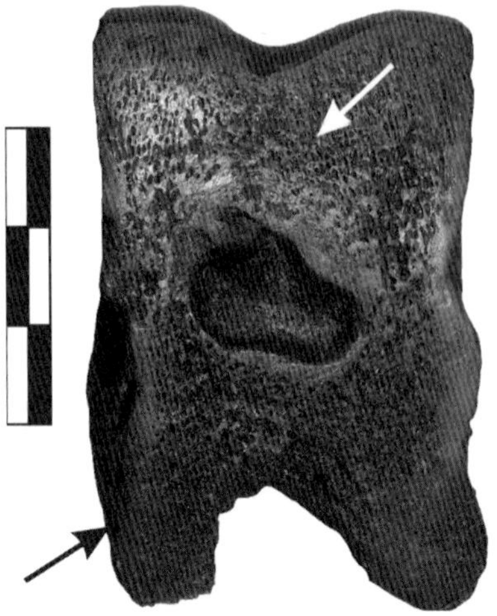

Figure 5.14: *Burnished cattle astragalus from Desyatinny-1. Photograph by Andrei Zinoviev*

Carcass processing summary (Mark Maltby)

As expected, the detailed analysis of the butchery marks on domestic mammal bones from Novgorod and the hinterland sites confirmed the intensity of carcass processing of the major food species indicated by the frequency of marks observed in the general scan of the material. Generally similar methods of carcass processing were employed at all these sites. Heavy blades and fine knives were both commonly used in processing and there were consistent methods of butchery that seem to have had longevity. Carcass divisions were strikingly similar in the cattle and pig assemblages in particular. These include the practice of severing the shafts of limb bones, some of which would have been associated with pot-boiling as well as primary segmentation. Scapula blades were commonly severed and skulls were split in similar ways to access the brain.

There were some variations, however. For example, relatively more cattle metapodials were longitudinally split to remove marrow in the lowest spits of Troitsky XI, where they comprised 22% of the butchery observations, compared with 9% in Spits 17 and above. Over 55% of the butchery marks on cattle metapodials at Gorodishche consisted of split bones. This implies that this practice became less popular in the later phases when transverse breakage became more common (Tables 5.24–5.25).

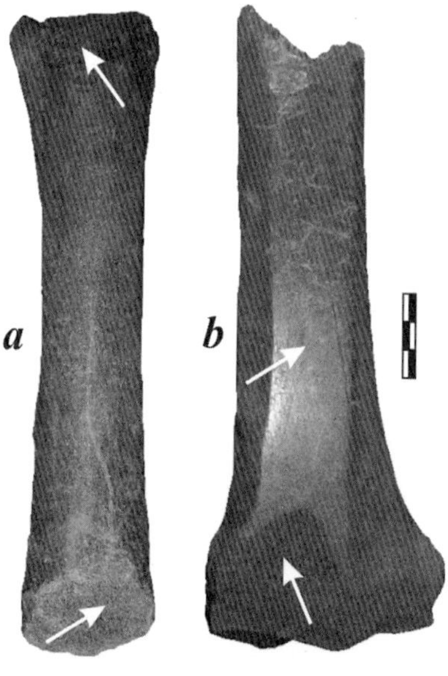

Figure 5.15: *Examples of polished horse metapodials and radius from Desyatinny-1. Photograph by Andrei Zinoviev*

There was little evidence in the Troitsky assemblage for accumulations of bones from different stages of processing, although this may emerge with more detailed recording of spatial variations of bones within properties. The impression gained is that most of the processing was carried out within the properties. Whether this involved specialist butchers is uncertain. This contrasts with Gorodishche where there is some evidence for mass processing of cattle carcasses in the South bank area.

-6-

METRICAL ANALYSES OF DOMESTIC MAMMAL BONES FROM NOVGOROD AND SITES IN ITS IMMEDIATE HINTERLAND

Mark Maltby

INTRODUCTION

This chapter will focus on the metrical data of domestic mammals from the sites studied in this project. Measurements were routinely recorded on bones from the scanned assemblages from the Troitsky sites in Novgorod as well as on the bones recorded in detail from the hinterland sites. Cattle bones obtained from the Fedorovsky site were also measured and these included a large group of metapodials that had been selected for bone-working. Most of the measurements follow those recommended by von den Driesch (1976). The individual measurements are stored in the archive. Summaries of the metrical data for each of the major domestic mammals are provided in the tables. Graphs of the most common measurements are included in this chapter. The assemblages from the Troitsky sites were divided up for analysis in the same way as for the analyses of species representation and ageing (Chapter 4). This allowed for comparisons between the assemblages from the lower levels of Troitsky IX and X and the lowest, middle and upper spits of Troitsky XI. For some species, the samples from the middle and upper spits were combined because of small sample size.

CATTLE

Despite high fragmentation rates and damage caused by gnawing, the assemblages produced large numbers of measurable cattle bones. The following sections will examine the results of the metrical analysis of different elements.

Horn core

Fragmentation restricted the number of measurements of the outer curvature lengths of horn cores. However, 43 specimens from the Troitsky sites were sufficiently complete for this measurement to be taken. Their average length was 140.3 mm but horn cores varied greatly in size from very small (96 mm) to large (205 mm). The coefficient of variation was much larger for all horn core measurements compared with those from limb bones (Table 6.1). Measurements of the minimum and maximum diameters at the base of the horn cores were taken much more frequently (Figures 6.1–6.2). There was a strong linear relationship between the two measurements, indicating that the horn core shape, at least at the base, was fairly consistent. Shapes of horn cores can vary between bulls, oxen and cows, although there is a

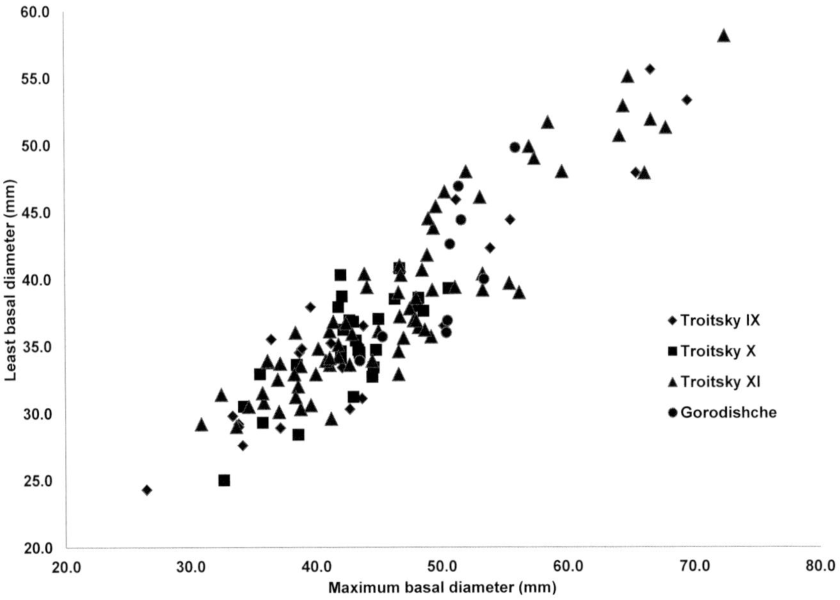

Figure 6.1: Troitsky and Gorodishche cattle horn core measurements

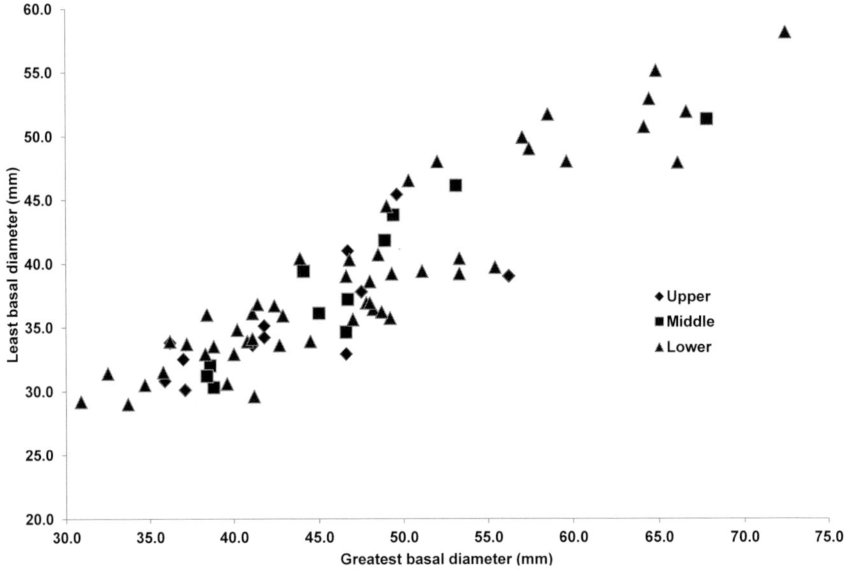

Figure 6.2: Troitsky XI cattle horn core measurements

lot of overlap (Armitage and Clutton-Brock 1976; Grigson 1982a). No consistent attempt was made to assign sex to the cattle horn cores during scanning, although the impression was gained that the majority probably belonged to cows. Horn cores could have been moved around with skins or collected for horn-working. However, there were no large accumulations of horn cores discovered in the Troitsky properties investigated. Within Troitsky XI, most of the largest horn cores were found in the lowest spits but most horn cores from these spits were no larger than those from the upper spits (Figure 6.2). The average size of both basal measurements decreased in the sample from the middle spits of Troitsky XI and horn cores from the topmost spits were on average 9% smaller than those from the lowest spits (Table 6.1). The means of horn core measurements at Gorodishche were slightly larger than those from Novgorod but these results were based on a small sample (Figure 6.1; Table 6.1).

Scapula

The scapula fuses during the first few months of the animal's life (Grigson 1982b), which means that most glenoids were fused and potentially available for measurement. Given the likelihood of post-fusion growth, slightly porous scapulae of immature cattle were not measured. Four measurements were routinely taken following von den Driesch (1976): the greatest length of the glenoid process (GLP); the length of the glenoid cavity (LG); the breadth of the glenoid process (BG) and the smallest length of the neck (SLC) (Table 6.2). SLC measurements were not taken as systematically as the others. SLC measurements are therefore summarised in Table 6.2 but not further analysed.

From a total of 148 specimens from the three Troitsky sites, the mean GLP was 60.4 mm. They were on average slightly larger on Troitsky X (61.3 mm) and Troitsky IX (60.7 mm) than in the lowest spits from Troitsky XI (60.5 mm). The average size of the GLP from Troitsky XI decreased by about 4% to 58 mm in the upper spits (Table 6.2). The mean size of the GLP from Gorodishche was 60.1 mm. Therefore there was no significant variation in size between the small sample of scapulae from that settlement and the much more substantial assemblages from the earliest layers of the three Troitsky sites. LG measurements, unsurprisingly as they are taken in the same plane as the GLP, produced similar results. The average size of all the Troitsky specimens was 51.3 mm. On Troitsky XI, the mean size decreased from 51.6 mm to 50.0 mm between the lowest and highest spits, a reduction of about 3%. Again the average size of the LG measurements from Gorodishche (51.8 mm) was closely similar to the mean obtained from the lowest spits on Troitsky XI (Table 6.2). The LG means from Troitsky IX and X were 51.9 mm and 50.7 mm respectively.

Scapula BG measurements, taken in the opposing plane to the LG, showed broadly similar trends. The overall mean from the Troitsky sites was 43.3 mm. The BG measurements from the middle and upper spits of Troitsky XI were generally 3–4% smaller (means of 41.6 mm and 41.9 mm respectively) than those from the lowest

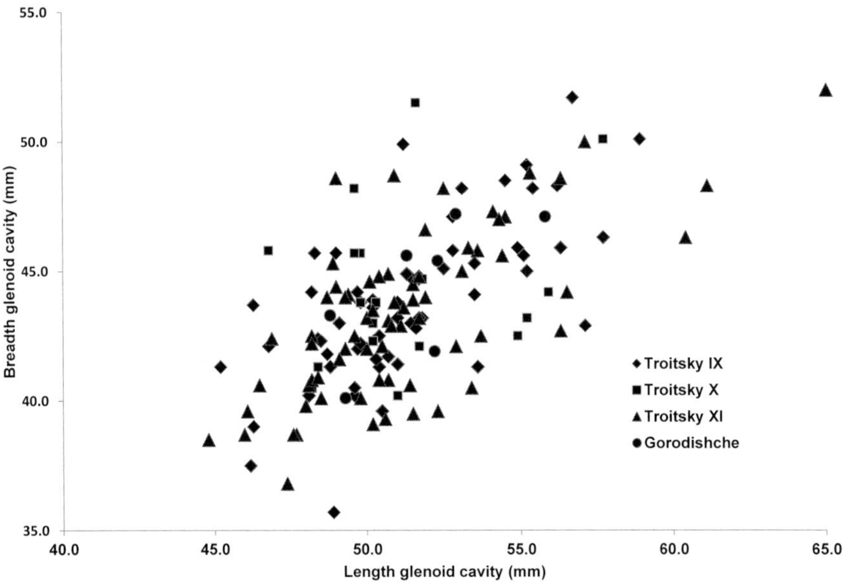

Figure 6.3: *Troitsky and Gorodishche cattle scapula measurements*

spits (mean 43.3 mm). In this case, the sample from Gorodishche produced the highest mean of 45.1 mm (Table 6.2).

Measurements comparing the length (LG) and breadth (BG) of the glenoid cavity of scapulae from all the Troitsky sites and Gorodishche are plotted in Figure 6.3. There was a fairly high positive correlation between the two measurements (correlation coefficient r = 0.68; coefficient of determination R = 0.82). The large sample from Troitsky XI produced the greatest range of measurements. There is, however, a denser cluster of specimens in the lower ends of the range. Figure 6.4 compares the results from the three spit groups from Troitsky XI. The lowest spits produced several specimens that were larger than any of those found in later deposits. The absence of such large specimens largely accounts for the decrease in mean size of the scapulae in the middle and upper spits. Therefore there may not have been a general decrease in cattle sizes but a greater focus on the slaughter of small cattle. It is tempting to attribute this to an increase in the relative abundance of (smaller) cows in the later levels, although sexual dimorphism in cattle scapulae is less marked than in many limb bones. Other factors such as age-related increases and/or abnormal lipping of the glenoid cavity can extend the size of LG measurements in particular.

Humerus

The distal humerus is another early-fusing epiphysis (Grigson 1982b), which allowed a large number of fused bones to be available for metrical analysis, even excluding

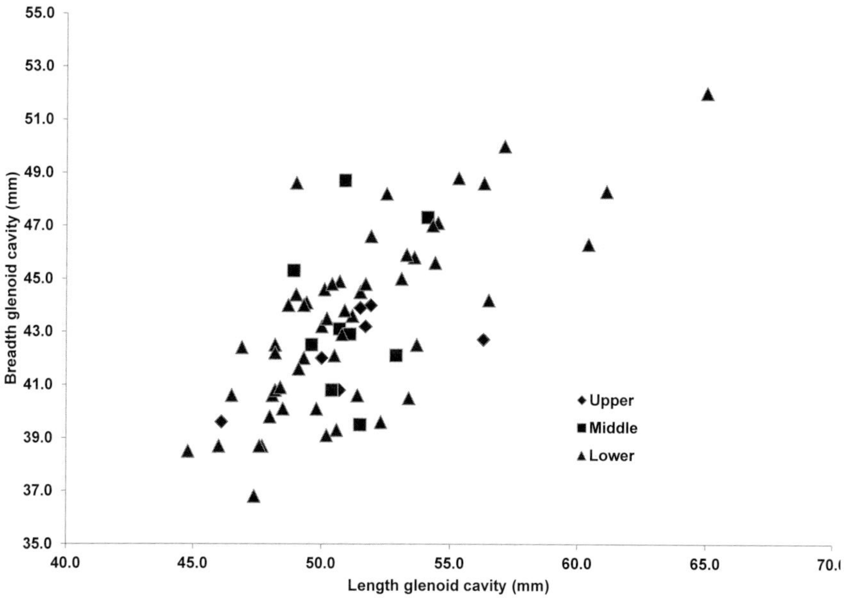

Figure 6.4: *Troitsky XI cattle scapula measurements*

the many that had been damaged by gnawing and butchery. Three measurements were taken on the distal humerus: the greatest breadth (Bd); the breadth of the trochlea (BT); and the greatest height of the trochlea (HT). The humerus Bd is a difficult measurement to take consistently and gnawing damage in particular prevented this measurement being taken in many cases. The Bd measurements are summarised in Table 6.3 but are not further analysed.

BT measurements were taken on 89 humeri from the three Troitsky sites, producing a mean of 66.5 mm. The specimens from Troitsky, IX, X and the lowest spits in XI had similar ranges, means and coefficients of variation (Table 6.3). Specimens from higher spits in Troitsky XI had a mean of only 64.2 mm, representing a decrease of 4% compared with the specimens from the lowest spits on this site (66.7 mm). This mirrors the trend observed on many other measurements but is based on only eight specimens from the higher spits. Again, it was the absence of the largest specimens that causes the decrease in the BT mean rather than a general decrease across the range of measurements. The 11 specimens from Gorodishche again produced the highest mean of 68.1 mm.

HT measurements were taken on 73 humeri from the Novgorod sites, mainly from Troitsky IX and XI. The mean size was 38.7 mm (Table 6.3). The largest specimens from Troitsky XI were again found in the lowest spits, where the mean size was 38.2 mm, compared with 37.6 mm in the small sample from higher levels. The six specimens from Gorodishche produced a slightly higher mean of 39.8 mm.

Radius

The proximal epiphysis of the radius is another that fuses during the first few months of life (Grigson 1982b). Therefore substantial numbers of fused proximal radii survived in a complete enough state to be measured. Three measurements were taken: the greatest proximal breadth (Bp); the greatest breadth of the proximal articular surface (BFp) and the greatest proximal depth (Dp). The mean size of the Bp from Troitsky IX, X and the lowest spits of XI all fell between 71 mm and 72 mm (Table 6.4). In the Troitsky XI assemblage, there is again a decrease in mean size from 71.5 mm in the lowest spits to 69.0 mm in the middle spits and 67.1 mm in the topmost spits, representing a substantial decrease of 9% in mean size, although the later samples are quite small. Again, it is the largest specimens that disappear from the later assemblages rather than a decrease throughout the range (Table 6.4). Some large specimens were also found at Gorodishche, where the mean size was higher than in Novgorod at 74.6 mm.

The radius BFp, as expected, shows the same trend as the Bp, as it is a measurement taken in the same plane. The mean size of the BFp from Troitsky IX and X and the lowest spits of XI were closely similar (65.3–66.3 mm – Table 6.4). At Troitsky XI the mean BFp decreased from 66.2 mm to 62.8 mm in the topmost spits – a decrease of over 5%. Again, all the largest specimens were found in the earliest assemblages (Figure 6.6). The mean BFp size at Gorodishche was 68.2 mm based on a small sample of nine specimens.

Very similar trends were observed in the radius Dp measurements (Table 6.4). Mean sizes ranged between 36.2 mm and 38.2 mm in the lowest (10th–12th century) spits from all three Troitsky sites. This decreased to 34.9 mm in the upper (mid-13th to 15th century) spits on Troitsky XI – a fall of 6%. Radii from the contemporary Fedorovsky site tended to be even smaller (Table 6.3).

Comparisons between the BFp and Dp measurements (Figures 6.5–6.6) show that the proximal breadth and depth measurements have quite a strong positive correlation ($r = 0.84$; $R = 0.92$), indicating that there were no major variations in the shape of the articular surface. The two smallest outliers from Troitsky X may have been from young animals, as the proximal end of the radius has substantial post-fusion growth. Most of the largest specimens were found on Troitsky IX and XI (Figure 6.5), with a group of five particularly large specimens coming from the lowest spits of Troitsky XI (Figure 6.6). These are probably from large male cattle, which were not represented in later deposits.

The distal epiphysis of the radius fuses later than the proximal end (Grigson 1982b), which reduced the number of fused specimens available to measure. Distal breadth (Bd) measurements were taken on only 36 specimens from the Troitsky sites. Summary results are provided in Table 6.4 but the data have not been analysed in detail.

Tibia

Distal tibiae survived well in these assemblages. This epiphysis fuses later than the scapula, distal humerus and proximal radius. Fused distal epiphyses therefore are

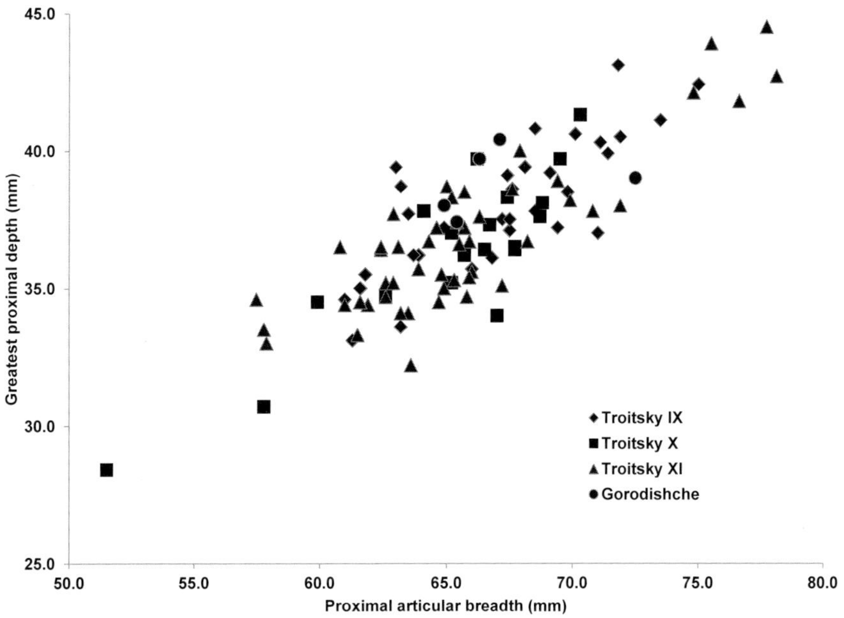

Figure 6.5: Troitsky and Gorodishche cattle radius measurements

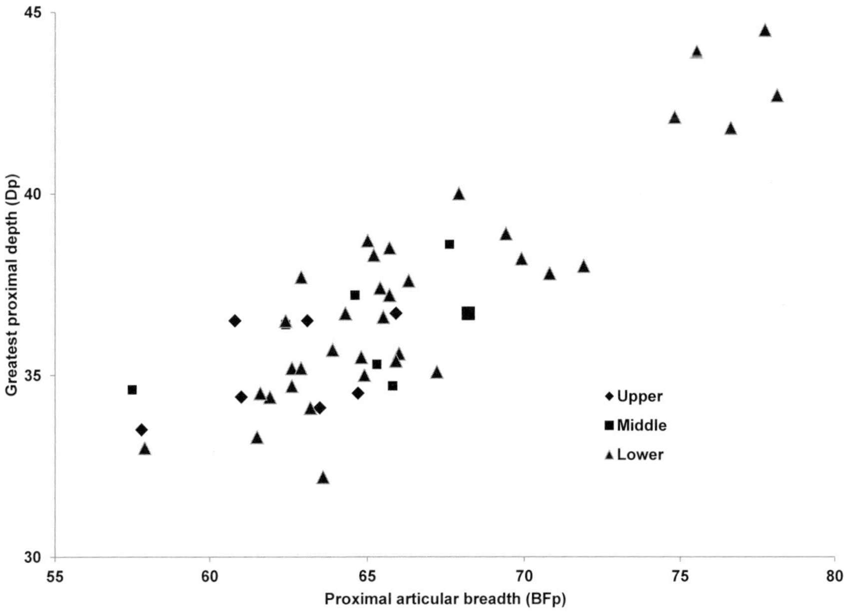

Figure 6.6: *Troitsky XI cattle radius measurements*

likely to belong to cattle over 18 months of age (Grigson 1982b). Two measurements were routinely taken: the distal breadth (Bd) and the distal depth (Dd). The Troitsky sites produced 235 Bd measurements. The means from Troitsky IX, X and the lowest spits of XI were almost identical (54.2–54.3 mm). The means from later levels from Troitsky XI decreased to 52.5 mm in the middle (mid-12th to early 13th century) spits and declined further to 51.7 mm in the upper (mid-13th to 15th century) deposits (Table 6.5). This therefore mirrors the decreases observed in the younger-fusing bones discussed above, representing a fall of around 5% in the mean measurement through the archaeological sequence. Again, it is the largest specimens found in the lowest (10th–12th century) spits from all three Troitsky sites that were absent from the later levels on Troitsky XI (Table 6.5), suggesting that it is mainly the largest adult males that have disappeared from the assemblage. The Bd mean of tibiae from Gorodishche was 55.5 mm, again indicating that the cattle were generally slightly larger than those from Novgorod.

The tibia distal depth measurements display parallel trends (Table 6.5). The three substantial samples from the lower levels of Troitsky IX–XI produced similar Dd means of 39.6 mm, 39.9 mm and 40.2 mm respectively. This figure decreased to 37.8 mm in the sample from the middle spits from Troitsky XI and to 37.5 mm in the sample from the highest levels. The means of the tibia Bd and Dd measurements from Gorodishche were 55.5 mm and 41.2 mm respectively, again indicating that the cattle were generally slightly larger than those from Novgorod (Table 6.5; Figures 6.7–6.8).

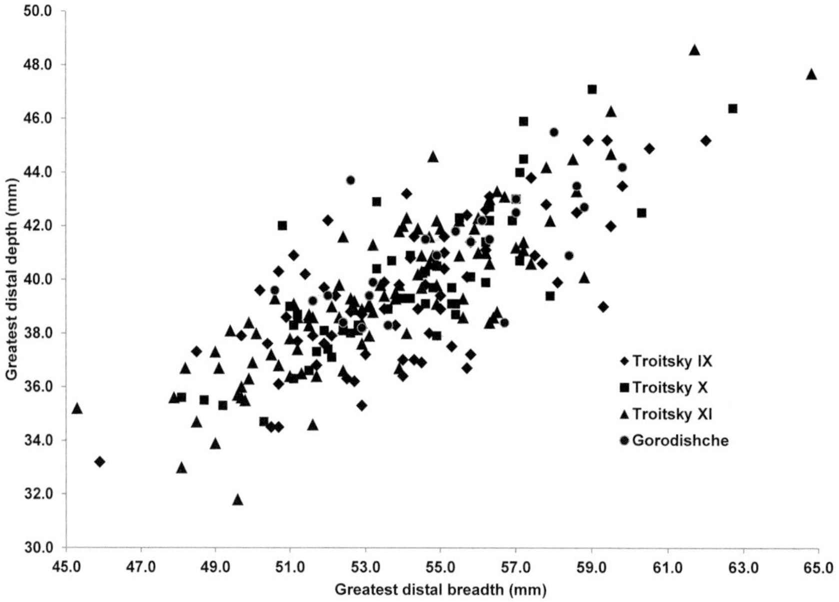

***Figure 6.7:** Troitsky and Gorodishche cattle tibia measurements*

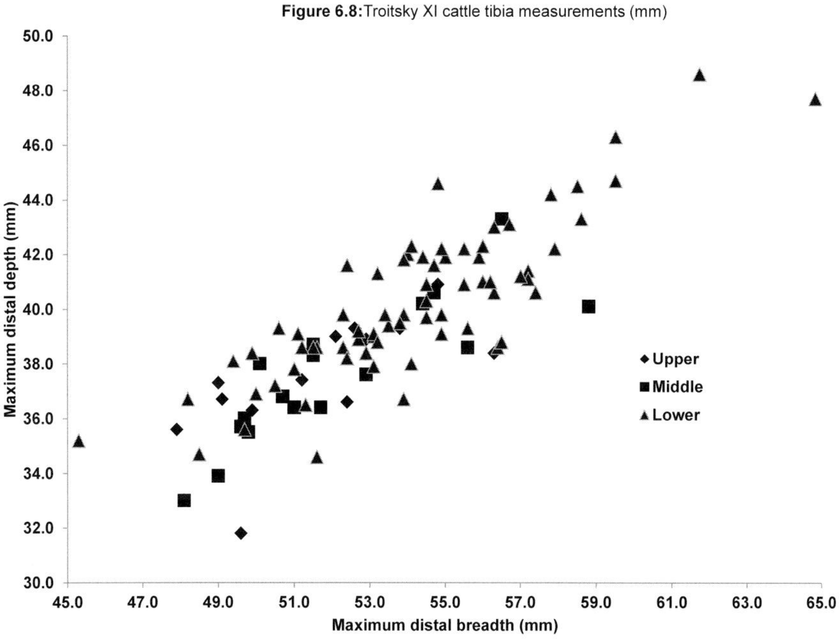

Figure 6.8: Troitsky XI cattle tibia measurements

Astragalus

Three measurements were taken on cattle astragali: the greatest length of the lateral half (GLl); the greatest length of the medial half (GLm) and the greatest distal breadth (Bd) (Table 6.6). The GLl and Bd measurements are compared in Figures 6.9–6.10. Both the length and breadth measurements showed similar trends to those obtained on the upper limb bones. The means of the GLl measurements from Troitsky IX, X and the lowest spits of XI were almost identical (59.8 mm; 60.1 mm and 60.0 mm respectively), whereas the specimens from later levels of Troitsky XI were on average 1% (middle spits) to 3% (upper spits) smaller (59.4 mm and 58.4 mm respectively). Most of the largest specimens came from the lowest spits of Troitsky XI (Figure 6.10). The mean GLl from Gorodishche was 61.3 mm. Unsurprisingly, as they are measured in the same plane, the GLm measurements produced very similar patterns. (Table 6.6).

The Bd means from the lower layers of the three Troitsky sites were also very similar (37.1–37.5 mm), again slightly smaller than the average size of those from Gorodishche (38.5 mm). The mean Bd measurements from Troitsky XI declined from 37.2 mm to 36.2 mm in the uppermost layers, a fall of around 3%.

Metacarpal

Six measurements were routinely taken on cattle metacarpals: the greatest length (GL); maximum proximal breadth (Bp); maximum proximal depth (Dp); maximum

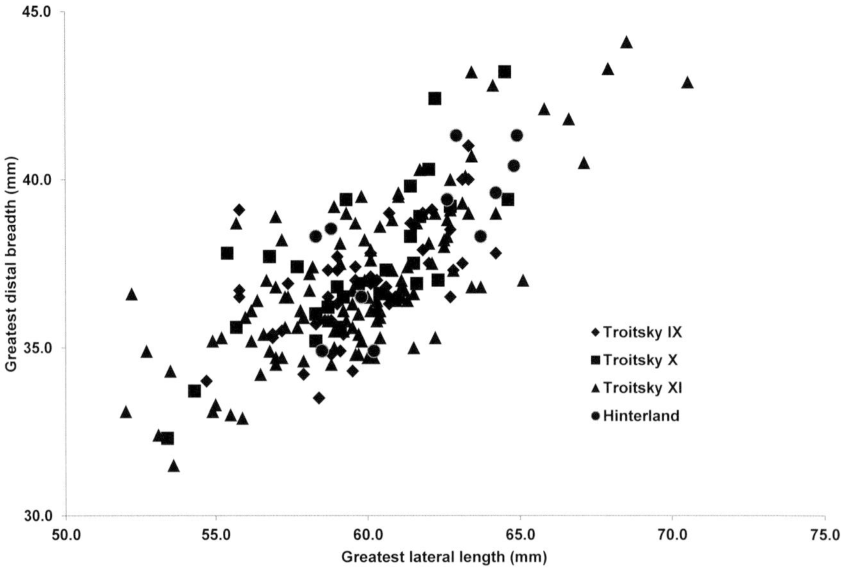

Figure 6.9: *Troitsky and hinterland sites cattle astragalus measurements*

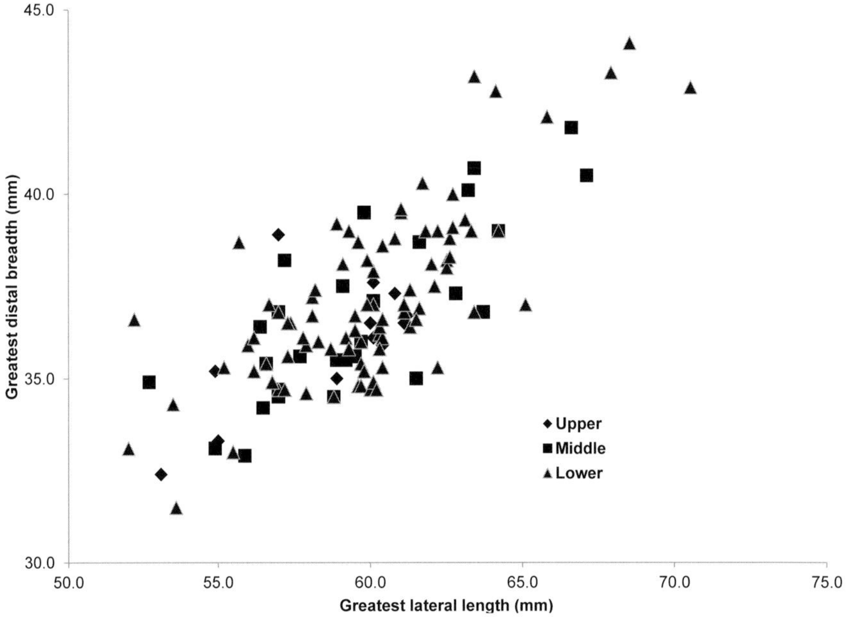

Figure 6.10: *Troitsky XI cattle astragalus measurements*

distal breadth (Bd); breadth at distal fusion point (Bdf) and depth at distal fusion point (Ddf) (Tables 6.8–6.9). The smallest breadth of the shaft (SD) was also recorded on complete specimens. The GL, Bd and Ddf measurements have already been discussed in relation to sexual dimorphism at Gorodishche (Figures 3.13–3.14) and the three

Troitsky sites (Figures 4.24–4.25). Length/distal breadth indices suggested that the majority of adult cattle at both settlements were females but that, although still in the minority, more metacarpals of male cattle were recovered from Troitsky IX and X than from Troitsky XI. The percentage of specimens assigned to males also decreased in the upper levels of Troitsky XI. This section will incorporate further metrical data into this discussion. The summary of the Bd measurements (Table 6.8) shows again that there was little variation in average size in the lower layers of the three Troitsky sites (52.3–52.9 mm) but the mean size decreased in the middle (50.1 mm) and upper spits of Troitsky XI (49.4 mm). The distal metacarpals from the Fedorovsky and Deysatinny-1 sites in Novgorod were of similar size but those from Gorodishche were on average slightly larger (53.9 mm). Comparisons with the greatest lengths indicated that most specimens smaller than 55.0 mm were probably females (Figures 6.11–6.12). All the samples were positively skewed indicating that more smaller (mainly female) adult cattle were represented throughout and the analysis confirms the virtual absence of metacarpals of males in the middle and upper levels of Troitsky XI. No metacarpals with a Bd of over 60 mm were recorded in those spits (Table 6.8).

Bdf measurements, as expected, showed virtually identical trends to the Bd measurements, as they are taken in the same plane. The Ddf measurements also showed that the same size variations were apparent (Table 6.8). The mean Ddf decreased by 7% from 25.7 mm in the lowest spits to 24.0 mm in the upper spits of Troitsky XI. The

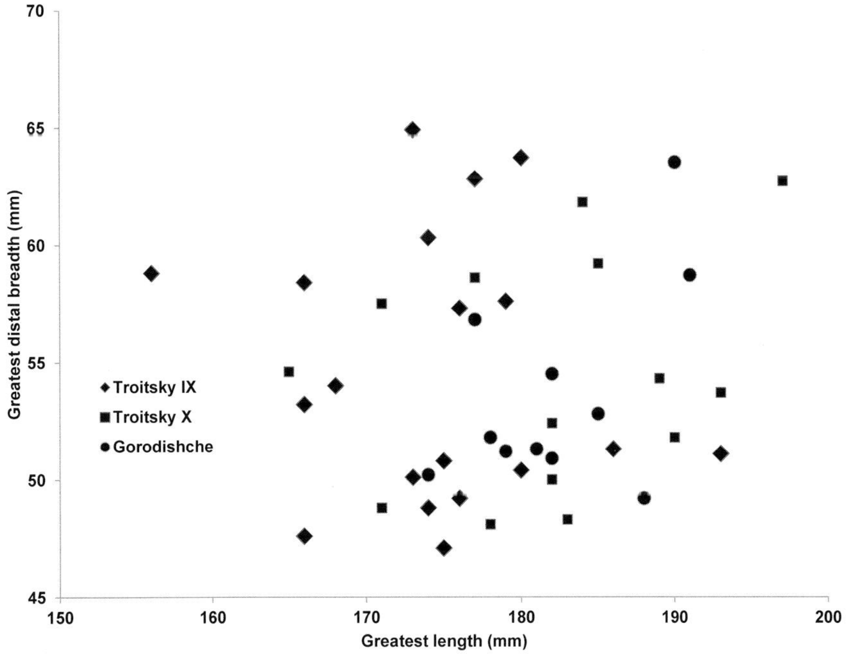

Figure 6.11: Troitsky IX–X and Gorodishche cattle metacarpal length and distal breadth measurements

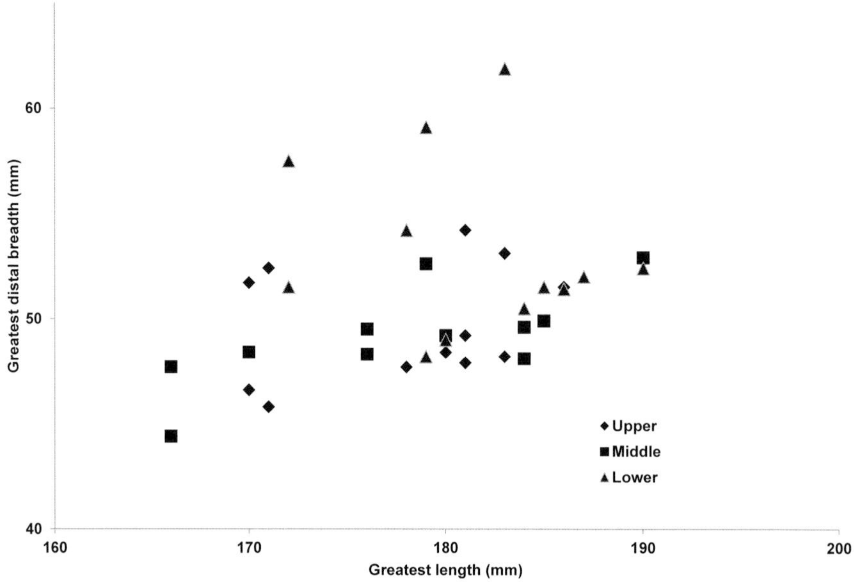

Figure 6.12: Troitsky XI cattle metacarpal length and distal breadth measurements

depth measurements from Gorodishche were similar to those from the lower layers of the Troitsky sites.

The proximal metacarpal provided large samples of Bp and Dp measurements (Table 6.9). Although porous specimens from calves that had not grown to full size were not measured, the sample will undoubtedly include some specimens with unfused distal epiphyses from immature animals. Although sexual dimorphism of the metacarpal Bp is not as marked as distal breadth measurements, it is likely that the majority of the smaller specimens belonged to cows. This is supported by comparisons of the Bp with the GL measurements of complete specimens, where it was possible to assign the specimens into probable male (Bp/GL > 0.302) or probable female categories (Figure 6.13). Employing this index, there were roughly equal numbers of male and females represented by complete metacarpals from Troitsky IX and X, whereas in the lowest spits of Troitsky XI and at Gorodishche around 75% of the complete metacarpals belonged to cows. In the middle and upper spits, there were no complete metacarpals that had Bp/GL indices of >0.302, whereas 19 had values lower than 0.302, implying that no males were represented by complete metacarpals (Figure 6.14; see also Figures 4.24–4.25).

By including broken specimens, on which the Bp could be measured, the sample sizes increased markedly. The Bp measurements displayed the familiar trends observed for other breadth measurements. The means from the lower layers of the three Troitsky sites were very similar (50.9–51.3 mm) and all had a positively skewed distribution (Table 6.9). In all three cases the modes lie between 49.0–50.9 mm, as

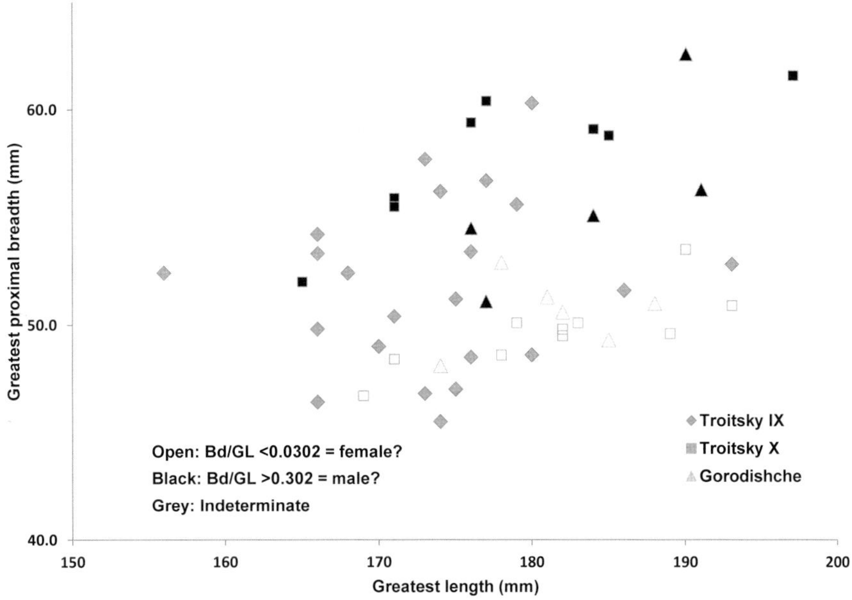

Figure 6.13: *Troitsky IX–X and Gorodishche cattle metacarpal length and proximal breadth measurements*

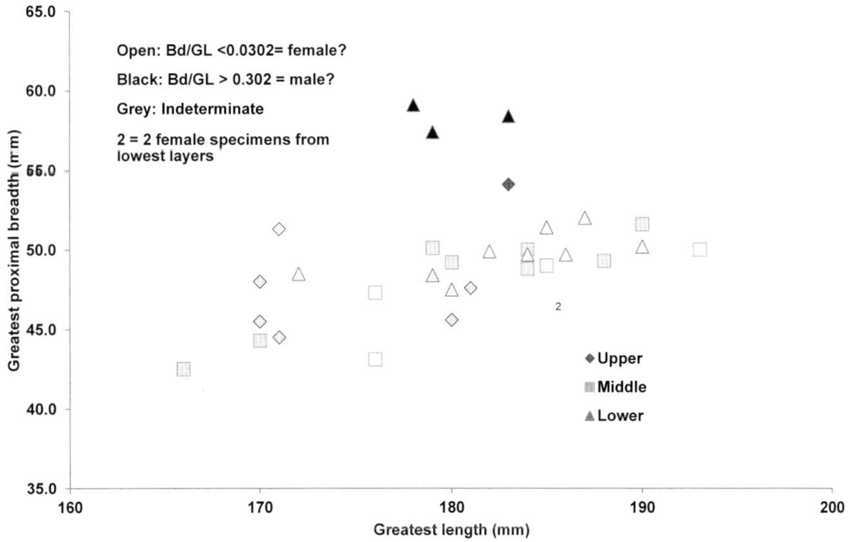

Figure 6.14: *Troitsky XI cattle metacarpal length and proximal breadth measurements*

illustrated in Figures 6.15–6.17, in which measurements from Troitsky IX, X and the lowest layers of Troitsky XI respectively are plotted. Specimens provisionally assigned as males or females are indicated. All bar one of the complete specimens assigned as

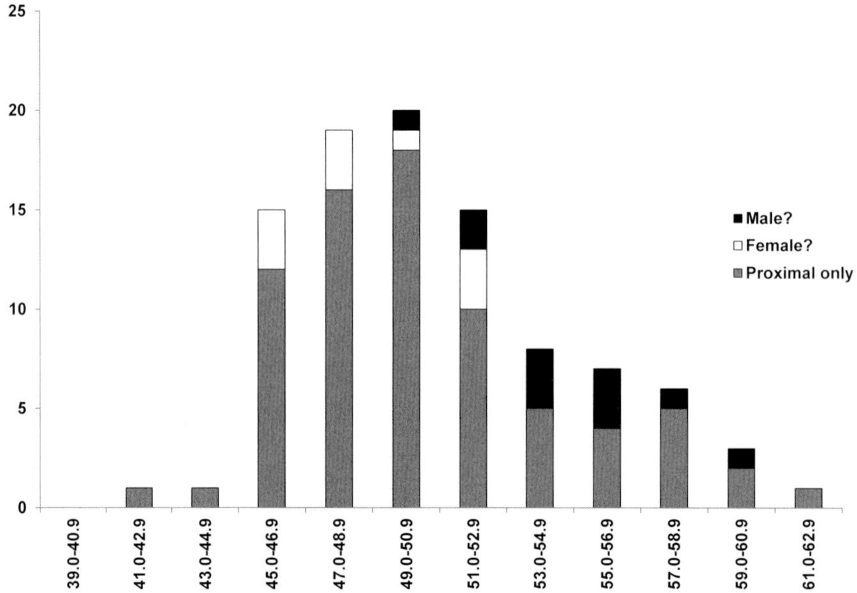

Figure 6.15: *Troitsky IX cattle metacarpal greatest proximal breadth measurements (mm)*

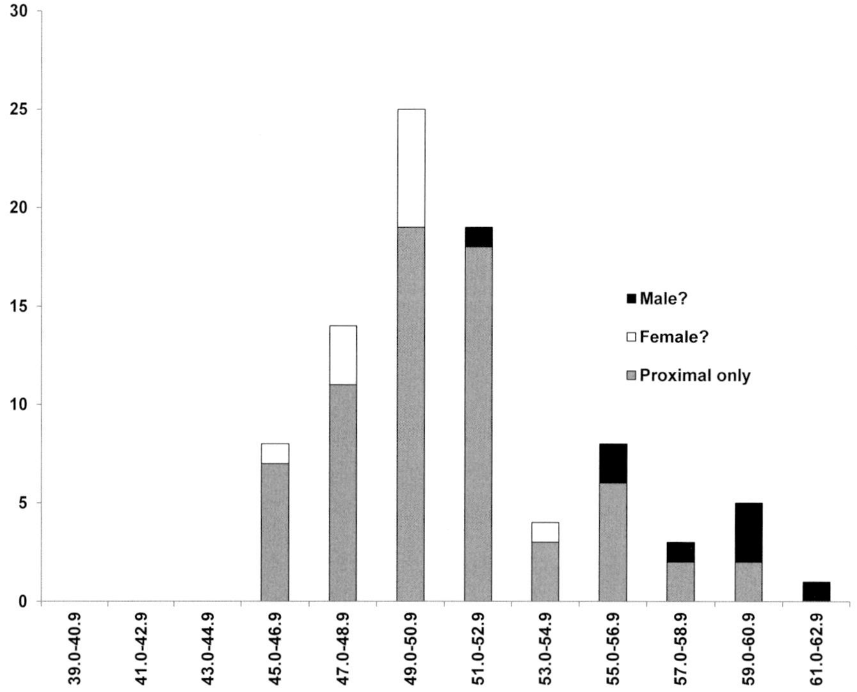

Figure 6.16: *Troitsky X cattle metacarpal greatest proximal breadth measurements (mm)*

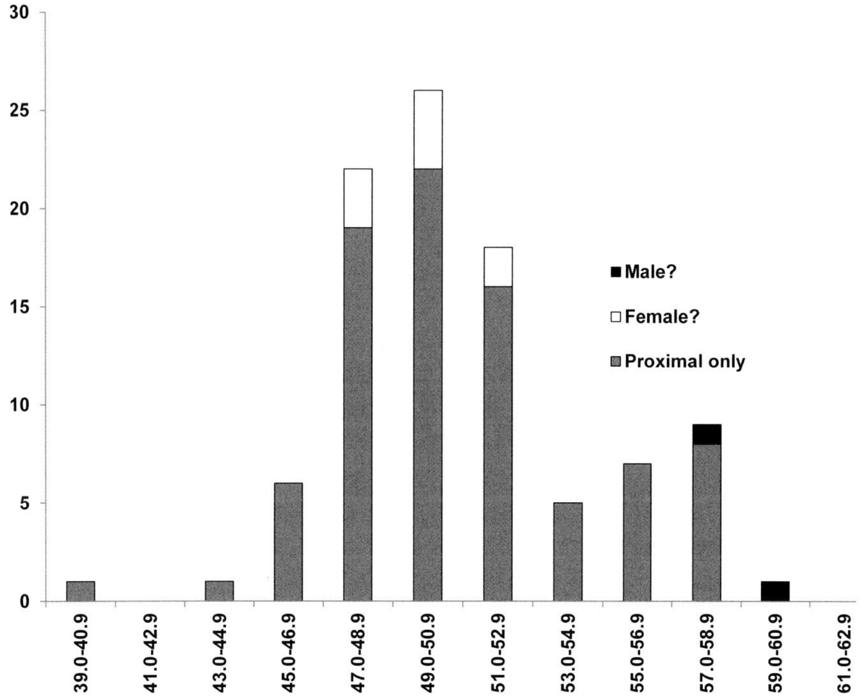

Figure 6.17: *Troitsky XI lower spits cattle metacarpal greatest proximal breadth measurements (mm)*

female had Bp measurements of <53 mm, whereas most of the probable male specimens measured >53 mm. If this was the case then around 75–80% of the measured proximal metacarpals in these assemblages belonged to cows. The percentage of cows rose to around 90% in the middle and upper spits of Troitsky XI (Figure 6.18). This was much higher than the results obtained from the 9th–10th-century deposits at Gorodishche, where proximal metacarpals of bulls and oxen provided up to 45% of the measured specimens (Figure 6.19). These results would support the interpretation that a significant factor in the size reduction in other cattle limb bone breadth measurements in the later medieval period was the decrease in the number of (mainly) adult males being consumed in Novgorod.

However, this does not explain all of the decrease in cattle size. The mode for the metacarpal Bp measurements at Gorodishche lay between 51.0–52.9 mm (Figure 6.19). This fell to 49.0–50.9 mm in the three assemblages from the 10th–12th-century Troitsky sites (Figures 6.15–6.17) and further still to 47.0–48.9 mm in the middle and upper spits on Troitsky XI (Figure 6.18). This indicates that the average stature of cows gradually decreased during the medieval period.

Metacarpal Dp measurements reflected the same trends observed in the Bp measurements (Table 6.9). There was a decrease in mean size from 32.3 mm on the 9th–10th-century hinterland sites to 30.7–30.9 mm in the 10th–early 12th-century

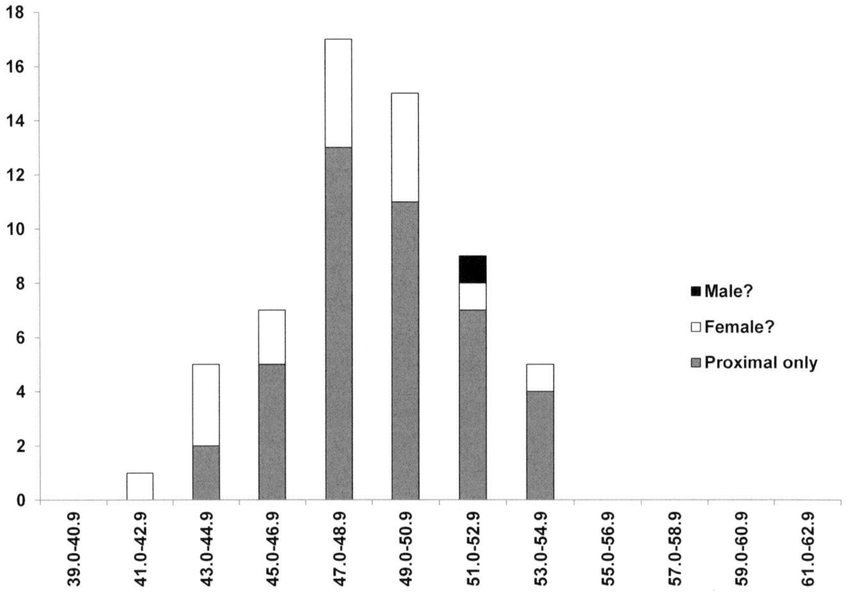

Figure 6.18: *Troitsky XI middle and upper spits cattle metacarpal greatest proximal breadth measurements (mm)*

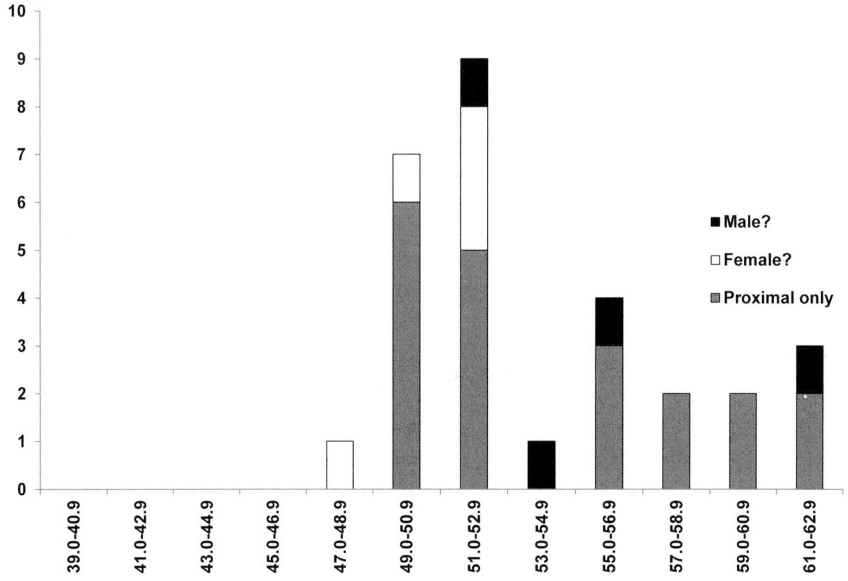

Figure 6.19: *Gorodishche and Georgii cattle metacarpal greatest proximal breadth measurements (mm)*

assemblages from the three Troitsky sites followed by a further decrease to 29.5 mm in the mid 13th–early 15th-century levels of the Troitsky XI site.

Because they were less commonly broken, substantial numbers of complete metacarpals provided GL measurements (Table 6.9). Seventy-eight complete specimens were measured from the three Troitsky sites. These will be discussed further in relation to withers height estimates of other bones below, but it should be noted that, compared with breadth and depth measurements, there was not as marked a chronological decrease in length measurements (2%) in the specimens from Troitsky XI and that the Troitsky IX site had more very small specimens (Figures 6.11 and 6.13) resulting in an average size (173.6 mm) being markedly smaller than those from contemporary assemblages from Troitsky X (180.1 mm) and XI (181.3 mm).

Metatarsal

Proximal breadth (Bp), distal breadth (Bd) and greatest length (GL) measurements were taken where possible on cattle metatarsals from all the Novgorod and hinterland sites (Table 6.10). Analysis of the Bp measurements produced similar trends to those of other breadth measurements. The slightly earlier hinterland sites produced the highest mean (44.3 mm); this decreased slightly in the 10th–early 12th-century assemblages from the three Troitsky sites (41.9–42.3 mm) (Table 6.10; Figures 6.20–6.21).

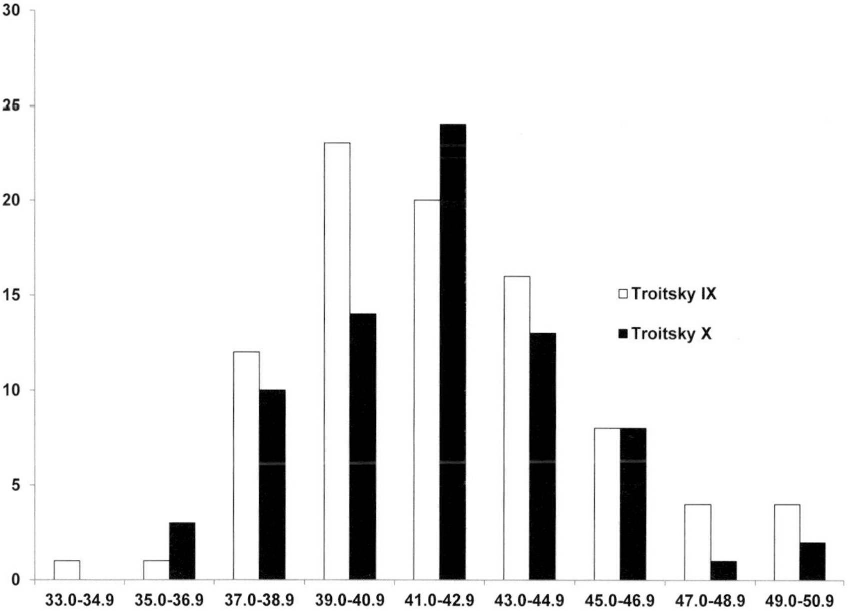

Figure 6.20: Troitsky IX–X cattle metatarsal greatest proximal breadth measurements (mm)

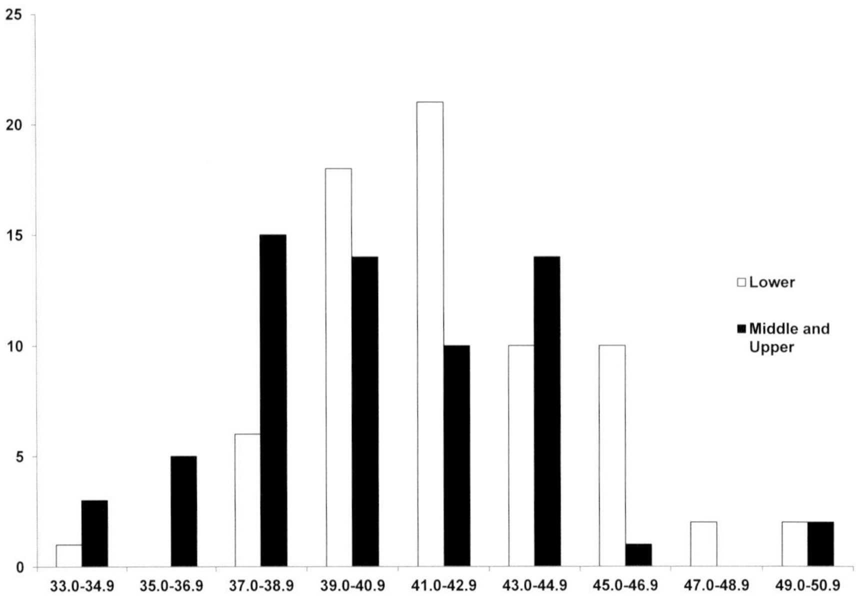

Figure 6.21 *Troitsky IX–X cattle metatarsal greatest proximal breadth measurements (mm)*

The average size fell further in the assemblages from the middle and upper spits on Troitsky XI to 40.9 mm and 39.7 mm respectively (Figure 6.21). Most of the distributions were positively skewed, reflecting the predominance of specimens in the lower part of the size range. Although sexual dimorphism is less marked in the metatarsal than in the metacarpal, this distribution probably largely reflects the predominance of females in the assemblages from Novgorod (Figures 6.20–6.21). On the other hand, the assemblage from Gorodishche, is negatively skewed and displays a clear bimodal distribution (Figure 6.22), which suggests that a more even proportion of males and females may be represented.

Variations in the metatarsal Bd measurements followed the same pattern as the Bp (Table 6.10). Gorodishche produced the highest mean (50.6 mm). The means from the three Troitsky 10th–early 12th-century assemblages were all slightly smaller (47.8–48.7 mm) and decreased further in the middle and upper spits of Troitsky XI to 47.5 mm and 46.7 mm respectively, despite the presence of an unusually large specimen measuring 67.2 mm from Spit 8. Although not noted during recording, the breadth of this specimen may have been expanded by splaying and lipping, a pathological condition that is more prevalent in working oxen (Bartosiewicz *et al.* 1997).

The Fedorovsky excavations produced a large dump of cattle metatarsals, which consisted of the sawn-off proximal and distal ends of the bones after the removal of the central part of the shaft for working. These were dumped in the 13th–14th century. Bp and Bd measurements were obtained from these offcuts. Their average

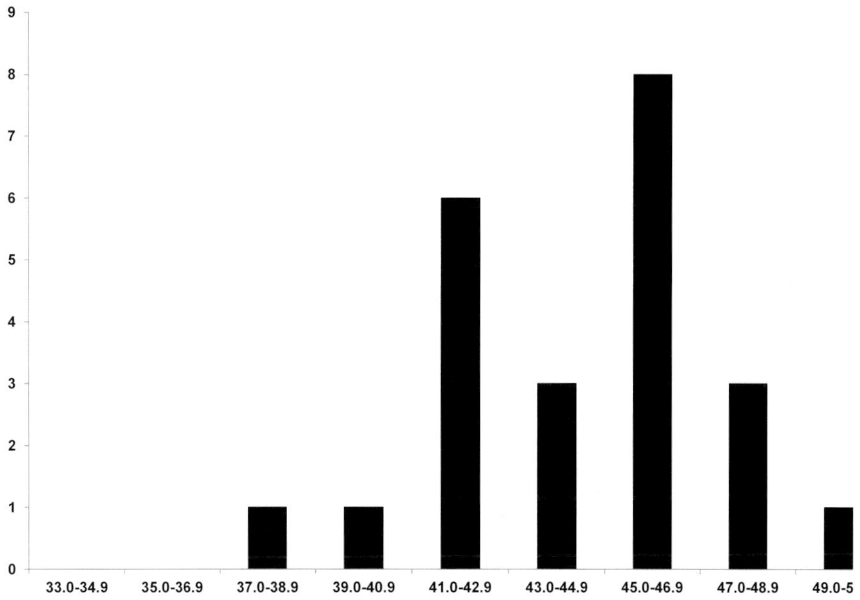

Figure 6.22: *Gorodishche (22 specimens) and Georgii (1 specimen) cattle metatarsal greatest proximal breadth measurements (mm)*

sizes (41.2 mm and 46.1 mm respectively) were similar to those from contemporary deposits from the middle and upper spits of Troitsky XI (Table 6.10). The worked distal ends were generally slightly smaller than other (non-worked) metatarsals from contemporary levels on the Fedorovsky site, which could imply that smaller metatarsals from cows were either more readily available and/or preferentially selected for working.

The GL of 14 and 64 complete metatarsals were measured from Gorodishche and the three Troitsky sites respectively (Table 6.10). The mean size of the former, as in the case of the metacarpals (Table 6.9) was slightly larger (209.1 mm) than in most of the Troitsky samples. However, the means of the metatarsals from Troitsky X were higher (215.4 mm) than those from Gorodishche as well as those from Troitsky IX (204.5 mm) and Troitsky XI (200.0 mm). In contrast to most other measurements, the means of metatarsal GL increased in the middle spits of Troitsky XI (207.1 mm) before decreasing to 197.2 mm in the upper fills. However, these calculations included an unusually large specimen (234 mm) from Spit 8. This specimen was 18 mm longer than any other metatarsal from Troitsky XI (although only 5 mm larger than the largest specimen from Troitsky X). It also had, by far, the largest distal breadth measurement (67.2 mm), as noted above, and the second largest proximal breadth (50.2 mm). This was therefore a large male specimen, possibly a large ox.

Withers heights

The large assemblage of cattle bones recovered from the Troitsky excavations, particularly from the 10th–early 12th-century levels has allowed detailed analysis to be carried out on individual bones. These assemblages have been compared with the smaller samples from Gorodishche and other hinterland sites. Generally, the cattle consumed in Novgorod and in its immediate hinterland were small by modern standards but they were not unusually small for the medieval period. They generally confirm the results obtained from other excavations in Novgorod (Tsalkin 1956).

The greatest lengths of complete limb bones provided withers height estimates. The vast majority of the complete bones were metapodials (Tables 6.9–6.10). Withers height conversion factors for these followed those recommended by Foch; namely GL*6.125 for metacarpals and GL*5.45 for metatarsals. Those for radius (GL*4.30) and tibia (GL*3.45) followed those of Matolsci (von den Driesch and Boessneck 1974).

A total of 155 withers heights estimates were obtained from the Troitsky assemblages and a further 29 were recorded from Gorodishche (Table 6.11). Overall estimates ranged between 95.6 cm and 127.5 cm. The assemblage from Gorodishche provided a mean of 112.5 cm with a broadly unimodal distribution (Figure 6.23). Very similar results were obtained from later seasons of excavations of this site (see Sablin below). The mean withers heights from the three Troitsky 10th–early 12th-century assemblages were all slightly smaller (Table 6.11). Both the Troitsky IX and X assemblages had unimodal distributions but the Troitsky IX withers height estimates were on average smaller (mean=108.4 cm) than those from Troitsky X (mean=111.9 cm)

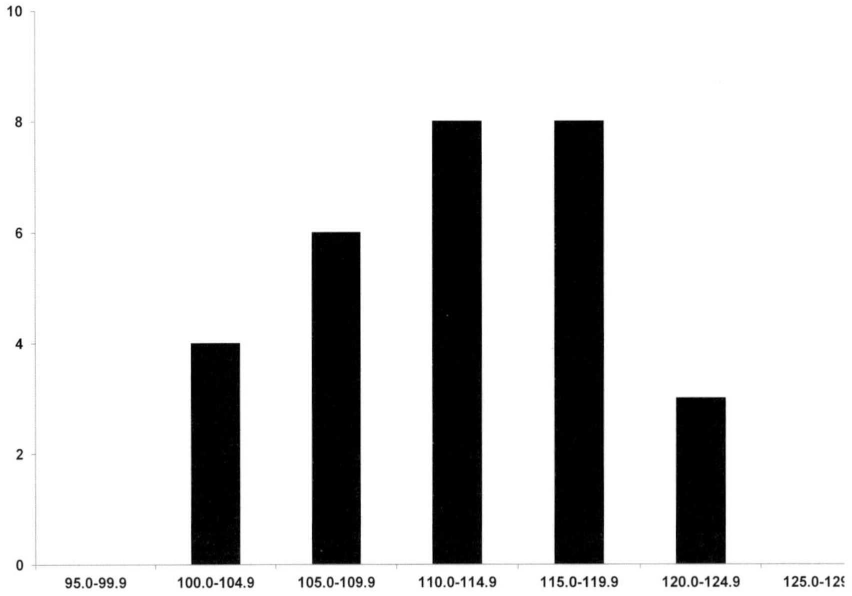

Figure 6.23: *Gorodishche cattle estimated withers heights (cm)*

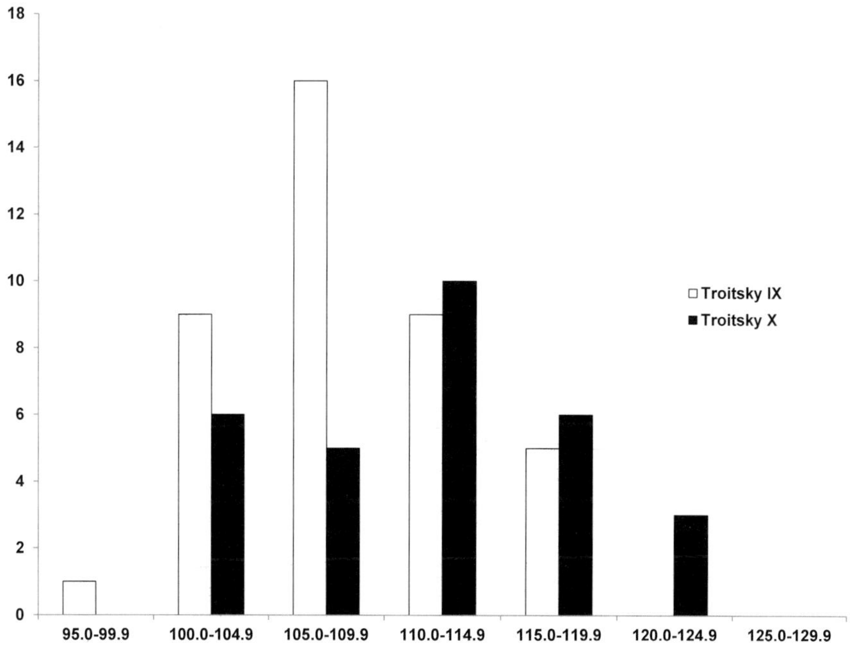

Figure 6.24: *Troitsky IX and X cattle withers height estimates (cm)*

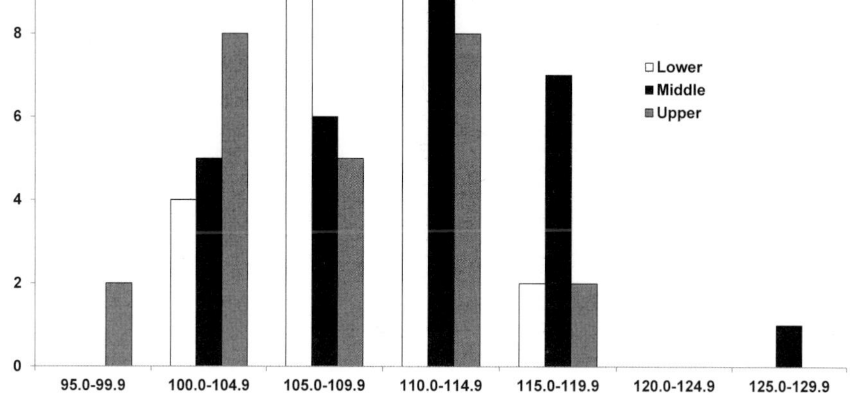

Figure 6.25: *Troitsky XI cattle withers height estimates (cm)*

(Table 6.11; Figure 6.24). In the Troitsky XI sample, the average withers heights initially increased from 109.7 cm in the 10th–early 12th-century levels (S22–S14) to 111.3 cm in the middle spits (S13–S8) before decreasing to 107.7 cm in the upper spits (S7–S1) (Table 6.11; Figure 6.25). The largest withers height estimate (127.5 cm) from the middle spits was derived from the unusually large metatarsal from Troitsky XI Spit 8 (discussed above). Excluding this specimen, the mean withers height in the middle spits fell to 110.7 cm. The distribution from the upper spits, in contrast to all the other samples, was bimodal, reflecting a significant increase in bones of cattle smaller than 105 cm at the shoulder, which accounted for 40% of the specimens compared with 14%–25% in the samples from earlier deposits from all the Troitsky sites and Gorodishche (Figures 6.23–6.25).

Log ratio analysis of breadth measurements

To increase the sample sizes of breadth measurements, log-ratio analysis was employed. This enables the breadth measurements of different elements to be amalgamated by comparing them against a set of standard measurements. The use of log-ratio analysis of metrical data of animal bones from archaeological sites has increased substantially in recent years. This method of analysis involves comparing individual measurements again standards derived from a variety of elements from an individual skeleton (*e.g.* Thomas *et al.* 2013) or from the means of a large (usually archaeological) sample (*e.g.* Albarella *et al.* 2008). However, the relative proportion of male and female specimens in standards derived from archaeological samples can vary and, as has been discussed, the samples from Novgorod and its hinterland seem to contain uneven numbers of male and females. Unfortunately, no sets of standard measurements derived from animals of known sex have been established for Russian material. For this analysis, the measurements have been compared with those obtained from an Iron Age skeleton recovered from South Cadbury, Somerset, England (Hamilton-Dyer and Maltby 2000). Observations of the morphology of the pelvis, metapodials and skull all indicated that this skeleton belonged to a cow. Estimates indicated that it had a withers height of about 106 cm. Therefore this specimen can be used to represent a cow of similar size to many of those represented in Novgorod. The following breadth measurements from the South Cadbury cow were selected for comparison: humerus: BT (63.6 mm); radius: Bp (70.3 mm); tibia: Bd (53.1 mm); astragalus: Bd (36.6 mm); metacarpal: Bp (49.3 mm) and metatarsal Bp (40.4 mm). These standard measurements were converted into base-10 logarithms. The breadth measurements of all the specimens from the Gorodishche and Troitsky sites were also converted to base-10 logarithms and compared against the equivalent standards. A positive value indicates that the archaeological specimen was larger than the equivalent measurement of the South Cadbury cow, whereas a negative value indicates that it was smaller. The results are shown in Figures 6.26–6.31.

The 101 measured specimens from Gorodishche ranged between logs −0.03 and +0.09, with 76% being larger (≥0.01) than the standard, including 26% at ≥0.05

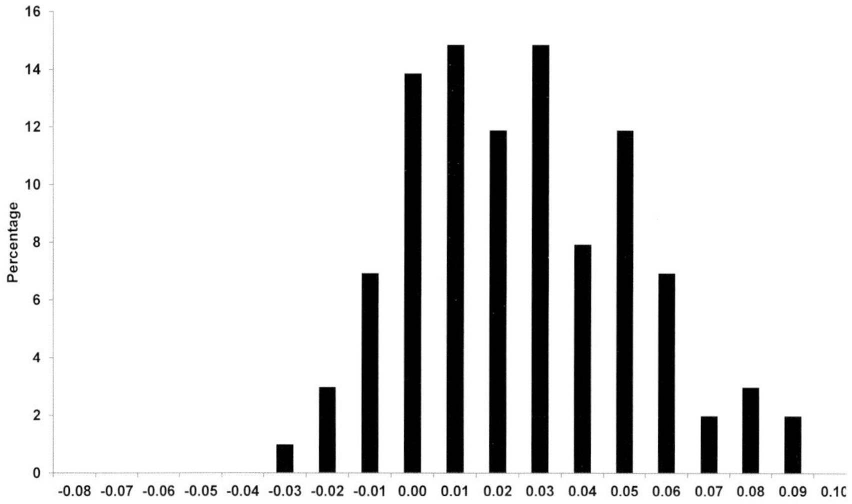

Figure 6.26: *Gorodishche log ratios of cattle breadth measurements (N=101)*

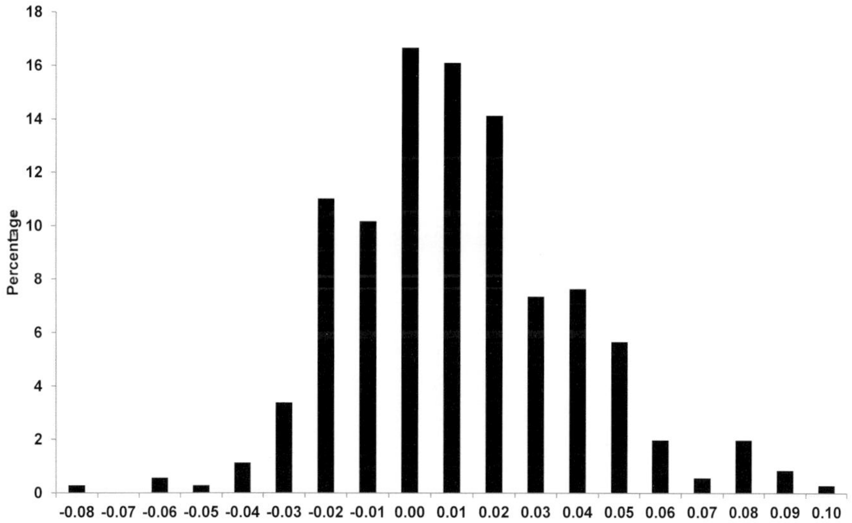

Figure 6.27: *Troitsky IX log ratios of cattle breadth measurements (N=354)*

(Figure 6.26). Given that the breadth measurements of the metacarpals indicated that around 25% of the measured specimens belonged to larger males (Figures 3.13–3.14, 6.11–6.12), it is reasonable to assume that most of the specimens at ≥0.05 belonged to males. However, the bulk of the specimens had log ratios of between 0.00 and + 0.03. Assuming these mainly belonged to cows, it shows that these were generally slightly larger than the Iron Age specimen from South Cadbury.

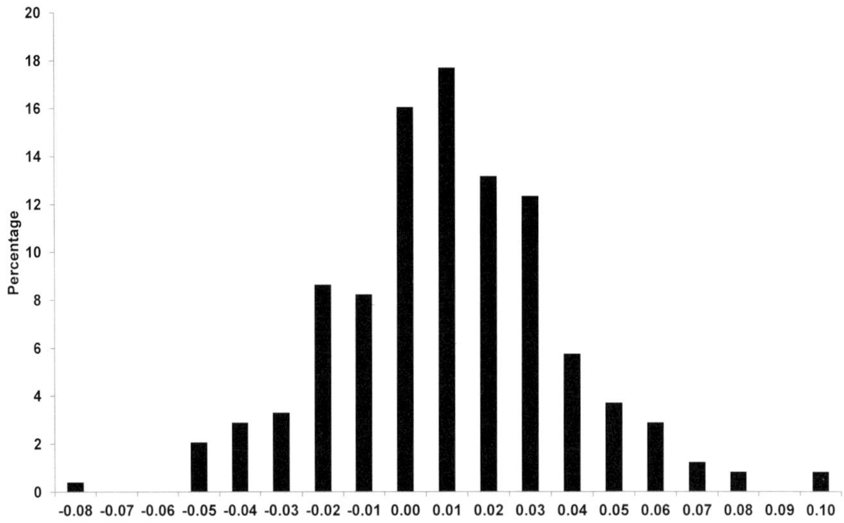

Figure 6.28: *Troitsky X log ratios of cattle breadth measurements (N=243)*

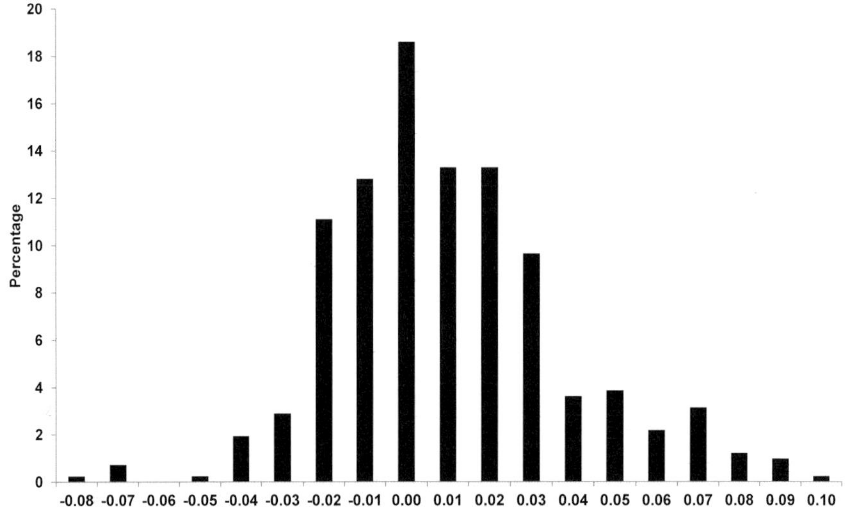

Figure 6.29: *Troitsky XI lower spits log ratios of cattle breadth measurements (N=414)*

The log-ratio distributions from specimens from the lower levels of Troitsky IX, X and XI showed a lot of similarities (Figures 6.27–6.29). Understandably, given that they were all derived from larger samples than from Gorodishche, the range in size was greater (log ratios −0.08 to +0.10) in all three samples. The modes of all three samples lay between 0.00 and + 0.01. However, compared with Gorodishche, slightly smaller percentages of specimens were larger (≥0.01) than the standard

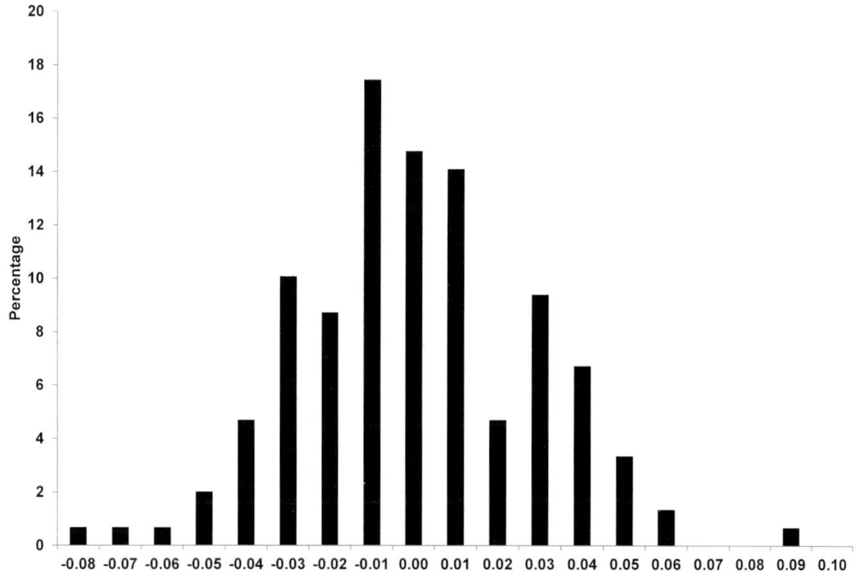

Figure 6.30: *Troitsky XI middle spits log ratios of cattle breadth measurements (N=149)*

(Troitsky IX 57%; Troitsky X 59%; Troitsky XI lower spits 51%) and there were also fewer specimens at ≥+0.05 (Troitsky IX 12%; Troitsky X 10%; Troitsky XI lower spits 11%). Instead, significantly more specimens fell below the standard measurements. Whereas at Gorodishche, only 11% of the specimens had values of ≤−0.01, specimens smaller than the standard formed 25% of the assemblage from Troitsky IX, 26% from Troitsky X and 30% from the Troitsky XI lower spits 14–22 (Figures 6.27–6.29). As at Gorodishche, the sexing data indicated that around 75–80% the metacarpals in these deposits were from females (Figures 4.24–4.25, 6.11–6.12). Therefore, it would appear that the slight decrease in the breadth of limb bones reflects a decrease in the overall size of both male and female cattle kept in and around Novgorod during the 11th and 12th centuries.

The assemblages from the middle and upper spits of Troitsky XI showed further slight decreases in the sizes of cattle. The range in size of the assemblage from the middle spits was similar to the ranges encountered in the lower spits (−0.08 to +0.09) but only 5% of the specimens had values of ≤−0.05, whereas 45% had values of ≤−0.01. The mode fell to −0.01 (Figure 6.30). In the upper fills, only one of the 78 specimens had a value above +0.05 (0.09). None of the others had values of over 0.03. Although the mode remained at −0.01 and 0.00, over half the specimens (58%) had values of ≤−0.01 (Figure 6.31). Although, analysis of sexual dimorphism of the metacarpals indicated a decrease in the percentage of male cattle (c.10%) in the upper spits (Figures 4.24–4.25, 6.11–6.12), it is clear that both female and male cattle became smaller during the later medieval phases. Although cattle provided

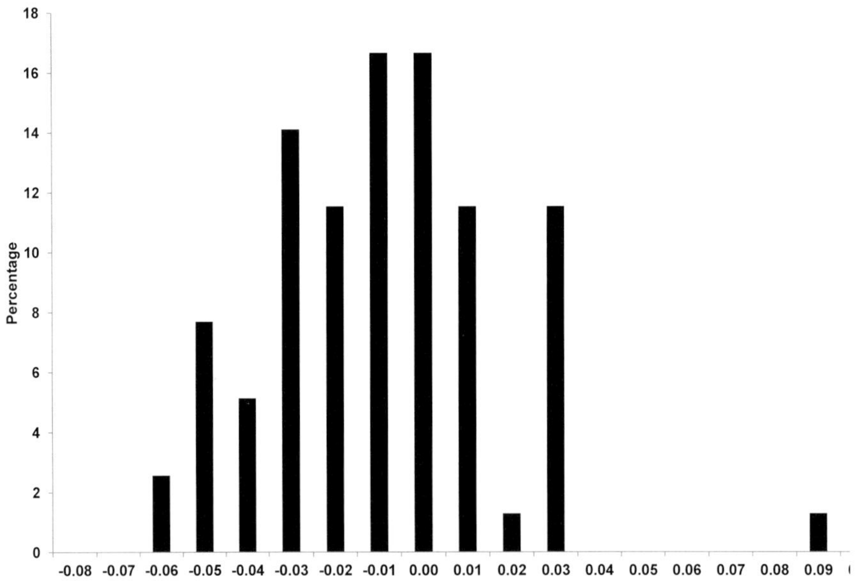

Figure 6.31: *Troitsky XI upper spits log ratios of cattle breadth measurements (N=76)*

the most meat for the Novgorodians, the methods of husbandry did not encourage improvements in productivity that is reflected in carcase size improvements, indeed the opposite seems to have occurred.

PIG

Metrical data obtained from pig bones was limited by the high percentage of unfused specimens. Because of the smaller sample sizes, specimens from the middle and upper spits of Troitsky XI (S13–S1) were considered together in the following comparisons. Unfortunately, the measurements of teeth were limited to those taken on third molars and not the broader range of measurements advocated by Albarella and Payne (2005). This omission remains to be addressed by future research.

Mandibular third molar

Because of the high level of immature slaughter, few third molars were fully erupted and thus measurable. Only seven specimens from the Troitsky sites and five from Gorodishche were measured (Table 6.12). The mean was slightly higher in the Gorodishche assemblage (31.5 mm) than on the Novgorod sites (30.5 mm) but the samples were too small to determine whether this represents a significant variation in size. The largest specimen (from Troitsky X, Spit 14) measured 36.5 mm, therefore falling below the threshold usually assigned to wild boar, which indicates that this and the remainder of the specimens all belonged to domestic pig.

Scapula

Three measurements were routinely taken: the greatest length of the glenoid process (GLP); the breadth of the glenoid process (BG) and (less frequently) the smallest length of the neck (SLC) (Table 6.13). Figure 6.32 plots the first two measurements on specimens where both measurements could be taken. The mean size of the GLP of 45 scapulae from all three Troitsky sites measured 32.9 mm, with those from Troitsky IX and X being on average slightly larger than those from Troitsky XI. The average size of the scapulae from Gorodishche was also slightly higher (34.0 mm) but only four specimens were measured (Table 6.13). The largest specimen from Troitsky IX (Spit 16) measured 39.7 mm (not plotted on Figure 6.32), which falls in the range of overlap between domestic pigs and wild boar. However, this is the only specimen that falls into that category. The GLP measurements are generally positively skewed. This may imply that there were more (smaller) females in the assemblages. However, the scapula fuses at a very early age and there is the possibility that post-fusion growth is a significant factor. Although porous specimens were not measured, it is possible that the smaller specimens may often have belonged to younger animals that were not fully grown.

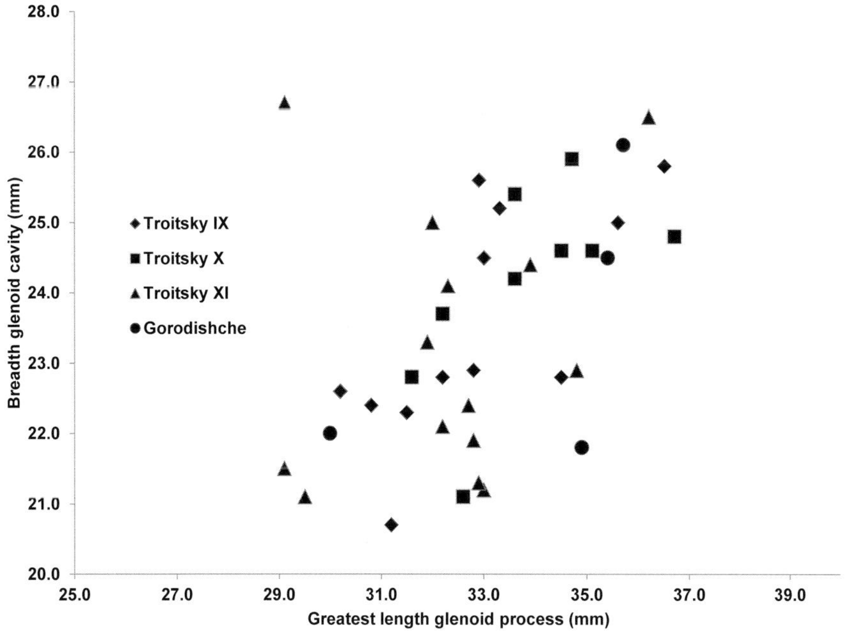

Figure 6.32: *Troitsky and Gorodishche pig scapula measurements*

The average size of the BG of 40 scapulae from the Troitsky sites measured 23.6 mm, again slightly smaller than the five specimens from Gorodishche (24.0 mm). The five specimens from the upper spits of Troitsky XI averaged only 22.9 mm. Again small sample sizes hinder interpretation of these variations.

Humerus

Three measurements were routinely taken: the greatest distal breadth (Bd); the breadth of the distal trochlea (BT) and (less systematically) the greatest height of the trochlea (HT) (Table 6.14). Again, the distal epiphysis fuses at an early age and some post-fusion growth is to be expected (Albarella and Payne 2005). Fifty specimens from the Troitsky sites had an average Bd of 37.0 mm, identical to the average of 17 specimens from Gorodishche. The 12 specimens from the lower spits on Troitsky XI were on average slightly larger (38.0 mm) than the ten examples from the upper layers (36.0 mm). Unsurprisingly, as the measurement is taken in the same plane as the Bd, the BT results showed very similar trends with the average size of 44 specimens from the Troitsky sites (30.3 mm) being almost identical to the 18 from Gorodishche (30.2 mm). The ten humeri from the upper spits of Troitsky XI were on average smaller (29.5 mm) than the 13 from the lower spits (31.5 mm). There is greater variation in the sizes of humeri from Troitsky XI. This is mainly due to the presence of one particularly large specimen (Bd = 41.8 mm; BT = 39.2 mm) from Spit 15. This falls within the range of wild boar (Rowley-Conwy and Owen 2011). However this and perhaps a handful of other large specimens are the only ones that could be considered to be from wild animals. The majority fell comfortably within the size range of domestic pigs.

Radius

The greatest breadth (Bp) and depth (Dp) of the proximal radius were routinely measured. Again, this early-fusing epiphysis is well known to have substantial post-fusion growth (Albarella and Payne 2005) and therefore size may partially reflect maturity as well as stature. Sixty-three specimens from the Troitsky sites had an average Bp of 26.9 mm compared with 27.7 mm from 20 specimens from Gorodishche (Table 6.15). Again, the radii from the lower levels at Troitsky XI tended to be on average slightly larger (27.3 mm) than those from the upper spits (26.4 mm). Similar trends were also reflected in the Dp measurements (Table 6.15). Figure 6.33 plots the two measurements and it can be seen that the majority of the Gorodishche measurements fell within the upper part of the size range. As none of the pigs fell within the size range usually attributed to wild boar (Rowley-Conwy and Owen 2011), it is possible there was a gradual decline in the average size of pigs during the medieval period in Novgorod and its hinterland. Significantly, four of the five smallest specimens from Troitsky XI plotted in Figure 6.33 were from upper layers dating to the mid-12th century or later.

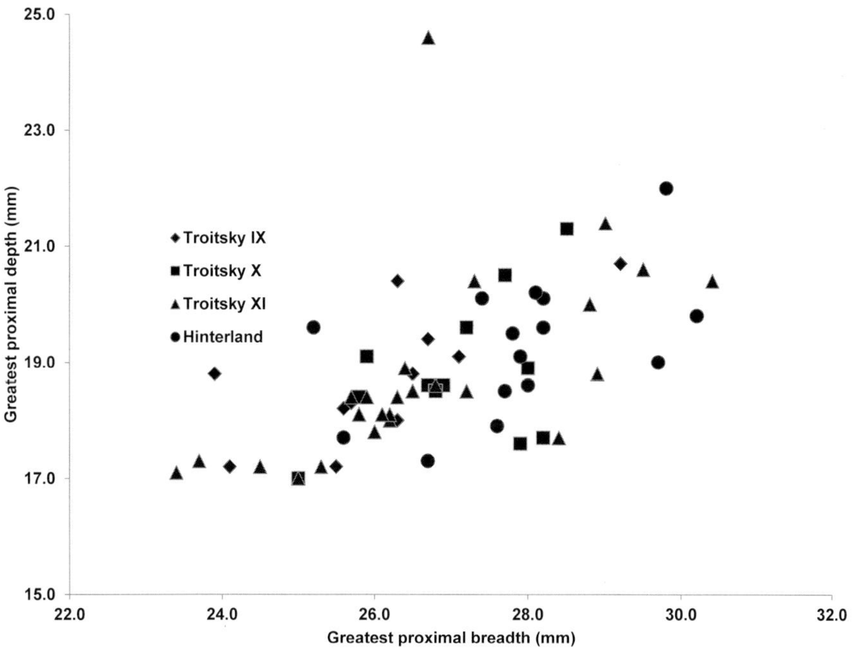

Figure 6.33: *Troitsky and hinterland sites pig radius measurements*

Tibia

The distal tibia is a later-fusing epiphysis and therefore is less likely to suffer from post-fusion growth. The greatest breadth (Bd) and depth (Dd) measurements were taken where possible on fused specimens (Table 6.16; Figure 6.34). A substantial sample of over 100 specimens was obtained from the Troitsky sites. The average Bd measured 28.2 mm, again slightly lower than the average from the small sample from Gorodishche (29.1 mm). The 21 specimens from the upper spits of Troitsky XI were also slightly smaller on average (27.4 mm) than the 32 from the lower layers (28.6 mm). The six smallest specimens from Troitsky XI are all from Spit 8 or above. One exceptionally large specimen from Troitsky XI, Spit 20, Property G (34.0 mm) is likely to have derived from wild boar (Figure 6.34) but the vast majority were again no larger than specimens recorded as domestic pigs on other medieval European sites. The Dd measurements follow the same trends as the Bd measurements (Table 6.16).

Astragalus

Lateral (GLl) and medial lengths (GLm) and (less frequently) distal breadth (Bd) measurements were recorded. Porous specimens were not routinely measured but

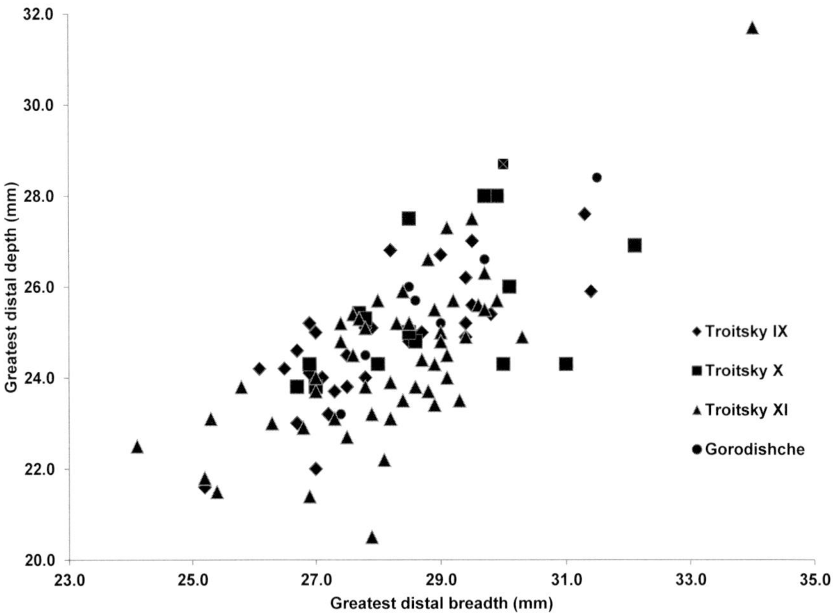

Figure 6.34: Troitsky and Gorodishche pig tibia measurements

the assemblage probably contains specimens from immature animals that had not attained their full size (Table 6.17). The analysis of these measurements showed similar trends to those observed on other pig bones. To take the GLl as an example, the average size of 90 specimens from the three Troitsky sites (38.7 mm) was slightly lower than the ten specimens from the hinterland sites (39.6 mm). The average size of the 39 specimens from the lower spits of Troitsky XI was in turn slightly higher (39.0 mm) than the 26 specimens from the upper spits (37.9 mm). Similar results were obtained from the analysis of the Glm and Bd measurements (Table 6.17). Six of the eight smallest specimens plotted in Figure 6.35 came from the upper fills of Troitsky XI.

Discussion of pig sizes

Although individual samples were often small, the analysis showed some consistent patterns. The first observation is that the vast majority of the measured specimens fell comfortably within the size range of domestic pigs. A few specimens were large enough to have potentially been from wild boar but it seems likely that the vast majority of the pigs found in Novgorod and Gorodishche were domestic. The second observation is that the pigs found at Gorodishche and perhaps the other hinterland sites (samples from Prost and Georgii were very small) tended to be slightly larger than those from Novgorod. At Troitsky XI, more small pigs were found in the later deposits. Decreases in stature can be the result of many factors.

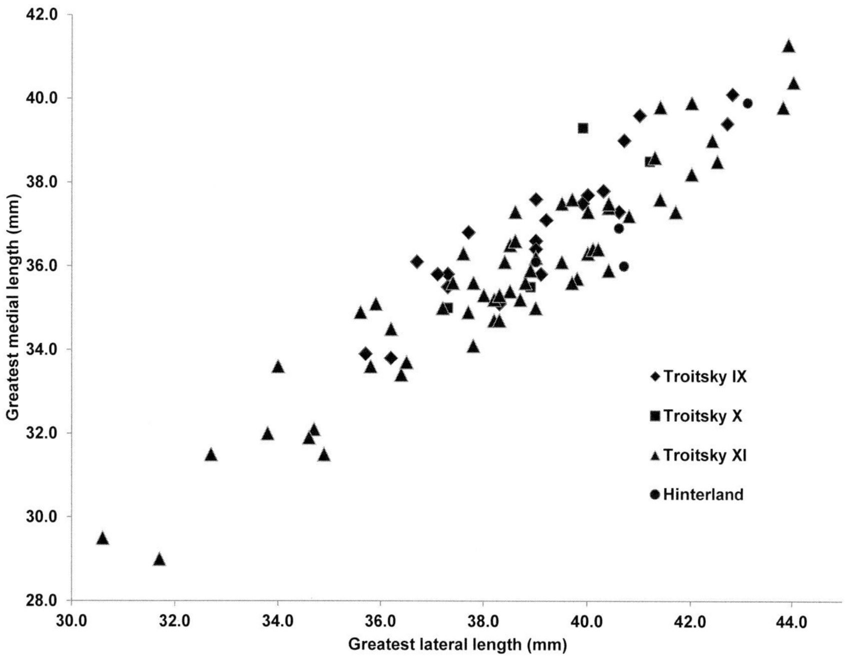

Figure 6.35: *Troitsky and hinterland sites pig astragalus measurements*

Given the possibility of post-fusion growth, a decrease in the average size may reflect the culling of younger animals. However, as discussed in Chapter 4, this was not the case in Novgorod, where there was a decrease in the relative number of juvenile pigs in the later deposits. The decrease may reflect the presence of a larger proportion of sows in the later periods. Unfortunately, it was not possible to analyse possible changes in the sex structure of pigs, because there was inconsistency in the recording of the presence of large (male) and small (female) canines in the jaws. Another possibility is that pigs in general became slightly smaller in the later medieval period. Decreases in woodland may have decreased the availability of suitable areas to fatten pigs for slaughter. Pigs kept in the town may also have been less well fed than in a more extensive woodland environment, although keeping pigs in sties in other periods and regions led to increases in size (*e.g.* in Roman Italy – MacKinnon 2001).

SHEEP AND GOAT

As discussed in Chapters 3–4, both sheep and goat were commonly represented in the Gorodishche and Troitsky assemblages. Distinguishing between the species can be difficult, even after the bones have fully developed and differentiation was not attempted on every element type. Many bones that were probably distinguishable on morphological grounds were not assigned to species during the rapid recording of metrical data from the Troitsky sites. It was estimated that around two-thirds of the sheep/goat assemblage from the Troitsky sites consisted of sheep. More sheep than goat bones were also recorded in the Gorodishche assemblage but goats were better represented than in the Troitsky assemblage. Generally, the sample sizes of measured bones from Gorodishche were too small for detailed analysis. Because of the small sample sizes, specimens from the upper and middle spits of Troitsky XI (S1–S13) were considered together.

Metacarpal

Some aspects of the metacarpal metrical data has been discussed in Chapter 4. Comparisons of the proximal breadth (Bp) against the greatest length (GL) from the three Troitsky sites indicated that both species were well represented with sheep narrowly outnumbering goat (Figure 4.37). Measurements of the proximal end only showed that the proximal breadth and depth (Dp) of goat bones were generally larger than those of sheep (Figure 4.39). Comparisons of the minimum shaft breadth and the distal breadth against the greatest length confirmed the slight dominance of sheep over goat in the metacarpal assemblage (Figures 6.36–6.37). The distal breadth measurements of specimens from the Troitsky sites that were assigned to sheep were rarely greater than 25 mm, whereas those of goats were nearly all larger than 25 mm (Figure 6.38). Accordingly, five of the six measured metacarpals originally unassigned to species were probably sheep, as none of the distal breadths were larger than 23.6 mm. One large specimen with a distal breadth of 28.4 mm probably belonged to a goat. On this basis, 15 (75%) of the fused (and measured) distal metacarpals belonged to sheep.

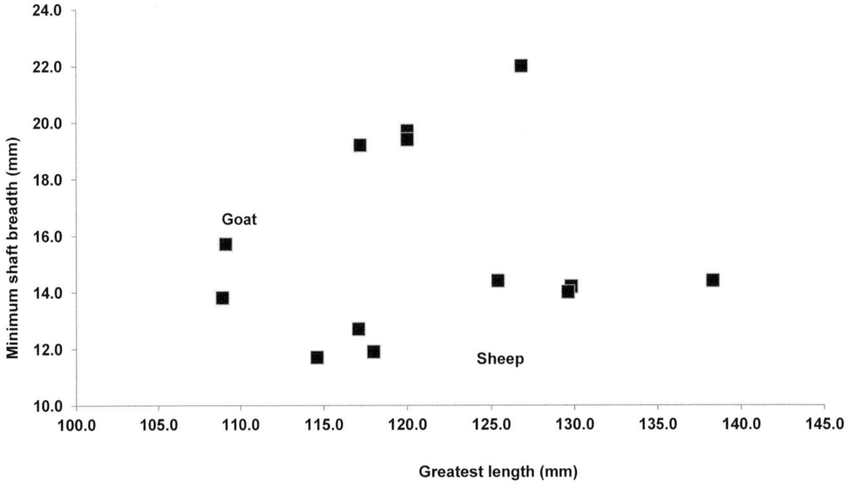

Figure 6.36: *Troitsky sheep and goat metacarpal length and minimum breadth measurements*

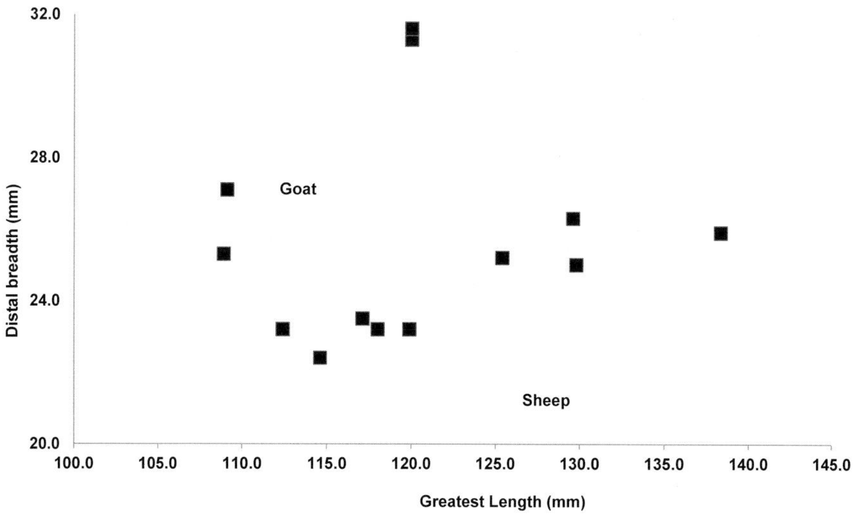

Figure 6.37: *Troitsky sheep and goat metacarpal length and distal breadth measurements*

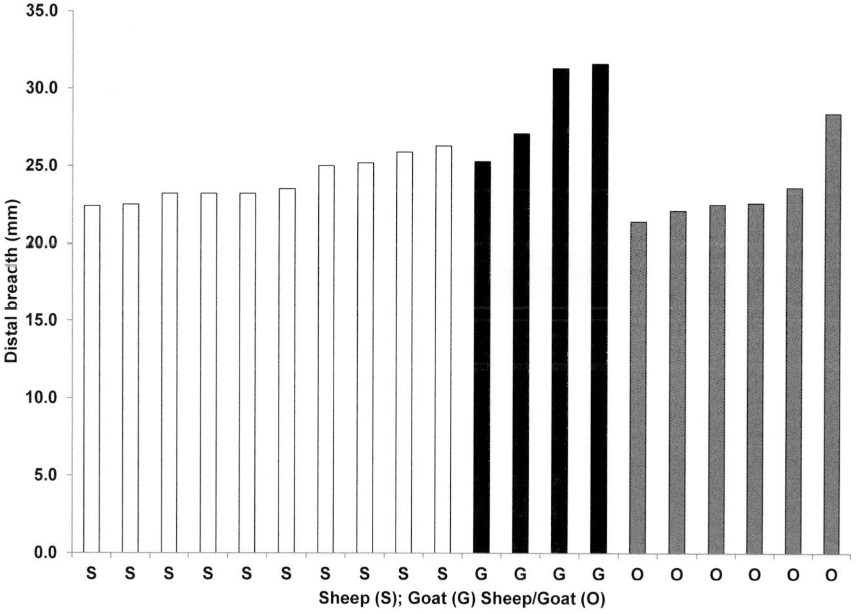

Figure 6.38: *Troitsky sheep and goat metacarpal distal breadth measurements*

Metatarsal

Measurements of the proximal breadth (Bp) showed no significant variations in the size of the specimens assigned to sheep and goat. The mean of all specimens from the three Troitsky sites was 20.1 mm (Table 6.19). There was a much clearer distinction in the size of the distal breadth (Bd). All four specimens assigned to goat were larger

(25.8–28.1 mm) than any of the 11 identified as sheep (21.9–25.3 mm). The four unassigned specimens all fell within the range of sheep. On this basis, 79% of the measured fused metatarsals probably belonged to sheep.

Horn core

The shapes of sheep and goat horn cores are quite easy to differentiate particularly in males, which grow significantly larger than those of females in both species (Boessneck 1969). The one complete horn core from Gorodishche belonged to a male goat with an outer curvature length of 158 mm. Only ten complete horn cores were measured from the Troitsky sites and most of these of both species were large male specimens (Table 6.20). Measurements of the greatest and minimum diameters of the horn core base confirmed that most of the horn cores of both sheep and goat came from males. Goats were better represented than in the case of other diagnostic bones (Figure 6.39). This indicates that larger horns were preferentially selected for horn-working. Some of these may have been brought into Novgorod after being detached from the carcasses of goats and sheep slaughtered elsewhere. The samples were too small to attempt assessments of chronological variations in the Troitsky XI assemblage.

Scapula

Differentiation between sheep and goat specimens based on morphological criteria was not attempted during the scan. Three measurements were routinely taken: the

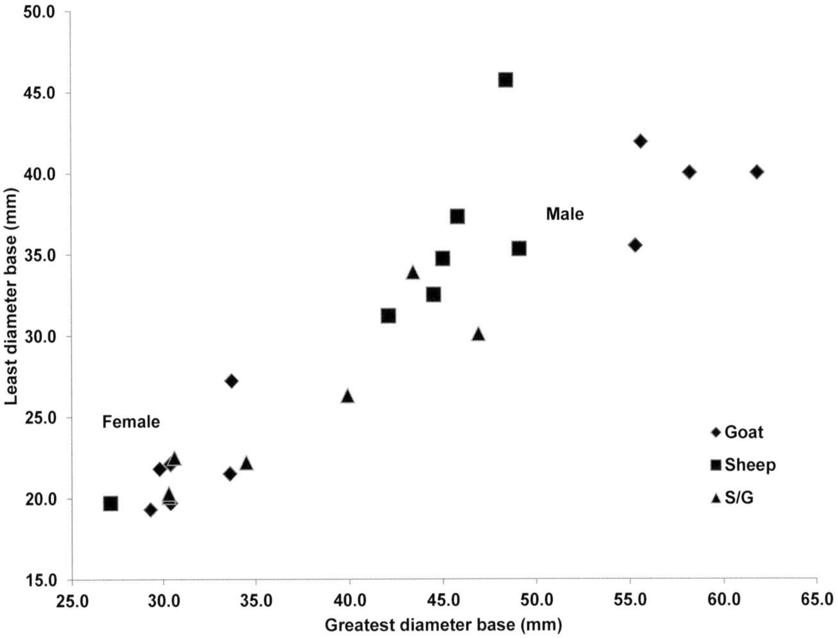

Figure 6.39: *Troitsky sheep and goat horn core measurements*

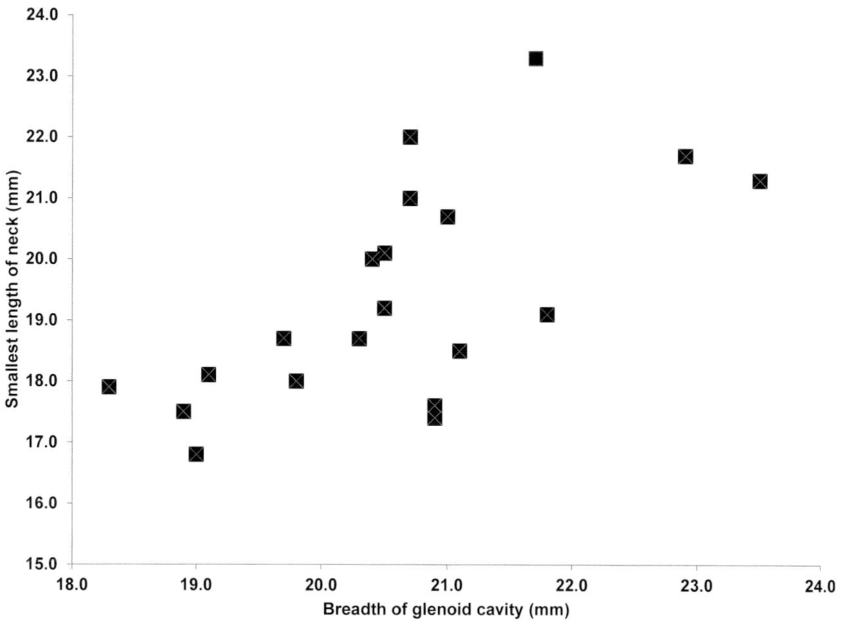

Figure 6.40: *Troitsky sheep and goat scapula measurements*

greatest length of the glenoid process (GLP); the breadth of the glenoid process (BG) and the smallest length of the neck (SLC). Very few scapulae from Gorodishche were measurable.

Although the Novgorod specimens are split by site and layer, only the samples from the lowest spits of Troitsky XI provided over ten specimens (Table 6.21). Given that the measurements are likely to be from both sheep and goats, analysis of size variations was not feasible. Comparisons of the BG and SLC measurements can be seen in Figure 6.40. On the basis of measurements from other bones, it is likely that most of the smaller specimens (BG <21 mm; SLC <20 mm) belonged to sheep whereas more of the larger ones were from goats.

Humerus

Differentiation between sheep and goat based on morphological criteria was not routinely attempted during the scan. Three measurements were taken: the greatest distal breadth (Bd); the breadth of the distal trochlea (BT) and the greatest height of the trochlea (HT) (Table 6.22). Given that the sample was comprised of indeterminate numbers of sheep and goats, analysis of size variations was not viable. Comparisons of the BT and HT measurements can be seen in Figure 6.41. It is again likely that most of the smaller specimens belonged to sheep, whereas most of the larger ones were from goats. However, although there are some interesting clusters, it was not possible to determine whether these were species-related.

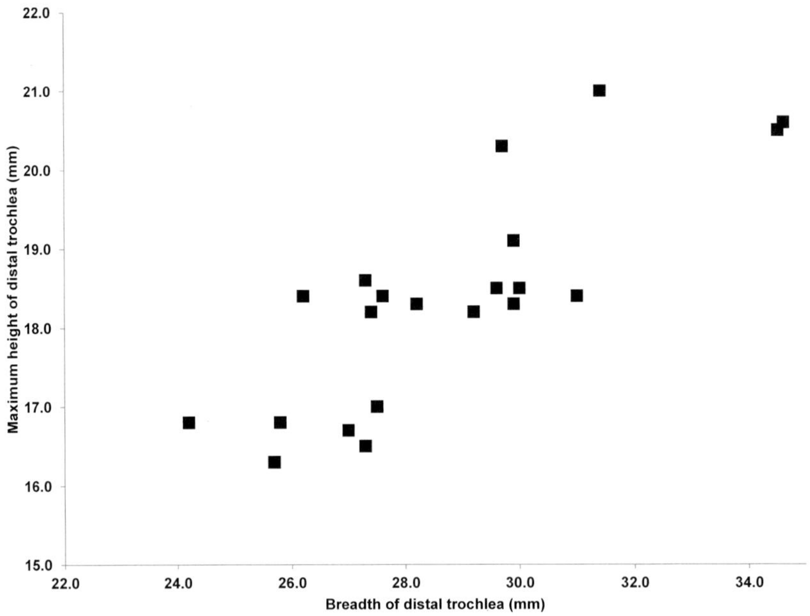

Figure 6.41: *Troitsky sheep and goat humerus measurements*

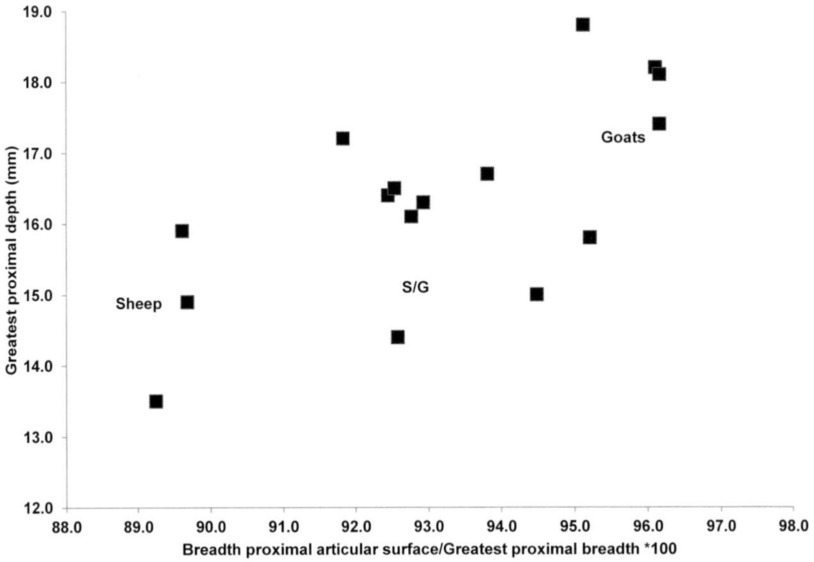

Figure 6.42: *Troitsky sheep and goat radius measurements*

Radius

There were no measurable radii from any of the hinterland sites. On the Troitsky sites three measurements were taken: the greatest proximal breadth (Bp); the greatest breadth of the proximal articular surface (BFp) and the greatest proximal depth (Dp).

Ranges and mean sizes were fairly similar throughout the deposits but most samples were very small and the presence of both sheep and goat makes interpretation of the results very difficult (Table 6.23). Although differentiation of sheep and goat specimens was not routinely attempted, there are some quite distinctive morphological variations at the proximal end of the bone (Boessneck 1969; Prummel and Frisch 1986). These morphological distinctions mean that the ratio between the BFp and Bp tends to be higher in goats than sheep, although there is some overlap (Salvagno and Albarella 2017). Following their analysis, this ratio is plotted against the Dp measurements in Figure 6.42. The seven specimens with BFp/Bp*100 of >93% were almost certainly from goats. The three with BFp/Bp*100 of <91% were from sheep. The six specimens with BFp/Bp*100 of 91%–93% could have been from either species. Because of this overlap, it is not possible to determine the exact ratio of sheep to goats, although goats may have been better represented by radii than by some of the other bones. The goat radii Dp tended to the larger than those of sheep (Figure 6.41) and this also applied to the Bp and BFp measurements.

Tibia

The distal ends of sheep/goat tibiae survived in larger numbers than most other bones and measurements of the greatest distal breadth (Bd) and greatest distal depth (Dd) were obtained quite frequently (Table 6.24). However, there was still the problem of not being able to distinguish between sheep and goat on morphological grounds and recently devised measurements to attempt to separate sheep from goats (Salvagno and Albarella 2017) were not taken. Comparisons of the two measurements from specimens from the Troitsky sites showed there was quite a wide range in sizes (Figure 6.43). Most of the larger specimens were probably goats and most of the smaller specimens were probably sheep but there was probably substantial overlap in the sizes.

Astragalus

Three measurements were taken on sheep/goat astragali: the greatest length of the lateral half (GLl); the greatest length of the medial half (GLm) and the greatest distal breadth (Bd) (Table 6.25). The GLm and Bd measurements from the Troitsky sites are compared in Figure 6.44. Measurements that have been created recently to successfully separate sheep and goat astragali (Salvagno and Albarella 2017) were not taken. Therefore, again it is only possible to note the comparatively wide range in sizes and postulate, based on measurements of other elements, that most of the larger specimens (*e.g.* GLm >28.0 mm) belonged to goats whereas most of the smaller astragali (*e.g.* GLm <26.00 mm) belonged to sheep (Figure 6.44). An additional problem is that some of the specimens may have belonged to young animals whose astragali had not attained full size, although porous specimens were not measured. There was much less variation in the sample from the lowest spits from Troitsky XI than in other

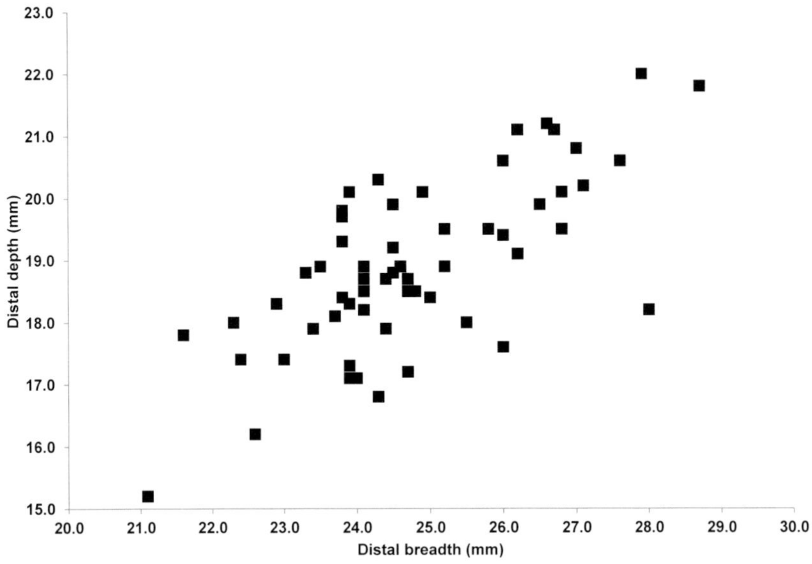

Figure 6.43: *Troitsky sheep and goat tibia measurements*

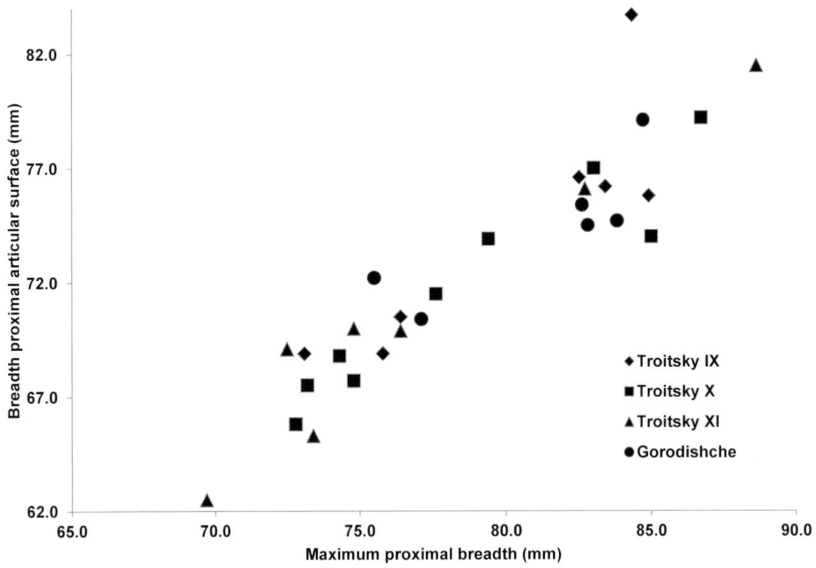

Figure 6.44: *Troitsky and hinterland sites horse radius proximal measurements*

samples (Table 6.25), perhaps suggesting that there was greater diversity in the sizes of sheep and goats in the later medieval period. However, further work needs to be undertaken to distinguish between the bones of the two species to determine whether this tentative interpretation can be sustained.

Withers heights

No complete limb bones of either sheep or goat were recovered from Gorodishche, so estimates of withers height were not possible. Estimates were obtained from 21 sheep and ten goat complete limb bones from the Troitsky sites using the conversion factors from greatest length measurements devised by Teichert (1975) and Schramm (1967) for sheep and goat respectively. Most of the withers height estimates were derived from the lengths of metacarpals and metatarsals (Tables 6.18–6.19) and these estimates are summarized in Table 6.26. The sheep were quite tall ranging between 55.0 cm and 67.2 cm, with an average withers height of 59.9 cm. The mean heights of the sheep from Troitsky XI (58.4 cm) were slightly smaller than those from Troitsky IX–X (60.7 cm), although both sample sizes were small and the differences were not statistically significant.

The goats tended to be a little taller than sheep, ranging in height between 60.8 cm and 72.9 cm and averaging 66.4 cm.

Discussion of sheep and goat sizes

Metrical analysis has assisted in separating sheep from goat. Amongst the measurable specimens, sheep consistently outnumbered goats with percentages usually in the range of 60–80% dependent upon which bones and measurements were under consideration. The main exceptions to this pattern were the horn cores which were preferentially selected by size. There was a focus on the acquisition of horns of male goat in particular. There was insufficient evidence from the Troitsky sites to investigate possible chronological changes in the stature of sheep and goats.

HORSE

Horse bones were generally less fragmented than those of cattle and as most of their bones belonged to adults, a higher proportion of the horse assemblage was measurable. The following discussion assumes that all equid bones belonged to horses rather than mules or donkeys. No attempt was made during the scan of the Troitsky material to differentiate between these species. At Gorodishche, only horse has been positively identified during this study and other discussions of metrical data from both Gorodishche (see Sablin below) and Novgorod (Tsalkin 1956) have not produced any evidence for mules. It is therefore assumed that all, or at least the vast majority, of the measured equid bones belonged to horse.

Scapula

Four measurements were routinely taken on horse scapulae following von den Driesch (1976): the greatest length of the glenoid process (GLP); the length of the glenoid cavity (LG); the breadth of the glenoid process (BG) and the smallest length of the neck (SLC). SLC measurements were not taken as systematically as the others and are less reliable in their accuracy.

Measurements were obtained from four horse scapulae from hinterland sites and 22 from the Troitsky excavations. Summaries are provided in Table 6.27. The average sizes of the Troitsky measurements were higher than their equivalents from Gorodishche and Georgii but the hinterland samples were too small for statistical comparisons.

Humerus

Measurements of the breadth of the trochlea (BT) were taken on nine specimens from Gorodishche and 14 from the Troitsky sites (Table 6.28). The average size of the Troitsky specimens (72.8 mm) was slightly larger than those from Gorodishche (71.7 mm), where the humeri were slightly more variable in size.

Radius

The following measurements were taken following von den Driesch (1976): the greatest proximal breadth (Bp); the greatest breadth of the proximal articular surface (BFp), the greatest proximal depth (Dp); the greatest distal breadth (Bd) and, on complete bones, the greatest length (GL) the lateral length (GLl) and the smallest breadth of the diaphysis (SD).

The mean proximal breadth measurements obtained from the Troitsky sites were in this instance slightly smaller (Bp 78.5 mm; BFp 72.6 mm) than their counterparts from Gorodishche (Bp 81.3 mm; BFp 74.4 mm), although the latter were again based on very small samples (Table 6.29). Comparisons of the two measurements (Figure 6.44) showed a bimodal distribution with one cluster having Bp measurements of <80 mm and another with values of >82 mm. Specimens from both Gorodishche and Novgorod fell into these two clusters. Whether the clusters are largely related to sexual dimorphism, the presence of more than one type of horse, or merely down to chance cannot as yet be determined.

The distal breadth measurements from the Troitsky sites were also broadly bimodal, with half the specimens measuring <72.0 mm, whereas larger specimens had a mode of 76-0-77.9 mm. All five measured distal radii from Gorodishche fell into the larger size group, having breadths of 74.0 mm and above (Figure 6.45).

Tibia

Measurements of the distal breadth (Bd) and distal depth (Dd) were routinely taken. In contrast to the radius, there was quite an even distribution across the range of the measured specimens from the Troitsky sites (Figure 6.46). The means of these measurements was very similar in the assemblages from the lower spits of the three Troitsky sites (Table 6.30). The distal tibiae from Gorodishche tended to be a little larger than those from Novgorod but again the sample was very small (Table 6.30; Figure 6.46).

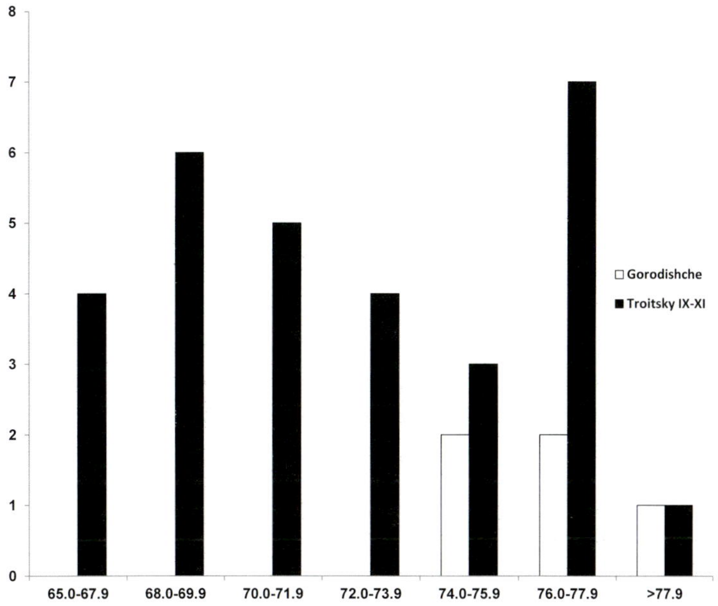

Figure 6.45: *Troitsky and hinterland sites horse radius distal breadth measurements (mm)*

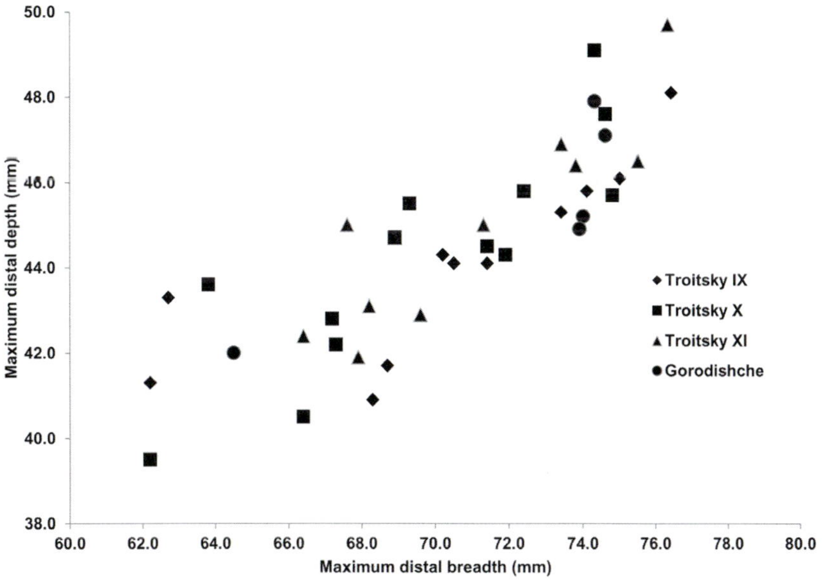

Figure 6.46: *Troitsky and Gorodishche horse tibia measurements*

Astragalus

The greatest heights (GH) of 24 astragali from the Troitsky sites were measured (Table 6.31). There was a unimodal distribution with a mean of 57.2 mm. The one specimen from Gorodishche was one of the largest measured (62.7 mm).

Calcaneus

Six specimens from the Troitsky sites had a mean greatest length (GL) of 105.5 mm (Table 6.31). Three calcanea from Gorodishche had a mean greatest length of 108.2 mm and included the largest specimen (111.8 mm).

Third metacarpal

The following measurements were taken: greatest length (GL); lateral length (GLl); maximum proximal breadth (Bp); maximum proximal depth (Dp); maximum distal breadth (Bd). The smallest breadth of the shaft (SD) was recorded on complete specimens.

The proximal breadth and depth measurements from the Troitsky sites produced means of 48.9 mm and 32.8 mm respectively (Table 6.32). A higher proportion of larger specimens were found on the Troitsky X site, including two bones which were significantly larger than the others. In contrast most of the smaller proximal metacarpals came from Troitsky IX (Figure 6.47). The mean sizes of the proximal breadth and depth measurements from Gorodishche were slightly larger than the combined Troitsky means (49.1 mm and 31.9 mm respectively) but these were based on small sample sizes (Table 6.32).

Distal breadth (Bd) measurements followed a similar pattern with the average size of those from Troitsky X (46.5 mm) being slightly smaller than those from the other sites. The mean size of all the specimens from the Troitsky sites (47.7 mm) was slightly smaller than those from Gorodishche (48.6 mm), which contained none of the smallest metacarpals (Table 6.32; Figure 6.48).

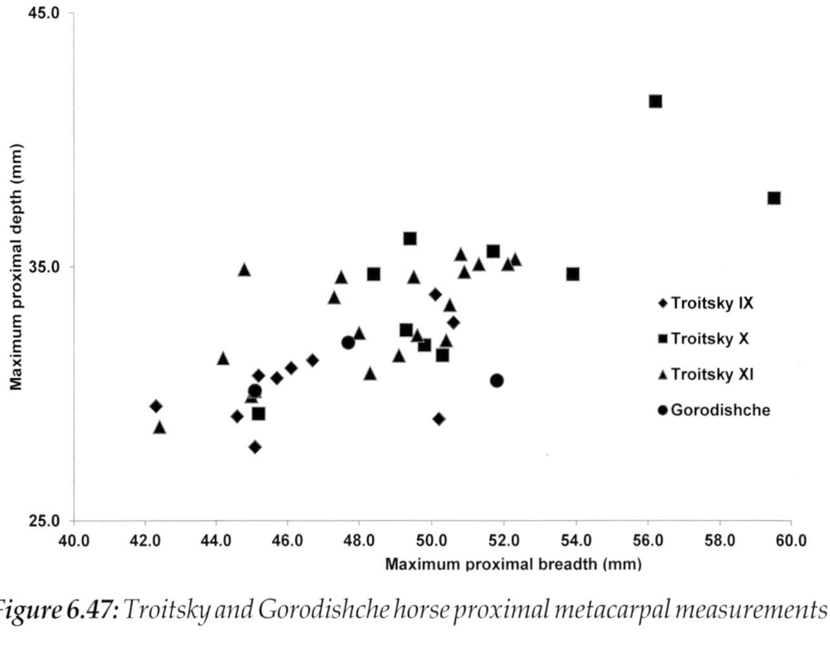

Figure 6.47: *Troitsky and Gorodishche horse proximal metacarpal measurements*

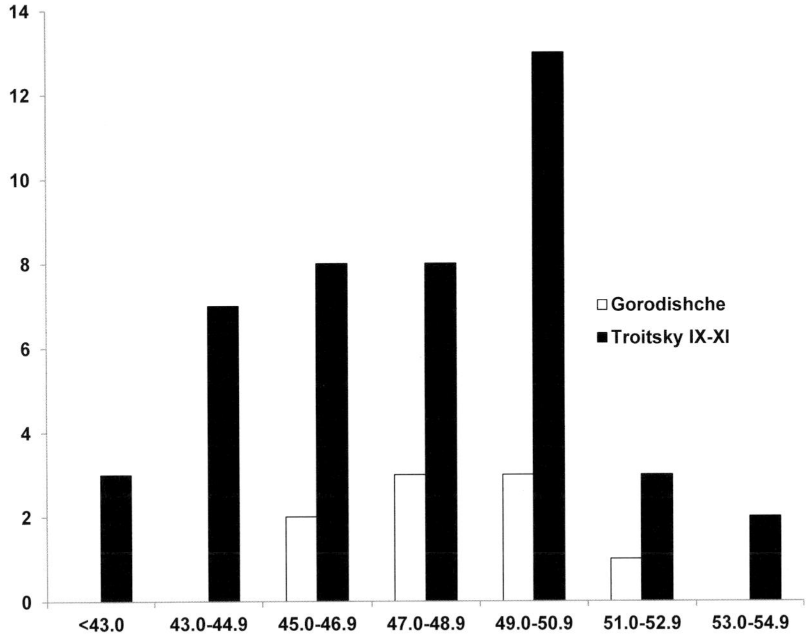

Figure 6.48: *Troitsky and Gorodishche horse metacarpal distal breadth measurements (mm)*

Many of the horse metapodials survived complete, which allowed comparisons of length and breadth measurements to be compared quite frequently. Plots of the greatest lengths (GL) and minimum shaft breadths (SD) of metacarpals are shown in Figure 6.49. These show some variations between the sites. Most of the metacarpals from Troitsky IX were small (GL<215 mm) and included some slender specimens. In contrast, most of the specimens from Troitsky X were larger (GL>215 mm), mirroring the variations observed in the proximal and distal breadth measurements of these and other specimens. The metacarpals from Troitsky XI were more evenly spread across the range but included most of the smallest specimens (GL<205 mm).

Third metatarsal

The following measurements were taken: greatest length (GL); lateral length (GLl); maximum proximal breadth (Bp); maximum distal breadth (Bd). The smallest breadth of the shaft (SD) was recorded on complete specimens.

The proximal breadth measurements from the Troitsky sites and Gorodishche produced means of 49.0 mm and 50.0 mm respectively, although the latter was based on only four measurements (Table 6.33). As with the metacarpal, the metatarsals from Troitsky IX (mean 48.1 mm) were generally slightly smaller than those from Troitsky X (mean 49.5 mm), partly because the latter included one very large specimen (58.5 mm). The metatarsals from Troitsky XI provided a mean of 49.1 mm (Table 6.33: Figure 6.50).

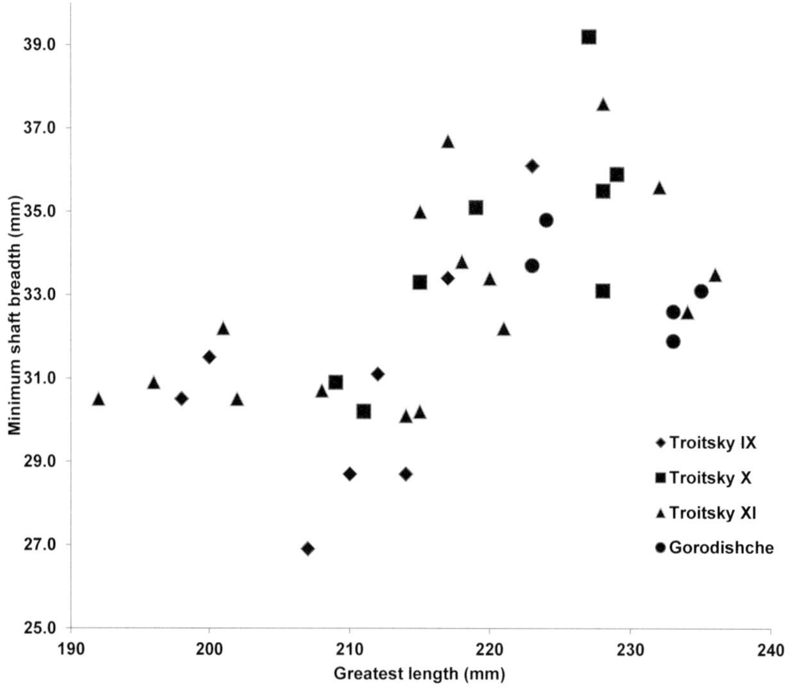

Figure 6.49: *Troitsky and Gorodishche horse metacarpal greatest length and minimum shaft breadth measurements*

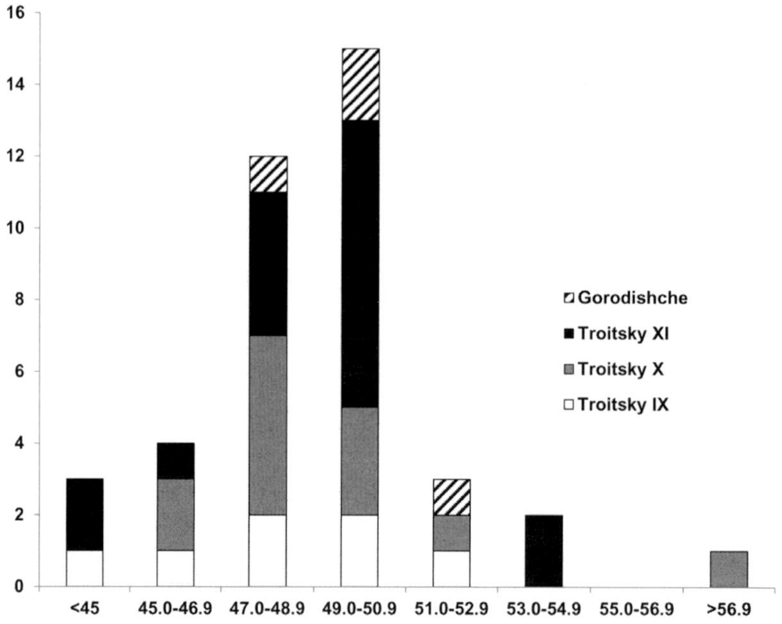

Figure 6.50: *Troitsky and Gorodishche horse metatarsal proximal breadth measurements (mm)*

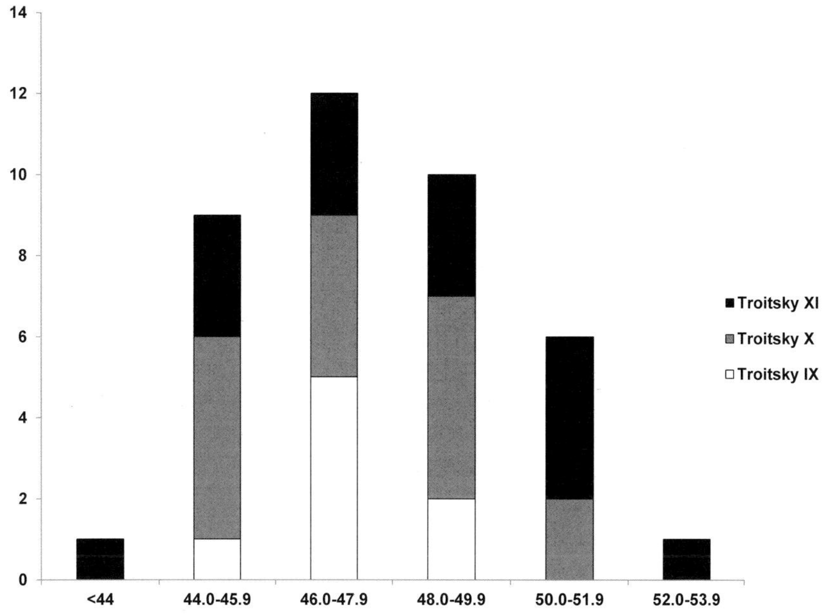

Figure 6.51: *Troitsky horse metatarsal distal breadth measurements (mm)*

Distal breadth measurements from the three Troitsky sites averaged 47.5 mm. The mean size of specimens from Troitsky XI (48.1 mm) was slightly larger than those from Troitsky IX (47.2 mm) and Troitsky X (47.1 mm) (Table 6.33; Figure 6.51). Gorodishche produced only two distal breadth measurements (Table 6.33).

The greatest lengths (GL) and minimum shaft breadths (SD) of complete metatarsals from the Troitsky sites are shown in Figure 6.52. Again there were some interesting clusters. Most had SD/GL indices of <0.12 but a group of broader specimens (SD/GL >0.12) can also be observed on some bones of medium length (GL 250–273 mm) from all three sites. Most of the smallest of the slender bones (GL<250 mm) came from Troitsky XI. Variations in the slenderness of horse limb bones have been observed on Russian sites (see Sablin below), which may be related to the presence of different types, although other factors such as sexual dimorphism also need to be considered.

Withers heights

Calculations of withers height based on the lengths of limb bones were made using the conversion factors for lateral lengths (Kiesewalter 1888) and greatest lengths (Vitt 1952) respectively. Measurements of greatest lengths are considered to have been more reliably recorded and therefore Vitt's conversion factors will be employed in this analysis (radius 4.111; tibia 3.947; metacarpal 6.102; metatarsal 5.239).

Withers height estimates were obtained from 93 bones from the Troitsky sites and 12 from Gorodishche. Height estimates ranged between 117 cm and 153 cm, roughly

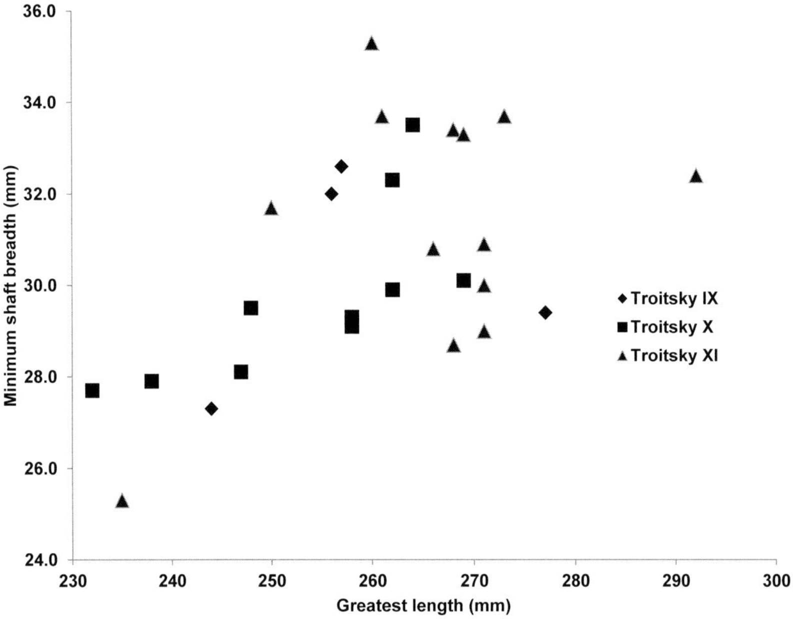

Figure 6.52: *Troitsky horse metatarsal greatest length and minimum shaft breadth measurements*

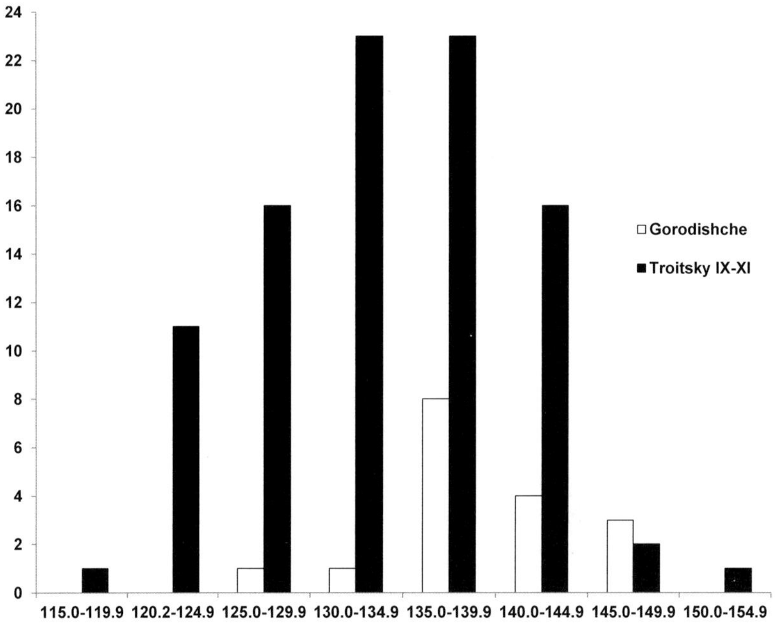

Figure 6.53: *Troitsky and Gorodishche horse withers height estimates (cm)*

equivalent to horses of 11.5 to 15.0 hands. The average height from the three Troitsky sites was 133.3 cm (13.1 hands). There was little variation in this estimate in the assemblages from the three sites (Table 6.34). The horses from Gorodishche were on average taller with 12 bones giving a mean withers height estimate of 140.0 cm (13.8 hands). Both assemblages had a unimodal distribution (Figure 6.53).

Log ratio analysis of breadth measurements

Log-ratio analysis was employed to increase the sample sizes of horse breadth measurements. These were compared against a set of standard measurements derived from the average of three modern Mongolian ponies (Johnstone 2004, 148).

The following breadth measurements from the Mongolian ponies were selected for comparison: humerus: BT (72.9 mm); radius: Bp (79.9 mm); tibia: Bd (71.6 mm); metacarpal: Bp (47.2 mm) and metatarsal Bp (48.7 mm). These measurements were converted into base-10 logarithms. The breadth measurements of all the specimens from Gorodishche and Troitsky sites were also converted to base-10 logarithms and compared against the equivalent standards. A positive value indicates that the archaeological specimen was larger than the equivalent measurement of the Mongolian ponies, whereas a negative value indicates that it was smaller.

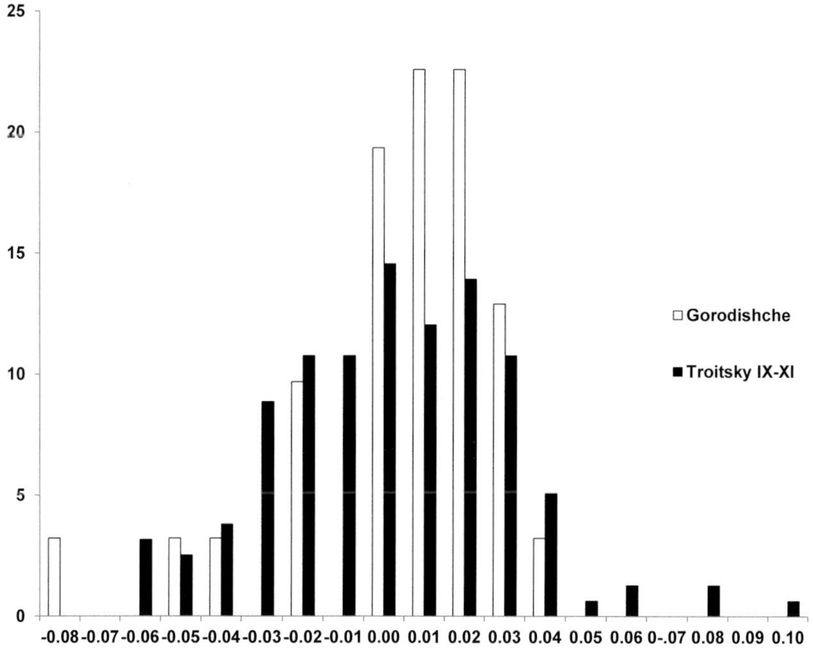

Figure 6.54: Troitsky (N=158) and Gorodishche (N=31) log ratios of horse breadth measurements

Comparisons of log ratios between Gorodishche and the three Troitsky sites (Figure 6.54) showed that although Gorodishche had the smallest specimen, 81% had log ratios equal or above those of the Mongolian ponies, compared with only 60% of those recorded on the Troitsky sites, indicating that a higher proportion of larger horses were represented at Gorodishche. However, Troitsky X produced six bones with values of ≥0.5, larger than any found on the other sites.

Discussion of horse measurements

Although there were some smaller and larger horses, most were the size of large ponies. The horses from Gorodishche were on average a little larger than those from Novgorod, although the largest specimens all came from Troitsky X. There were some variations in the sizes of horses from the Troitsky sites, with those from Troitsky IX tending to be a little smaller than those from Troitsky X. Some measurements formed clusters of specimens with similar dimensions, which would repay analysis on other material. In particular, the slenderness of metapodials varied markedly (Figures 6.49 and 6.52). It is likely that the horses kept and exploited in Novgorod were obtained from many different sources.

DOGS

A summary of the main measurements taken on dogs from Gorodishche and the Troitsky sites is provided in Table 6.35. In most cases the samples are too small to analyse in detail, particularly in respect of Gorodishche. The largest samples from the Troitsky sites were obtained from the humerus, radius and tibia. The standard deviations of both the length and breadth measurements of these bones showed a much larger amount of variation than encountered for any of the other domestic species, with coefficients of variation of 19% and above. This is not surprising as dog breeding has resulted in huge variations in their sizes. The samples are too small to investigate chronological variations.

Shoulder height estimates derived from conversion factors of the greatest length of limb bones (Harcourt 1974) also illustrates this diversity (Figure 6.55). The smallest dog from the Troitsky sites had an estimated shoulder height of 33.4 cm, about the height of a small modern beagle (Table 6.35). This was the only complete limb bone of a dog of under 40 cm high (Figure 6.55). Most of the dogs had estimated heights of 40–45 cm, roughly the height of a modern cocker spaniel. Bones of a few larger dogs were found at both Gorodishche and Troitsky. Four complete limb bones came from dogs with shoulder heights of 55–60 cm, in the size range of modern English setters. Estimates of the heights of dogs from later excavations at Gorodishche based on skull measurements suggested that larger dogs were more common there than on the Troitsky sites (see Sablin below). The largest specimen

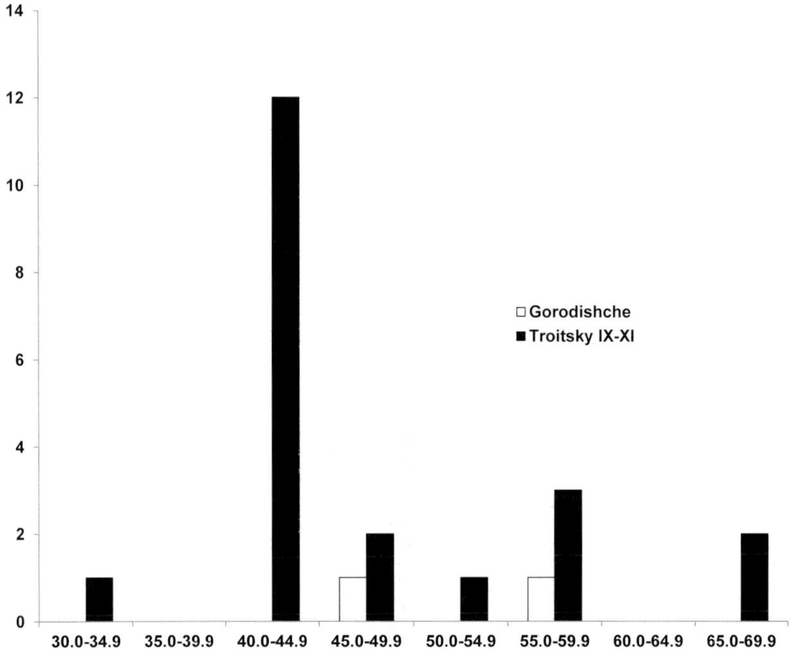

Figure 6.55: *Troitsky and Gorodishche dog shoulder height estimates (cm)*

from Troitsky had an estimated shoulder height of 69.3 cm, about the height of a greyhound. It should be emphasised that this metrical analysis can only estimate the stature and robustness of dogs rather than breed and it is likely that many of the dogs were mongrels.

Therefore most of the dogs found in Novgorod were of medium size and would have acted as companion animals and guard dogs to the inhabitants of the boyar properties. The high levels of gnawed bones indicated that they had extensive access to processing, cooking and table waste.

CATS

The most common measurements of cats from the Troitsky sites are shown in Table 6.56. These showed much less variation than dogs with coefficients of variation for length measurements being consistently around 4%. Breadth and depth measurements were a bit more variable with coefficients of variation ranging between 6% and 9%. All the measurements fell within the ranges of cats found on Viking and medieval sites in Northern Europe (O'Connor 2007). It is assumed that all the cats represented in the Novgorod assemblage were from domestic animals resident in the town.

COMMENTS ON METRICAL DATA FROM GORODISHCHE 2000–2004 (Mikhail Sablin)

The average withers height of cows calculated on the basis of the length of their metapodials (n=37), using Tsalkin's (1960) coefficients, is 112.4 cm, which accords well with the data collected by Maltby and Hamilton-Dyer (1995; 2001) from earlier excavations at Gorodishche (see above). The same parameter calculated using the Boessneck method (1956) came to 117.6 cm. It seems that the method used by Tsalkin gives the best estimate. The cattle from Gorodishche were taller than medieval cattle from the wooded zone of medieval Russia, the withers heights of which measured approximately 100 cm (Tsalkin, 1956) and were similar in size to medieval cattle from the middle reaches of the Dnieper – 114.1 cm (Timchenko, 1972).

The average withers height of horses calculated on the basis of the greatest length of the skulls found in the 9th-century ditch (Nehring 1884) came to 144.1 cm. The same parameter calculated on the basis of the length of the metapodials (n = 25) following Kiesewalter (1888) was 142 cm. The horses from Gorodishche fall into the category of medium-sized horses as classified by Vitt (1952). The withers height of the horses from Gorodishche was 10 cm greater than the average height of horses from the wooded zone of medieval Russia (Tsalkin 1956) and 7 cm greater than the average height of medieval horses from the wooded steppe zone (Timchenko, 1972). On the basis of the relative width of their foreheads (Gromova 1949; von den Driesch, 1976) three skulls from Gorodishche belonged to a group characterised by narrow foreheads, while the other should be categorised as belonging to a group characterised by foreheads of medium width. There was no sign of horses with wide foreheads (Asiatic steppe horses). On average the index for forehead width was 43.3% which coincides with the data provided by Tsalkin (1956) for domesticated horses from Novgorod and also with the data provided by Timchenko (1972) in relation to the middle reaches of the Dnieper. As regards the relative width of longbone shafts, three animals from Gorodishche should be classified as belonging to the group of slender-legged horses, nine to the group of semi-thin-legged horses and two to the group with legs of average size according to the classification by Brauner (1916). No semi-fat-legged horses (Asiatic steppe horses) were found. The average index for the width of the long-bone shaft for the group came to 14.9%, which coincides with the data from medieval Novgorod (Tsalkin, 1956) but is less than that for horses from the wooded steppe – 15.8% (Timchenko, 1972).

The withers height of dogs found at Gorodishche calculated according to the condylo-basal length of the skull (n=6) was approximately 60 cm, which would be similar to that of a large husky. Craniological features would also point to that type of dog.

– 7 –

THE EXPLOITATION OF WILD MAMMALS IN NOVGOROD AND SITES IN ITS IMMEDIATE HINTERLAND

Mark Maltby with Sheila Hamilton-Dyer, Andrei V. Zinoviev and Mikhail Sablin

INTRODUCTION

This chapter will be organised in the same manner as Chapter 5, looking first at the evidence of the presence and exploitation of wild mammals at Gorodishche and other sites in the immediate hinterland of Novgorod during the period slightly preceding its foundation. The second part of the chapter will be focused on evidence from Novgorod itself. As in previous chapters the evidence will be based principally on the bone evidence. Other evidence related to the exploitation of wild mammals will be incorporated into Chapter 11.

THE HINTERLAND SITES (Sheila Hamilton-Dyer, Mark Maltby, Ellen Hambleton and Mikhail Sablin)

Gorodishche 1979 and 1993–1998

Wild mammal bones were poorly represented in the assemblages obtained from excavations between 1979 and 1998. Only 94 bones were assigned to a total of nine species (Figure 7.1; Table 7.1). Wild mammal bones provided only 2% of the total mammal NISP counts (Figure 7.2). They were slightly better represented in assemblages from the South bank (4%) than in other assemblage groups (Table 7.2).

Elk (*Alces alces*) and bear (*Ursus arctos*) were the only large wild mammal species positively identified. The smaller mammals identified were hare (*Lepus* sp.), beaver (*Castor fiber*), pine marten (*Martes martes*), stoat (*Mustela ermina*), red squirrel (*Scirius vulgaris*), wolf (*Canis lupus*) and vole (*Microtus* sp). As discussed in Chapter 6, a few of the bones may have belonged to large wild boar but the vast majority were from domestic pigs.

Only seven elk bones were positively identified, forming 0.2% of the identified mammal assemblage. It is possible that a few of the unidentified large mammal bones also belonged to this species. All the positive and tentative identifications of elk in the pre-1998 assemblages were made on bones from the South bank. The 1998 excavations

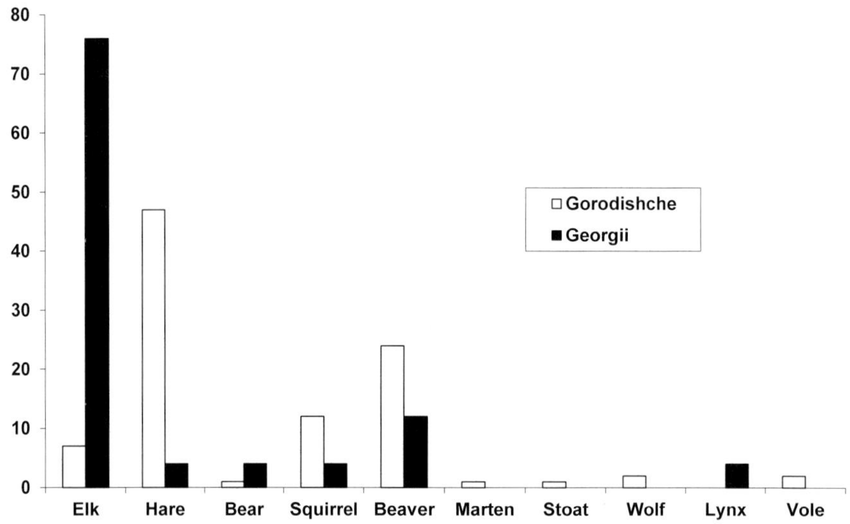

Figure 7.1: *Percentages of wild mammal species from Gorodishche (1979–1998) and Georgii*

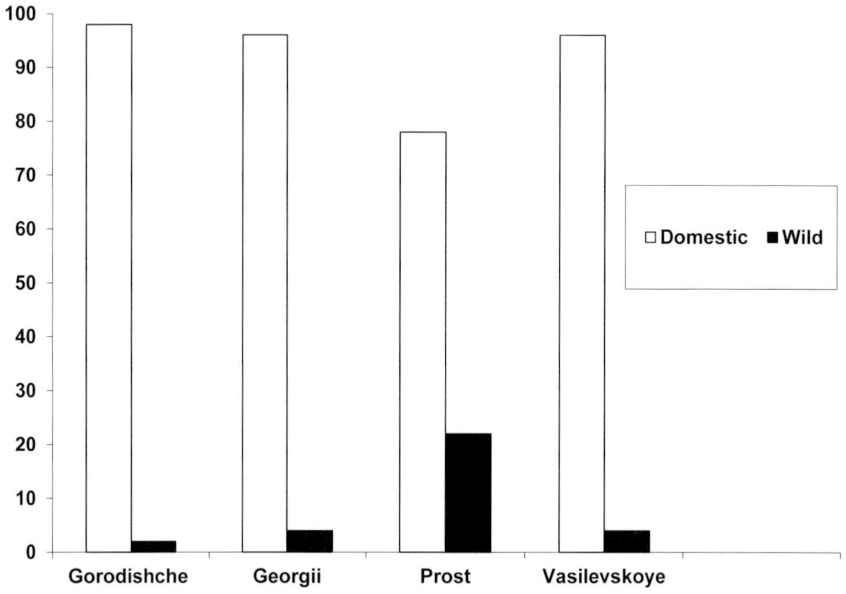

Figure 7.2: *Percentages of domestic and wild mammal bones from hinterland sites*

on the North bank produced three elk bones. There is little evidence that elk was exploited regularly at Gorodishche, although the bones of animals killed in the hunt need not necessarily have been brought back to the settlement, if field butchery took place.

The seven specimens positively identified as elk consisted of a maxilla, two other skull fragments, and fragments of two antlers, a pelvis and a metatarsal (Table 7.3).

Figure 7.3: Butchered elk skull from the 1998 excavations at Gorodishche. Photograph by Ellen Hambleton

In addition, seven fragments (2 tibiae; radius; skull; astragalus; calcaneus; centroquartal) recorded as unidentified large mammal may also have belonged to elk. Combs made out of elk antler have also been recovered from Gorodishche but are not considered in this analysis (Smirnova 2005). There is evidence for antler working. One of the elk antler fragments was a sawn offcut. One of the skulls found in 1998 included the base of the antler which had been partially severed through the pedicel prior to the removal of the antler stem. This skull and associated maxilla belonged to an adult male. Only the right side of the skull was present and it had been axially split (Figure 7.3). It is possible that this skull was butchered at the kill site. Splitting the skull in this manner would have made it easier to transport the antler whilst still attached to the skull. Splitting the skull would also have provided access to the brain. Butchery marks were also observed on the other elk skull fragment. This consisted of a zygomatic, which bore superficial chop marks on its lateral surface.

A skull of a wolf was found in the 1979 assemblage. A second skull and a tooth were found in 1998. Both skulls were from adults. Metrical data from the 1998 specimen are provided in Table 7.4. The size of the skull is significantly larger than any of the dog skulls from Novgorod and the lack of overcrowding of the cheek teeth is characteristic of wolf (Clark 1996). No evidence of butchery was found but the skulls may have been brought to the site with the pelts.

A third phalanx was the only bear bone identified from stratified deposits. It was found in the 1998 excavations. Other bones from bear claws have also been observed in mixed deposits of uncertain date (Nosov pers. comm.). These probably were attached to bearskins, although third phalanges could also have been used as pendants.

47% of the wild mammal remains belonged to hare, mostly from the South bank (Figure 7.1; Table 7.1). They still only formed 1% of the total mammal assemblage. There are two species which now occur in the region, brown hare (*Lepus europaeus*) and mountain hare (*Lepus timidus*). The sizes of the two species overlap and none of the diagnostic skull elements were present. Therefore the bones have not been identified beyond genus level. However, the brown hare is believed only to have been introduced to Russia in the 18th century to enrich royal hunting grounds (Thulin 2003; Neronov *et al.* 2008, 84). Therefore it is assumed that all these bones belonged to the native mountain hare. Six of the hare bones came from one skeleton located in Square 5 of the South bank excavations. A humerus had been cut during processing for meat.

Beaver provided 24% of the identified wild mammal bones (Figure 7.1), forming 0.5% of the total identified mammal assemblage (Table 7.1). The 23 fragments were retrieved from both the North and South banks, and from the 1979 excavations. Five of the bones bore butchery marks. A femur from the 1979 excavations had been chopped through the distal part of the shaft and an ulna also found during that season had been scraped down the medial side of the shaft. The 1998 excavations produced a beaver femur that had been chopped through both the proximal and distal ends. It also bore superficial blade marks on its shaft. Another femur had finer knife cuts running obliquely on its shaft. A mandible had been chopped through the ramus and had been cut on the outer (buccal) aspect below the cheek teeth. Only the cuts on the mandible are likely to have been associated with skinning. The other butchery marks resulted from segmenting the carcass and removing meat. The marks provide clear evidence that beaver meat was occasionally consumed at Gorodishche.

All 11 squirrel fragments were recovered from the South bank, both from the (lower) brown and the (upper) black layers (Table 7.1). Five of the squirrel bones from the South bank were from the same animal. The single bone of stoat was found in the North bank black layer. A mandible from an adult pine marten was recovered from the 1998 excavations. No butchery marks were recorded on the bones of any of these species. However, these animals may all have been brought to the site for skinning.

The two rodent bones belonged to voles and came from the South bank black layer. Several vole species are now present in the area but, of the smaller ones, only the short-tailed field vole (*Microtus agrestis*) is common and the two jaws are likely to be from this species.

Wild mammals from Gorodishche 2000–2004 (Mikhail Sablin with Mark Maltby)

A total of 91 fragments from wild mammals were recorded for the three periods involved (Table 7.5). The percentage of bones of wild mammals increased between the 9th (0.2%) and the 12th centuries (3.6%), which is possibly a result of the gradual increase in the role of hunting in the lives of the high status residents of the settlement in the later period. The main animal represented (33% of the total wild mammal assemblage) was elk (*Alces alces*), which was hunted for its meat. Some elk bones bore traces of butchering, which show that the carcasses, seemingly brought to Gorodishche intact, were then processed there.

All the hare was identified as mountain hare. Most of the 27 hare bones were found in 10th-century deposits, in contrast with elk which was found mainly in 11th–12th-century layers (Table 7.5). Four bones from the 11th–12th-century deposits were classified as wild boar. Four bones of bear were also from deposits of

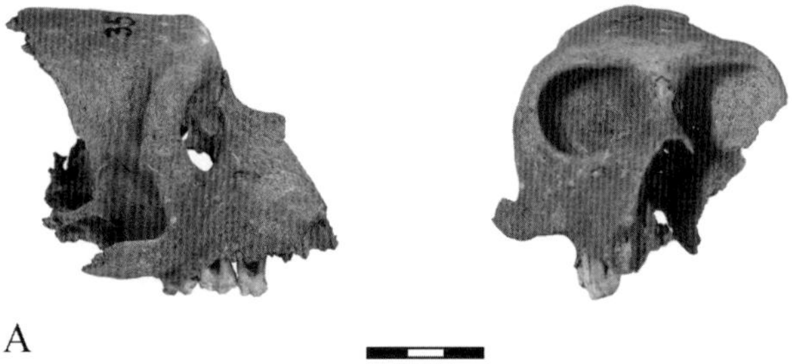

A

Figure 7.4: *Macaque skull from Gorodishche. Photograph supplied by Mikhail Sablin*

this date. All ten bones of beaver, by contrast, were from 10th-century layers, as were both the bones identified as pine marten. Ten of the 11 bones identified as fox (*Vulpes vulpes*) were also from 10th-century deposits. No bones were identified as squirrel. The species mainly associated with exploitation for their skins and pelts were more common in the 10th-century assemblage than in later deposits, in which (apart from bear) species mainly hunted for their meat formed the majority of the wild mammal bones.

During the 2003 excavations at Gorodishche a substantial portion of a skull was found belonging to an adult female Barbary ape or tail-less macaque (*Macaca sylvanus*) (Figure 7.4). The skull was subsequently radiocarbon-dated to between AD 1160 and AD 1220. Although the macaque was presumably tame, it has been included here amongst the wild species. Today this species is widespread in the mountain regions of north-west Africa and Gibraltar whence this monkey could have been brought live to the Novgorod area, perhaps as a gift for the Prince. It is also possible that this exotic animal could have been brought in from the West, particularly in view of the fact that the remains of Barbary apes have been found at a number of medieval sites in Europe. It is essential to note that the find of macaque at Gorodishche is the earliest in the continent of Europe for this era. Further discussions of the significance of this find can be found in Brisbane *et al.* (2007a; 2007b).

Georgii

The proportion of wild to domestic mammal bones was slightly higher at Georgii, forming 4% of the identified mammal bones (Figure 7.2). More bones of smaller species are likely to have been recovered had sieving been undertaken. Georgii produced a larger number of elk bones (18) than Gorodishche, despite producing a much

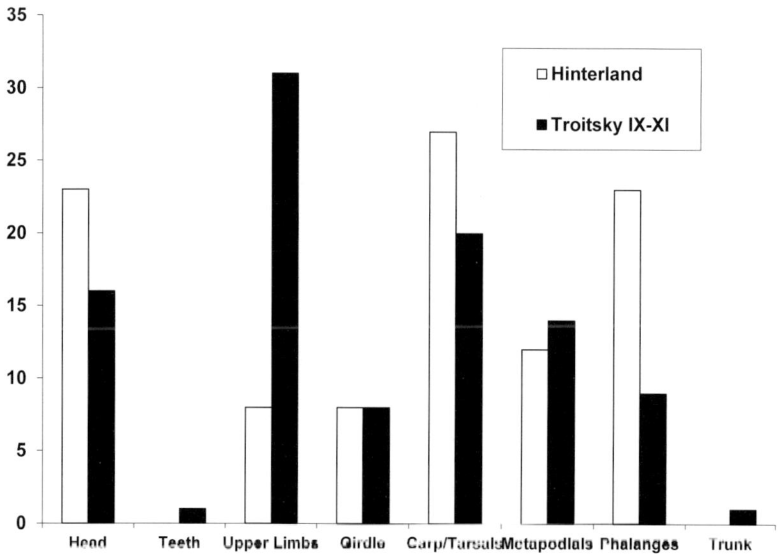

Figure 7.5: NISP percentages of elk body areas from Troitsky and hinterland sites

smaller assemblage. Elk provided over 3% of the total identified mammal bones, a much higher percentage than in any of the assemblages from other hinterland sites or Novgorod itself (Tables 7.6 and 7.8). Most of the elk bones consisted of tarsals and phalanges (Figure 7.5; Table 7.3). The bias towards foot bones could indicate that elk skins were brought to Georgii for processing. The low representation of upper limb bones could imply that meat was filleted from elk carcasses at the kill sites. Two second phalanges bore superficial oblique blade marks on their shafts that probably resulted from skinning. Knife cuts on the anterior and medial aspects of an astragalus were made during the removal of the feet. A pelvis bore superficial chop marks near the caudal end of the ischium. A maxilla with only the first two molars in wear came from a sub-adult animal. The proximal epiphyses of five phalanges, a radius and ulna and the distal epiphyses of a metacarpal and metatarsal had fused. Most of the elk represented were therefore probably adults. Two tarsals were pathologically fused with the centroquartal, which is a condition that probably developed in an older adult.

A squirrel femur and a beaver mandible, rib and tooth were also recovered. The squirrel femur and beaver mandible came from adults. A fifth metacarpal of a bear was also recovered. Knife cuts on its shaft were made during skinning. A second phalanx of a lynx (*Lynx lynx*) was also identified and probably was also brought to the site attached to a skin.

The higher percentage of wild mammal bones and the wider range of species suggest that the inhabitants of Georgii may have exploited wild mammal resources more frequently than the residents of Gorodishche. The evidence also suggests that most of these animals were exploited for their skins, which were brought back to Georgii for processing.

Prost

The much higher percentage (22%) of bones of wild mammals in the Prost assemblage (Figure 7.2; Table 7.2) can be largely explained by the discovery of a deposit of skin processing waste. An accumulation of 54 foot bones of pine marten were found. These consisted solely of tarsals, metacarpals, metatarsals and phalanges. A minimum of four animals were represented by some of the metatarsals (Figure 7.6; Table 7.7). Although no butchery marks

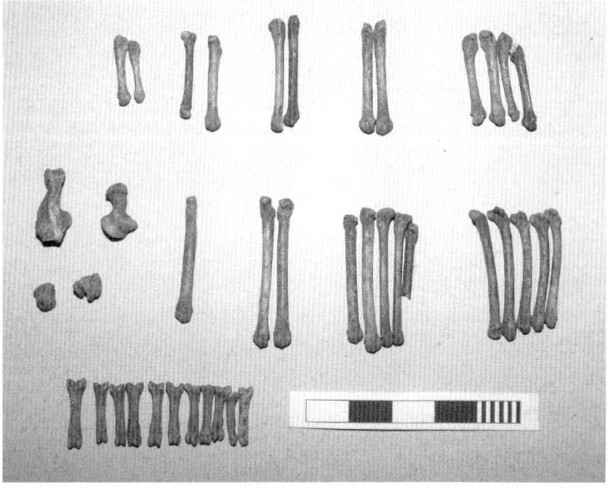

Figure 7.6: *Foot bones of pine marten from Prost. Photograph by Ellen Hambleton*

were observed, these remains must represent skinning waste. Presumably the marten skins were brought to the site with the feet still attached. These were removed during preparing the skins at the settlement.

A second phalanx of a large wild bovid was also recovered. This probably belonged to a bison (*Bison bonasus*) or possibly an aurochs (*Bos primigenius*). Further identification was hindered by gnawing damage. Again the presence of a foot bone may indicate the importation of a hide. The incisor of a beaver was also found.

Vasilievskoye

The only bone of a wild mammal identified consisted of a carpal of an elk. Again, this may indicate that skins were being imported.

NOVGOROD

Wild mammal bones from Troitsky IX–XI (Mark Maltby and Sheila Hamilton-Dyer)

The wild mammal assemblages were subdivided into the same general groups as the domestic mammals. The proportion of wild to domestic mammals was very low on all three Troitsky sites, forming only around 1% of the total (Figure 7.7; Table 7.8). Even acknowledging that many bones of small mammals such as hare and squirrel would have been overlooked using normal retrieval methods, it does not appear that meat from wild mammals was a common component of the diet. The small assemblage of

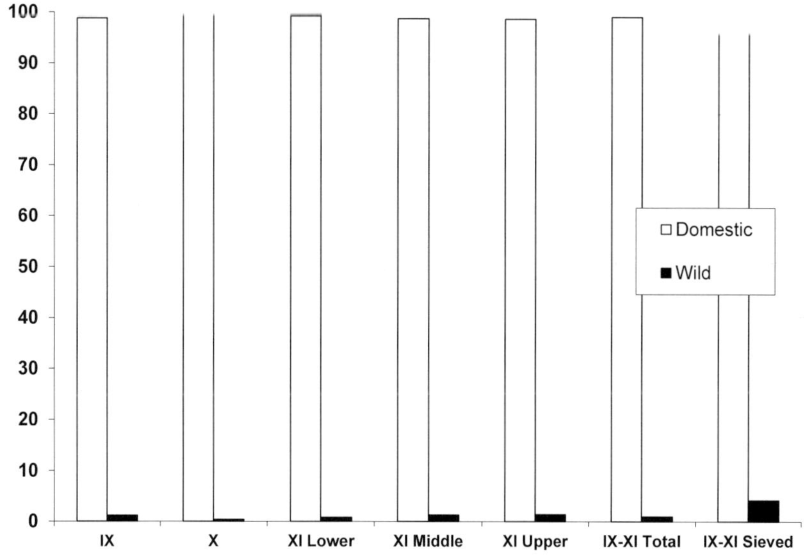

Figure 7.7: *Percentages of domestic and wild mammal bones from Troitsky IX–XI*

sieved material did produce a higher percentage of bones of wild mammals but still formed less than 4% of the overall assemblage.

Elk

Elk provided around a quarter of all the wild mammal bones from the Troitsky sites, although none were retrieved from sieved samples (Figure 7.8; Table 7.9). Overall, elk provided only 0.2% of the identified mammal bones, showing the dominance of the bones of domestic species in the Troitsky assemblage. This may underestimate the true contribution of elk. It is conceivable that they could have been commonly butchered and filleted at the kill sites. This would have eased the problem of transporting the heavy carcasses of this large mammal. However, on present evidence, it does not seem that elk was a common addition to the diets of the inhabitants of this part of the town.

There was no significant variation in the abundance of elk bones between the three Troitsky sites. On Troitsky XI, elk bones were encountered rather more frequently in spits assigned to the 12th–13th centuries than in earlier or later levels, providing over half of the wild mammal bones but still only 0.6% of the identified mammal assemblage (Figure 7.8; Table 7.9).

The types of elk bone represented (Tables 7.10–7.13) differed to some extent from the assemblages recovered from the hinterland sites (Figure 7.4). Upper limb bones were much better represented (31%), perhaps implying that joints of elk meat were being imported. However, bones from the lower limbs were generally better represented than their cattle counterparts, with 43% of the elk assemblage being comprised of carpals, tarsals, metapodials and phalanges. This would suggest either that complete carcasses were commonly imported, or, perhaps more likely, some of the foot bones were brought in still attached to skins.

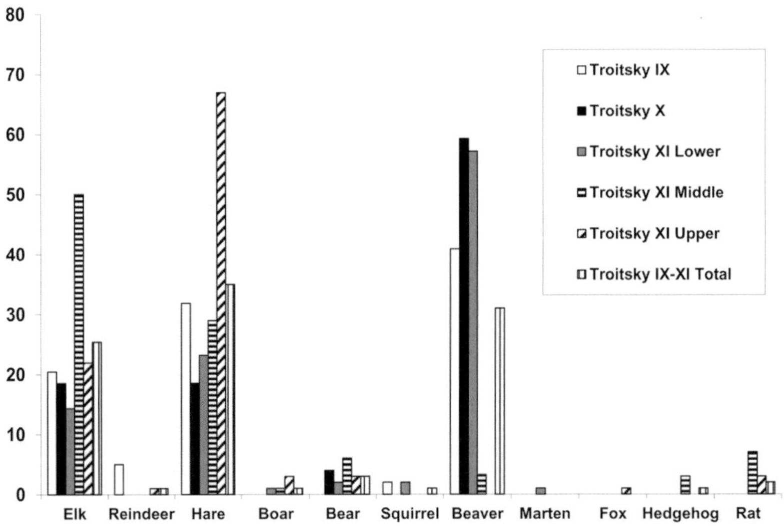

Figure 7.8: *Percentages of wild mammal species from Troitsky IX–XI (excluding sieved samples)*

Twenty-six elk bones were recorded as butchered, representing 33% of the assemblage, indicating that elk carcasses were being processed as intensively as cattle. Twelve of the butchered bones came from the Troitsky XI spits chosen for detailed recording of butchery marks. In all but one case the marks had been inflicted by a cleaver or a heavy blade. Three marks were found on one portion of a single skull (from Troitsky XI Spit 21). The specimen consisted of the right frontal and parietal. A fine incision on the top of the frontal indicated skinning. The skull had been split in half, separating the right from the left side and providing access to the brain. Finally the base of the antler had been chopped off through the pedicel. This may have been a portion of skull brought in with the antler still attached after it had been skinned and segmented at the kill site. A very similar example was found at Gorodishche (Figure 7.3). Evidence for the processing of elk antler itself was very limited in this assemblage, as only two pieces of antler were recovered, one of which had evidence of sawing. However, the retrieval policy militated against the retention of antler within a 'normal' faunal assemblage. All worked bone and antler that was recognised during excavation were routinely removed for separate study. These included not only finished objects, particularly composite combs, but also worked antler waste and comb blanks (Smirnova 2005, 198–212). It is clear that large numbers of elk antlers were brought to Novgorod for working by itinerant or resident specialist manufacturers. There is evidence that sometime in the middle of the 11th century, a comb-maker manufactured antler and wooden combs on Property G in Troitsky XI (Smirnova 2005).

A summary of some of the most common measurements taken on elk bones can be found in Table 7.14. There are, however, too few bones to merit discussion of the stature of the elks.

Reindeer

Only three bones were identified as reindeer, two from Troitsky IX and one from the upper layers of Troitsky XI (Figure 7.6; Table 7.9). Combs were occasionally made out of reindeer antler (Smirnova 2005).

Hare

Hare was the most common wild mammal species recovered from the Troitsky excavations providing 33% of the wild mammal bones, in both the hand-collected and sieved samples (Figure 7.8; Table 7.9). Overall, however, hare provided only 0.3% of the identified mammal bones. Normal excavation methods handicapped the recovery of the smaller hare bones in particular. On the other hand, it is more likely that any hares exploited for meat would have been brought to the town as complete carcasses in contrast to larger mammals such as elk that were probably often butchered in the field. Hare bones were outnumbered by beaver in the lower levels of all three Troitsky sites and ranked second behind elk in the middle spits of Troitsky XI. Hare became the dominant wild mammal in the assemblage from the upper spits of Troitsky XI, providing 69% of the wild mammal NISP counts. This increase partially reflects the disappearance of beaver from the later assemblages and the decline of elk from its high levels in the middle spits.

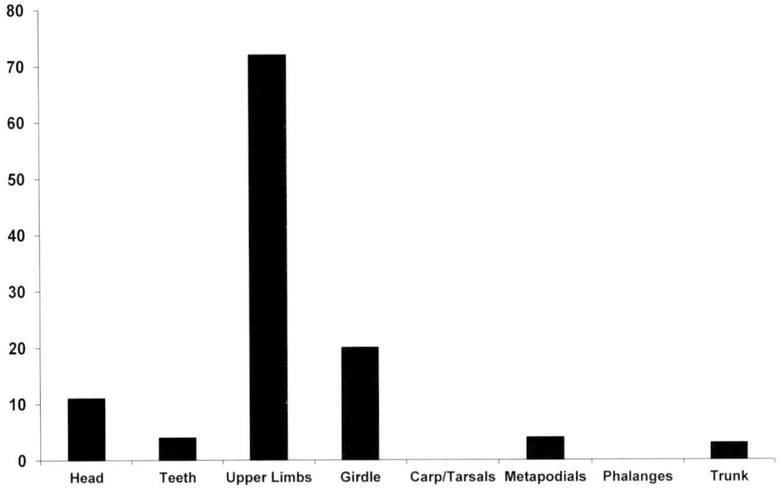

Figure 7.9: *NISP percentages of hare body areas from Troitsky IX–XI (excluding sieved samples)*

The types of hare bone represented were heavily biased towards upper limb bones (Figure 7.9; Tables 7.15–7.18). This largely reflects the relative sizes of the bones. The small bones of the feet were unlikely to be recovered by normal excavation methods. The relative abundance of bones from different areas of the body is quite similar to that of cat (Figure 4.51), which has a skeleton of very similar size to hare. This observation plus the presence of substantial numbers of cranial elements supports the contention that most hares were brought to Novgorod as complete carcasses. Butchery marks were observed on ten hare bones. Nearly all of these consisted of knife cuts located on various parts of the upper limbs, scapula and pelvis and were associated with both dismemberment and filleting.

Analysis of epiphyseal fusion data revealed that most of the hares were fully grown adults, with 76% of the latest-fusing epiphyses having fused (Table 7.19). Both preservation and retrieval biases favoured the recovery of fused bones and there is sufficient evidence to indicate that substantial numbers of immature as well as adult hares were caught and brought back to the town. Analysis of the limited metrical data (Table 7.20) showed that all the measured bones fell within the overlap range between mountain and brown hare but it is assumed that only the native mountain hare was represented during this period.

Wild boar

Positive identifications of wild boar were very limited. There are problems of differentiating between wild boar and domestic pigs because of size overlap. A conservative policy of identification was adopted in which only exceptionally large bones were assigned to wild boar (see also Chapter 6). Only four bones from Troitsky XI were thus classified as wild boar. Although, it is conceivable that other bones in the pig assemblage were from wild boar, it is believed that the vast majority in this assemblage would have been from domestic pigs.

Brown bear

Nine bones of brown bear were recorded, all bar one from Troitsky XI (Table 7.9). They formed only 3% of the wild mammal bones from the Troitsky sites and were most commonly found in the middle spits of Troitsky XI (Figure 7.8). All the bear bones were third phalanges. These may have been imported to the town attached to skins, which would suggest that bearskins were used by the inhabitants of these properties. However, one third phalanx had been perforated in order to be used as an ornament of some sort. Two other phalanges bore disarticulation marks near their proximal end, indicating that they had been separated from the rest of the foot and the skin. Therefore it may only have been the third phalanges themselves that were brought to the properties rather than the skins.

Red squirrel

Only four bones of squirrel were recovered including one from a sieved sample (Table 7.9). They formed only 1% of the wild mammal assemblage. Their rarity in the faunal assemblage stands in stark contrast to the documentary evidence that indicates trade in squirrel fur formed an integral part of the Novgorod economy (Martin 1985; Makarov 2012a; 2012b).

Although, the low frequency of squirrel bones could to some extent reflect the lack of sieving, it is much more likely that their paucity is a consequence of their bones never being brought to the town. It is much more likely that squirrels hunted for their furs would have been skinned either at the kill sites or in camps or larger settlements, such as Minino, inhabited by their hunters throughout the Novgorod territory. If any bones were still attached to the skins, they are likely to have been foot bones that would be difficult to observe during normal excavation. More likely, however, is that the foot bones were also removed prior to the importation of the furs. The squirrel bones found in Novgorod are most likely to have belonged to animals caught in the vicinity of the town and brought in for processing. There is no evidence that squirrel meat was eaten.

Beaver

Beaver were the second most common wild mammal in the Troitsky assemblage. A hundred bones were identified from normal excavation and six were retrieved from sieved samples (Table 7.9). They formed 32% of the identified wild mammal bones from the Troitsky sites (Figure 7.8). However, these overall figures mask a significant decline and eventually the disappearance of beaver bones in the assemblage through the stratigraphic sequence. In the spits dated to the 10th to 12th centuries from all three sites, beaver provided between 41% and 59% of the wild mammal assemblage and between 0.3% and 0.5% of the identified mammal assemblage (Table 6.9). They formed only 3% of the wild mammal bones from the middle spits of Troitsky XI and

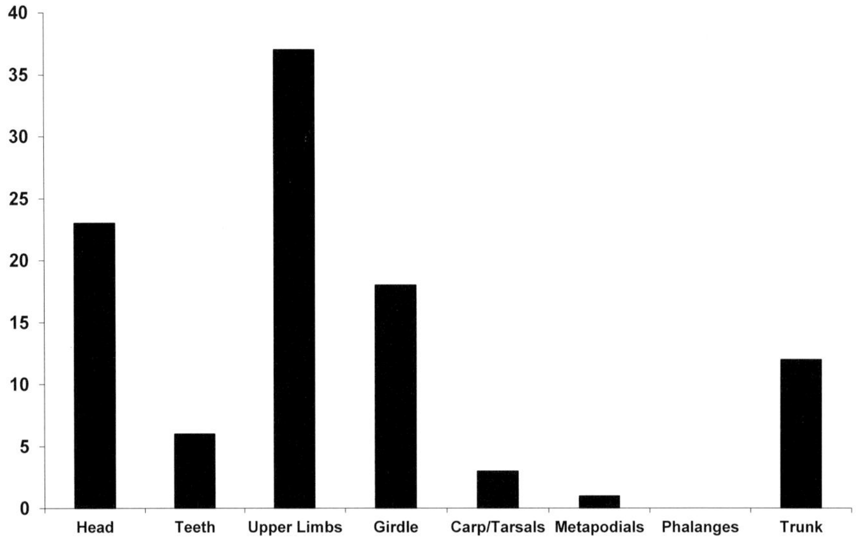

Figure 7.10: *NISP percentages of beaver body areas from Troitsky IX–XI (excluding sieved samples)*

were absent from the upper layers (Figure 7.8). The latest occurrence of a beaver bone was in Spit 9 (Property G). However, only three bones of beaver were found above Spit 17. Therefore, there is very little evidence for the presence of beaver from the end of the 11th century onwards and it disappeared entirely from deposits dated later than the early 13th century. This evidence broadly corresponds to documentary evidence that also indicates a marked decline in references to beaver furs being brought to and traded from Novgorod beyond the 13th century (Makarov 2012b). It also coincides with the decrease in the relative abundance of beaver bones at Minino (see Chapter 8).

The types of beaver bones recovered from the Troitsky sites, as expected, show a bias towards larger bones (Tables 7.21–7.22). Mandibles were the most common element represented but the assemblage also included a high proportion of upper limb and girdle bones (Figure 7.10). The presence of both low- and high-quality meat bones on the Troitsky sites indicates that whole carcasses of beavers were sometimes brought to the town and processed for meat as well as their skins. It is also possible that sometimes only joints of meat were brought to the town. Although, retrieval bias could partially explain the low representation of carpals, tarsals and metapodials and the absence of phalanges, it does not entirely account for their rarity. Comparing the element representation of beaver (Figure 7.8) with those of cat and dog (Figure 4.51), the percentages of metapodials of the latter are substantially higher. One would have expected the larger metapodials of beaver to have been recovered in at least similar quantities. This would suggest that, unlike the carcasses of the resident cats and dogs, which were usually deposited as complete carcasses, beaver carcasses were more frequently dismembered. Metapodials were probably often removed with the pelts either before the meat-bearing bones were imported to the town, or removed

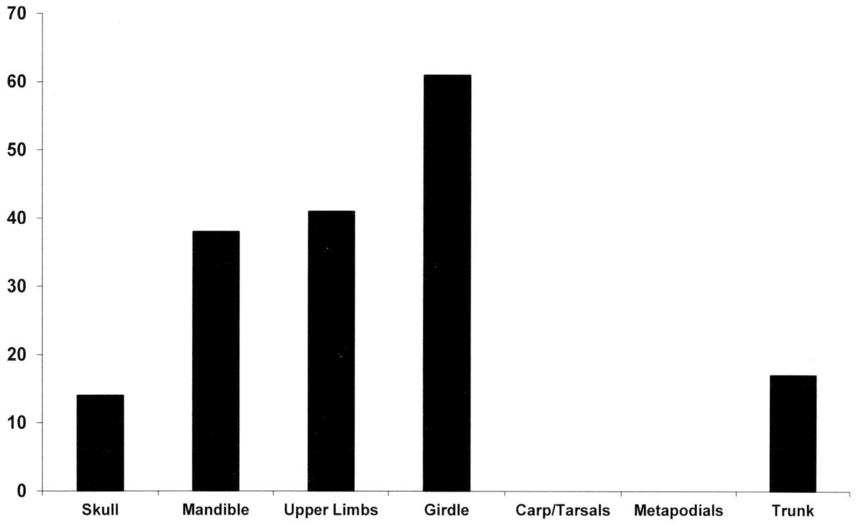

Figure 7.11: *Percentages of butchered beaver bones from Troitsky IX–XI (excluding sieved samples)*

from the site after processing in Novgorod itself. Like squirrels, most of the beaver skins would have been obtained from areas distant from the town in the Novgorod territory (see Chapter 8). It is likely that most of the beaver remains recovered from the excavations were from local populations that could be exploited for their meat as well as their skins by the urban residents.

Evidence that beaver meat was consumed was clearly supported by the frequency and location of the butchery marks encountered. Processing marks were observed on 35% of the beaver elements. Marks were found very frequently on upper limb bones, scapula and pelvis (Figure 7.11). Skinning marks were observed near the orbits of one skull and incisions on the buccal aspect (cheek) of a mandible were probably made during skinning. However, most of the remaining butchery marks were the result of dismemberment and filleting. Several of the upper limb bones had been chopped, sometimes severing the shaft (Table 7.23). Most of the butchery on the scapula and pelvis consisted of fine incisions inflicted during filleting.

Epiphyseal fusion data were obtained from 44 beaver bones (Table 7.24). 43% of the latest-fusing epiphyses were unfused indicating that immature beavers were frequently captured. Very few measurements of beaver bones could be taken and there was insufficient evidence to carry out metrical analysis.

Marten

A femur of an immature marten was found in Troitsky XI Spit 16. It is likely that it belonged to pine marten. As in the case of squirrel, unless complete carcasses were brought in for skinning, it is unlikely that the bones of marten would have been brought to the town. Undoubtedly trade in their pelts was an important item in the economy but evidence for this is unlikely to be manifest in the zooarchaeological record from the town of Novgorod itself.

Fox and wolf

A fairly complete skull of a fox was recovered from Troitsky XI Spit 5 (Property I). All other canid bones were identified as dog. No positive identification of wolf was made.

Small mammals

Bones of small mammals were unlikely to have been recovered during routine excavation. However, a few were retrieved by hand and confirmed the presence of several species within the town. The most commonly identified bones belonged to rats, all seven of their bones coming from Troitsky XI. Specific identifications were not possible in every case but all the rat bones are likely to have belonged to the black rat (*Rattus rattus*). Rats would have been attracted to the abundant quantities of kitchen and carcass processing waste along with other rubbish that was deposited within and around the properties. However, the earliest occurrences of rat bones in Troitsky XI come from spits and street levels dated to the late 12th century. The date of this appearance coincides well with the evidence from Minino (Savinetsky and Krylovich 2011) where black rats were identified in 12th- and 13th-century deposits. Savinetsky and Krylovich noted that there were possible rat bones dating to the 7th to 10th centuries from the 'Ves settlement near Vladimir but they also observed that the bones come from mixed deposits and they could more feasibly be of 12th- or 13th-century date. They argue that it is likely that the black rat spread to north-west Russia as stowaways on trading vessels. There is currently no strong evidence for a heavy infestation of rats in medieval Novgorod. However, more extensive sieving could provide evidence for this.

A mouse mandible was recovered in sieved deposits but it could not be identified to species. Several other small mammal bones were also found in the sieved deposits. These were the size of mice or smaller voles.

Two bones of hedgehog were recovered from Troitsky XI Property I in Spits 8 and 9. These consisted of a skull and mandible. Hedgehogs would have found favourable habitats within parts of the town.

The wild mammal bones from the Desyatinny-1 Site, Novgorod (Andrei V. Zinoviev – edited by Mark Maltby)

The bones of wild mammals are listed in Tables 7.25–7.26. The totals exclude possible wild boar. Some of the porcine mandibles approached the size and form of wild boar but further study of morphometry of pigs from the region of medieval Novgorod is necessary to obtain more precise data on separation of wild and domestic forms. All of the pig bones have been assigned to domestic pigs (Figure 4.52; Table 4.74). Excluding possible wild boar, the overall percentage of wild mammals is 0.2%, ranging between 0% and 0.4% in the different spits (Table 7.25). This compares closely with the 1% from the unsieved Troitsky assemblage (Table 7.8).

Elk was represented by a single bone dated to the second half of the 11th century. Other wild mammals, such as roe deer (*Capreolus capreolus*), hare, bear, beaver, fox and wolf were also each represented by only a few bones and obviously did not constitute significant source of meat for citizens in this part of medieval Novgorod (Table 7.26). Only one wolf bone has been found in a late 11th century deposit. This species was not identified on the Troitsky sites. As in the case of beaver and fox, wolves might have been used in fur production. Only two beaver bones were recovered from Desyatinny-1, both belonging to the early period (10th and 11th centuries). This supports the evidence from the Troitsky sites (Figure 7.8) for a marked decline in the beaver population in the Novgorod region prior to the 13th century due to over-exploitation. This decline eventually led to the complete disappearance of the species from most of European Russia by the end of 19th century. The population was successfully restored only in the 20th century and beaver is now again abundant in the Novgorod region, although its popularity as a fur animal has drastically declined.

Wild mammal bones from the Kremlin Excavations 1996 (Mark Maltby)

Bones from some of the later medieval layers (Spits 4–7) produced just three bones of wild mammal, all of hare. A complete radius (greatest length 107.3 mm) was found in Spit 5. Spit 4 produced two humeri of adult hares. This included a complete specimen (greatest length 110.6 mm; distal breadth 12.8 mm). This was therefore slightly larger than the specimens measured from the Troitsky sites (Table 7.20). Wild mammal bones accounted for 3% of the identified mammal counts (see also Table 4.77) but the assemblage of 92 specimens is too small to determine whether hares played a more significant part of the diet of the inhabitants within the Kremlin than those who were resident in other parts of Novgorod.

– 8 –

ARCHAEOZOOLOGICAL MATERIALS FROM MININO AND CHANGES IN POPULATIONS IN UTILISED MAMMALS FROM THE NORTH OF EUROPEAN RUSSIA FROM THE MESOLITHIC TO THE MEDIEVAL PERIOD

Olga Krylovich and Arkady Savinetsky

(Translated by Kathy Judelson)

INTRODUCTION

Archaeozoological research in the north-west of Russia goes back a fairly long way: A.A. Inostrantsev (1882), V.I. Tsalkin (1956), N.K. Vereshchagin (1979), E.G. Andreyeva (1971; 1991) and various others carried out work in this region. This has meant that the range of the main species of mammals hunted by the ancient inhabitants of this region is for the most part known to us. However, little attention has so far been paid to changes in the hunting of mammals in this territory and to changes in their numbers. This results, first and foremost, from the fact that collections of faunal remains from sites in the north-west of Russia are usually very small and it is also because most of the materials collected are from excavations of urban sites (Novgorod, Pskov, Staraya Ladoga), where bones of wild animals only make up a small part of the whole. Secondly, most of the sites relate to only one period, which makes it extremely difficult to resolve questions connected with diachronic changes in hunting on the basis of the material collected. Without finding answers to questions regarding the way the hunting of mammals evolved and the size of their populations, it is impossible to assess the degree to which humans interacted with their environment in the territory under discussion.

The archaeological site at Minino is in this respect highly unusual. Apart from the fact that large amounts of osteological material have been found there during excavations, most of the faunal remains are those of wild animals and, most important of all, this settlement existed in one place over a long period – Mesolithic, Neolithic, Early Iron Age and the Medieval Period (Makarov 2001). At the beginning of the 13th century this particular settlement ceased to exist. No signs have been found of violent events or natural catastrophes, which might have put an end to the existence of the settlement, whose inhabitants were engaged primarily in hunting and fishing.

Because of this, it has been suggested (Makarov *et al.* 2000) that the reason might have been over-exploitation of the populations of the main mammals hunted for food and fur, in particular beavers, the most valuable fur-bearing animals in the hunting grounds of this settlement. Determining the age structure of the beavers caught could serve to confirm or undermine this hypothesis, since changes in that structure could indicate the state of the population and the degree to which hunting impacted on it (Naumov 1955).

A separate and most important task is to determine changes in the relative numbers of animals being hunted. The difficulty lies in the fact that the bones of these animals found during excavations testify only to the range of animals hunted and the proportion, which the bones of particular species account for in the whole osteological spectrum. The task of plotting changes in the numbers of animals caught and the share of different animals in the total demands a special approach (Dinesman *et al.* 1999; Dinesman and Savinetsky 2003).

MATERIALS AND METHODS

In the course of the excavation of the settlements making up the Minino archaeological complex, a total of 48,325 bones of mammals, birds and amphibians were retrieved. Fish bones were collected separately during the excavations and the analysis of a sample of these is discussed in Chapter 10. The material was collected from quadrants measuring 1 m by 1 m and the thickness of each spit from which bones were collected was 10 cm. Material from the occupation levels was carefully sieved through 3–4-mm mesh, which made it possible to obtain a more complete range of faunal remains. All the material was dated by artefacts found in each layer and, in a number of cases, by radiocarbon dating (Makarov 2001).

When counting the bones from the cultural layer we used the indicator 'number of individual specimens' (NISP). The likelihood of finding all or most of the bones of a single animal in the occupation levels is fairly low, since the accumulation of most of the faunal remains took place over a long period of time measured in hundreds and sometimes even thousands of years (Dinesman *et al.* 1999). This does not apply to the material from the rubbish pits. The in-fill of these pits built up over a very short period, so it is highly likely that the bones of a single individual might be found there. This applies both to burial sites and to other parts of the settlement in which material accumulated over a short period. When quantifying the bones in these areas, we used the index 'minimum number of individuals' (MNI). This index was calculated on the basis of the most commonly found skeletal element. At the same time the capacity for variation between individuals and age-related variation of skeletal elements was taken into account.

As has already been pointed out, the number of bones for each species in the occupation levels testifies to the frequency of occurrence of the species in question. We should, however, be cautious in making direct comparisons between the numbers of one species with those of another. This is due to differences in the methods used for procuring different species of animal, the way they were skinned, transported, processed, utilised and also to the quality of bone preservation, to the selectiveness of the hunters and to many other factors. However, changes in the relative abundance of the bones of any particular species in different phases of occupation can indicate change in how frequently the species was caught (provided other conditions remain the same). It is evident that the more thoroughly a site is excavated, the more precise the results will be.

It should be noted that change in the size of a catch depends on a number of factors: the size of the animal populations concerned, the number of hunters, the methods for catching the animals and the extent to which the hunt is selective. We have worked on the principle that hunting methods, particularly during the medieval period, did not undergo any major change, at least there is no evidence that changes took place. For most of the species, no changes were noted with regard to the degree of selectiveness on the part of the hunters. Various modifications of the method devised by DeLury (1947) have long been used for determining the size of recent populations of animals hunted for food or fur. These are now widely used for analysis of recent populations of cetaceans, pinnipeds and fish for determining quotas for their catches (Allen 1966; Bockstoce and Botkin 1983; Allen and Kirkwood 1988). In the modification of this method proposed by Chapman (1974) the average number of animals – N during the period – t, the catch – C and the effort – f spent on the catch can be calculated using the equation $N_t = C_t/qf_t$, in which q is the hunting coefficient (profitability, potential yield). The hunting coefficient is determined by the particular features of the hunting methods used. The ratio C_t/f_t which forms part of this equation is used to assess the size of the catch per unit of effort on the part of the hunters, or, in other words, by the effectiveness of the hunting (CPUE – catch per unit of effort). If the hunting coefficient (q) remains stable, the size of the CPUE changes in proportion to the size of the population being hunted, regardless of its replenishment and its natural mortality rate or its mortality rate at the hands of the hunter (Riker 1979). In this case it can itself serve as an index of the relative number of animals. It is obvious that for the reconstruction of the numbers of animals available within the hunting grounds of ancient communities, it is essential not only to know the size of their catch and the time during which bones from the hunted animals accumulated in a settlement, but also, based on observations of the osteological material, the amount of effort spent on catching that prey and the hunting coefficient (q) of the hunting methods used by the community involved.

A relative index for the size of hunters' catches in the past can be determined by the number of bones from each species (Nbs) collected within horizons of the cultural layer (Nbs = C). The time it took for them to accumulate (t) was established using archaeological methods. What proved far more complex was the assessment of changes in the number of hunters operating from the settlement under investigation. For such an assessment we used the thickness of the cultural layer. Our investigations in Chukotka (Dinesman *et al.* 1999) showed that changes in the thickness of the cultural layer of the settlement correlated well with the number of burials relating to that settlement, which, in its turn, is an accepted criterion for assessing human palaeo-populations (Ascádi and Nemeskéri 1970; Ubelacker 1978). This meant that we were regarding the thickness of the excavated horizons as an index for the size of the settlement's population and, accordingly, as an index for the hunting effort (f). The number of faunal remains in these horizons gave an indication of the size of the hunters' catch (C). In order to determine changes over the centuries in the numbers of animals hunted for food and fur, we calculated the relationship between these two indices, which in essence meant the concentration of bones per volume of the cultural layer. In this particular case we made calculations only for those quadrants, which had gradual accumulations of the cultural layer, *i.e.* we omitted quadrants containing pits, burials and various structures such as hearths. Another index for hunting effort could be the number of hunting weapons. Unfortunately not very many of these were discovered, certainly not enough for any quantitative analysis.

In order to determine possible changes in the age structure of the medieval beaver population in the Kubenskoye Lake area and the influence of hunting on their population in the Early Iron Age and medieval periods, it was necessary to determine the age structure of the animals whose remains were collected from the Minino archaeological complex. On the basis of the tooth eruption sequence, measurements of long bones, and epiphyseal fusion evidence, we were able to divide the beaver remains into four age groups: animals under one year old; individuals aged one to two years; individuals aged three to four years; and individuals aged five years and over (Krylovich 2004).

RESULTS

Characteristics of the zooarchaeological assemblage

The total number of specimens in the osteological collection placed at our disposal, both from 'secure' levels, *i.e.* relating to one or other specific period, and also from mixed ones came to 48,325 bone fragments (Table 8.1). Of these 47,598 fragments were from mammals (98.5% of the total number of animal bones excluding fish), 641 from birds (1.3%) and 86 from amphibians (0.2%). In the course of our work we did not identify the species to which bird bones belonged. The amphibian bones, without exception, belonged to representatives of the genus *Rana* (frog). As for the mammal bones, it was only possible to determine the species to which 8.5% of the fragments belonged, which serves to indicate the poor state of preservation of the material in question. This very low percentage of identified bones is, however, a result of the large quantity of very small bone fragments in the material. The presence of so many extremely small bone fragments can be explained by the very thorough collection of material by archaeologists during the excavations. For these reasons a comparatively large number of the bones in the material were those of small animals, such as squirrels, small rodents and amphibians.

Of the 4,089 identified mammal-bone fragments 1,185 (29%) belonged to six domestic species and 2,904 fragments (71%) belonged to 18 species of wild mammals. The list of domestic animals included all species typical for the region in question (Table 8.1).

Domestic cattle (Bos taurus L.)

A total of 456 of these bones were found. Nearly all the bones were broken and they were fragments from all parts of the skeleton. Teeth and mandible fragments were found in the largest numbers. Fragments of long bones were found less frequently. Some fragments bore traces of having been gnawed or butchered. Fragments of horn core found in this material confirm that the cattle bred in this area was horned (Tsalkin 1956). Cattle accounted for 38.5% of domesticated mammals in the total sample (Table 8.1).

Domestic sheep (Ovis aries L.) and goat (Capra hircus L.)

It was not always possible to determine to which of these two species the bones of very similar morphology belonged. For this reason most researchers link them together in a single category – sheep/goat. A total of 436 fragments had been found. All but two of the 21 elements specifically identified belonged to sheep (Table 8.1). The most numerous skeletal elements were teeth and mandible fragments. The sheep/goat bones were also badly damaged and bore traces of gnawing and butchery. Sheep/goat bones account for 36.8% of the total number of bones from domestic animals.

Pig (Sus scrofa domestica L.)

It is usually only possible to distinguish remains of domestic pig from those of wild boar on the basis of their smaller measurements. Since our material contained bones of both domestic pig and the larger wild boar, we were able to distinguish between the bones of these two animals. The number of domestic pig bone fragments came to 162 and the most common skeletal elements were teeth. Long bones were badly fragmented. Some of the bones bore traces of gnawing and chopping. The pig bones made up 13.7% of all the bones from domestic mammals (Table 8.1).

Horse (Equus caballus L.)

The fragments of horse bones recovered were difficult to distinguish from bones of wild equid species. However, at contemporary sites from the forest zone the presence of wild horse has not been recorded. For this reason we have assumed that, in the period we are studying in the Vologda region, there were no wild representatives of the *Equus* genus. Altogether, 79 of the bone fragments found came from horses. Among them there are fragments of long bones, phalanges and teeth, some of which also bear traces of gnawing and cutting. Horse bones account for 6.7% of all the faunal remains from domestic animals.

Dog (Canis familiaris L.)
Fifty-two fragments were collected. More often than not these were teeth and mandible fragments. Dog bones make up 4.4% of the total number of bones belonging to domestic animals (Table 8.1).

Beaver (Castor fiber L.)
Thanks to the unusual structure of the beaver skeleton and the absence of any closely related species in the region, bones of this animal were easy to identify. The beaver bones accounted for more than half of all those belonging to wild mammals – 56.6% of the total identified (Table 8.1). In total there were 1,643 bone fragments from this large rodent and they included bones from almost all parts of the skeleton. Most frequently teeth, long bones, pelvis fragments, metapodials and phalanges were encountered. Mandibles were also found, along with astragali and calcanei, vertebrae from various areas of the spine, fibulae, scapula and clavicle fragments. Skull including maxilla fragments were rare. Most of the bones found were fragmentary and some of them bore traces of cutting and charring.

Red squirrel (Sciurus vulgaris L.)
This species accounted for the second largest amount of faunal remains – 17.5% of the identified wild mammal bones. A total of 508 fragments of squirrel were found (Table 8.1). Almost all parts of the skeleton were represented: skull fragments, mandibles with teeth, some incisors, all long bones, scapula and pelvis fragments, astragali and calcanei, metacarpals, metatarsals and vertebrae.

Pine marten (Martes martes L.)
All identifiable bones belonging to the species *Martes* were assigned to pine marten. The fragments of mandibles in our material lacked the characteristic features that enable the distinction between the two kinds of *Martes* (pine marten and sable). Isolated finds of sable bones have been recorded in north-west Russia (Vereshchagin and Rusakov 1979; Tsyganova 1999). However, because only pine marten bones have been recorded at the majority of archaeological sites, we have assumed that it was the pine marten which inhabited the territory of interest to us. These bones make up the third largest group of wild mammals (6.9%). Mandibles were the bones encountered most frequently. Fragments of limb bones were also noted.

Elk (Alces alces L.)
When it comes to the wild ungulates represented in our material, elk make up 5.4% of the total number of bones from the identified wild mammals (Table 8.1). When the faunal remains from this species are identified, problems do not usually arise: elk bones differ significantly in size from the bones of all other species of deer (*Cervidae*),

which might have inhabited this territory. Most frequently it was the teeth of this species which were encountered and also fragments of metapodials and phalanges. Many of the fragments of long bones bear traces of cutting and gnawing.

Hare (Lepus sp.)
Forty-four of the bone fragments (consisting of mandibles, loose teeth and fragments of limb bones) belonged to hares, which amount to 1.5% of all the identified wild mammal bones (Table 8.1). The absence of any skulls or complete mandibles in this material makes it difficult to determine exactly to which species these bones belong.

Common fox (Vulpes vulpes L.)
Fox bones account for 1% (29 specimens) of the faunal remains from wild mammals. For the most part fragments of mandibles and shaft fragments of long bones were encountered. These bones do not bear any traces of gnawing or butchery.

Other wild mammal species
The remaining species are represented in our collection by only a very small quantity of bones (Table 8.1). Bones of other ungulates, roe deer (*Capreolus capreolus* L.) and reindeer (*Rangifer tarandus* L.) were only found in small numbers in this collection. Seven fragments of roe-deer bones (0.2% of the identified wild mammal bones) and 22 fragments of reindeer (0.8%) were identified. Seven specimens (0.2%) belonged to species of deer not further identifiable. It is possible that some of them were bones of red deer (*Cervus elaphus* L.), remains of which have been found in small quantities at sites of the same period in the region further to the west.

Fifteen bone fragments (0.5%) in our material were recorded as wild boar (*Sus scrofa ferus* L.)

Reference has already been made to the difficulties involved in distinguishing bones of wild boar from those of domestic pigs. Wolf (*Canis lupus* L.) is poorly represented in our collection – only six bones (0.2%). Bones of this large beast of prey from the family of canids (*Canidae*) can be confused with bones of large dogs.

Remains of small mustelids (*Mustela*) were also found – eight bones (excluding the bones in the polecat skeleton discussed below) represented 0.3% of the total of identified wild mammal bones. Two varieties of *Mustela* might be found in the territory adjacent to the Minino archaeological complex – the polecat and the European mink. On the basis of their post-cranial skeleton, these two species can, in theory, be distinguished from each other using the proportions of their long bones, but in this case not a single reliably-identified mink bone has so far been found, and therefore all the collected fragments were classified as belonging to polecats. In one of the medieval burials an almost complete skeleton of a small mustelid was found, which included a skull, thanks to which we were able to identify these remains reliably as belonging to a polecat (*Mustela* (*Putorius*) *putorius* L.).

The skeleton of a male polecat (Figure 8.1) found in this burial was of particular interest: firstly, on account of its position in the burial itself – the human skeleton

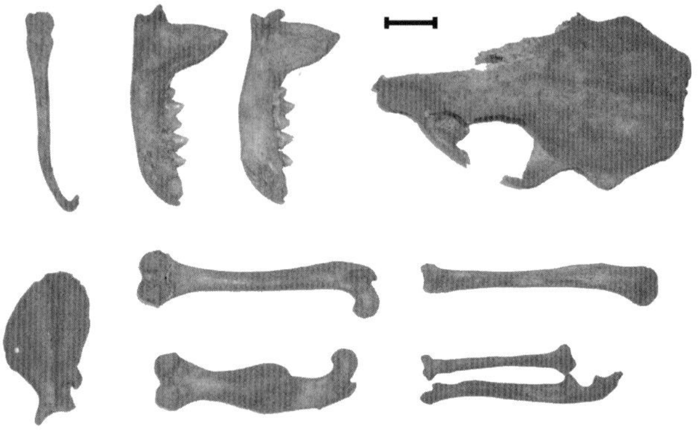

Figure 8.1: Bones of a male polecat skeleton buried with a woman in a grave at Minino

from this burial had been identified as that of a woman and the skeleton of this small animal had been placed near her head (it was assumed that it had actually been placed around her neck). Secondly, one of the femora bore traces of having grown together incorrectly after a break. As a result one leg was slightly shorter than the other. The skull on the other hand did not bear any traces of deterioration in the occipital area, as was typical for the skulls of small mustelids of that period. It is possible that this polecat had perhaps been tamed.

The identification of bones belonging to the hedgehog (*Erinaceous* sp.) left no room for doubt. Thirteen of the bones from our osteological collection (0.5%) belonged to this animal, in particular, fragments of mandibles and limb bones.

Five bones (0.2%) belonged to brown bear (*Ursus arctos* L.): one was the distal end of a radius, which bore traces of gnawing. One bone each of common badger (*Meles meles* L.) and wolverine (*Gulo gulo* L.) were identified, the latter from an Iron Age layer.

Bones from small rodents accounted for 7.7% of the total wild mammal bones. The bones of these species could have found their way into the occupation levels of the Minino archaeological complex at a later period via burrows, but they are more likely to have turned up in the pits before they were infilled.

Period comparisons of species abundance

It can thus be seen that in the Minino area all the typical species of domesticated mammals were exploited. The most significant of the animals hunted for their fur and meat were beavers, squirrels, elk and pine martens. Other mammals made up only an insignificant part of the osteological collection.

The quantities of identified animal bones, which can confidently be dated are not evenly spread over the various periods in the life of the settlement (Table 8.2). The material from the Mesolithic and Neolithic levels was of very limited quantity: all that we succeeded in identifying from those levels was a small quantity of beaver and dog bones, which have been included in the counts in Table 8.1.

There was a larger quantity of material from Early Iron Age levels. Bones of domestic animals make up only 4.2% (Figure 8.2; Table 8.2), but despite this all the main species of domestic mammals were already represented: cattle (4 fragments), sheep/goat (2), pig (3), horse (3) and dog (2). The vast majority of faunal remains are from wild animals (95.8%) and among these the remains of 11 species used for food and fur were found (Figure 8.3; Table 8.2). The largest quantities of bones were from beavers (75% of the wild mammal elements), pine marten (13.2%), squirrels (4.1%) and elk (3.1%).

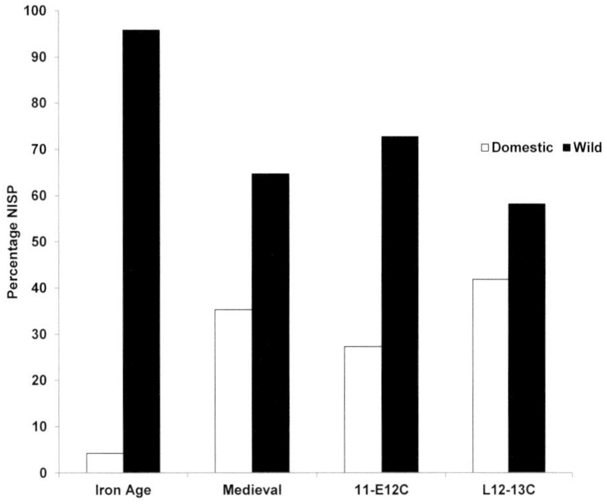

Figure 8.2: *Percentages of domestic and wild mammals at Minino*

By far the largest quantity of osteological material belonged to the Medieval period. It proved possible to identify 2,456 fragments of the mammals used for food and fur to species or genus (Table 8.2). A total of 35.3% of all the identified bones came from domestic animals (Figure 8.2) with cattle and sheep/goat bones being represented in very similar numbers: 36.5% and 39.8% of the domestic species respectively. A further 15.1% of the bones from domestic animals found in the medieval layers were from pig. The shares for horses and dogs were 6.9% and 1.6% respectively (Figure 8.4). Among wild animals hunted for food and fur, pride of place went to the beaver with 54% of the identified bones of wild mammals. In comparison with the figures for the Early Iron Age the percentage of beaver bones was slightly smaller. The species of wild mammal with the next largest number of bones was the squirrel with 26.5%. The amount of squirrel bones contained in the medieval levels increased sharply. There were also quite large proportions of elk (6.7%) and pine marten (6.5%) although the latter was significantly less well represented than in the Iron Age sample (Figure 8.3). The increase in domestic mammals in the medieval period compared to the Iron Age should be noted (Figure 8.2).

It proved possible to subdivide much of the medieval material into groups from two periods: (1) from the late 10th/early 11th century up until the first half of the 12th century and (2) from the second half of the 12th century and beginning of the 13th century (Table 8.2). This enabled us to chart changes in the ratio of bones from domestic mammals to those hunted for food and fur in the Medieval period.

In the levels of the first period the species and genera of 1,025 mammals bone fragments were identified and in the levels relating to the second period, the equivalent number was 820 bones (Table 8.2). The ratio of domestic to wild mammals in the earlier medieval period was 27.3%: 72.7%. In the second period the share of wild mammal bones decreased to 58.2% with bones from domestic mammals rising accordingly to 41.8% (Figure 8.2). As regards the range of domestic animals, cattle increased from

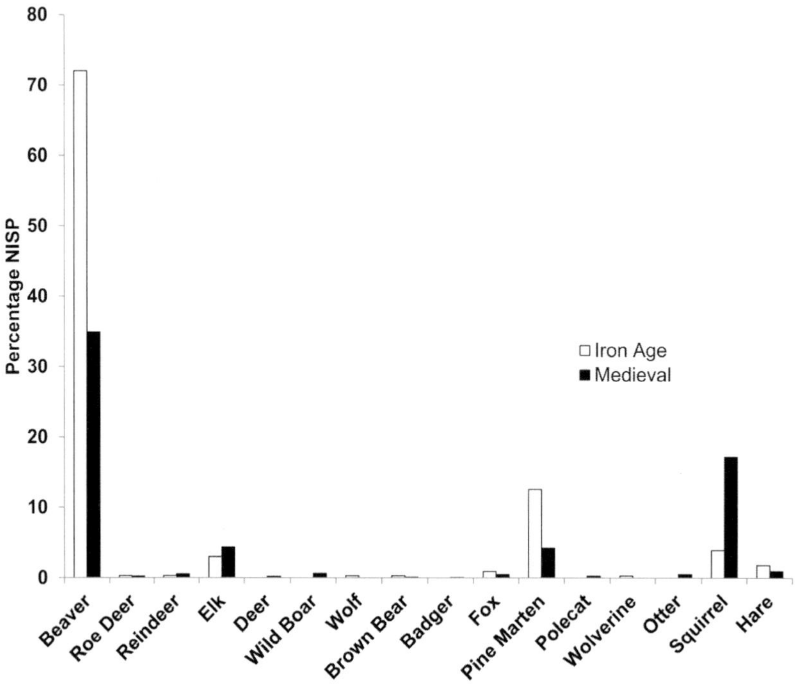

Figure 8.3: *Percentages of wild mammal species in Iron Age and medieval levels at Minino*

31.1% to 43.1%. The percentages of sheep/goat bones, pig and dog bones remained virtually the same. There was a sharp drop in the proportion of horse bones from 13.6 to 2.3% (Figure 8.4; Table 8.2).

Among the remains of animals hunted for food and fur, beaver predominated, but its share decreased from 56.8% to 38.4%. The share of bones from the second most plentiful species, the squirrel, increased slightly: from 28.7% to 32.1%, as did that of the pine marten from 6% to 8.4%. There was a significant increase in the share of elk from 4.2% to 11.7% (Figure 8.5).

As has already been noted in the methodology section, in order to calculate the relative numbers of animals, quadrants were used in which the cultural layer had gradually accumulated, *i.e.* without taking into account quadrants including pits, burials or hearths. There were 160 such quadrants containing Early Iron Age layers and 1,066 in the medieval period (Table 8.3). There were 322 quadrants relating to the earlier medieval period and 267 quadrants relating to the second period. The cultural layers in the remaining squares from the medieval period were of a mixed nature.

Characteristic for the levels of the medieval period was the considerable increase (by a multiple of 9.3) in bone concentration (*i.e.* density of bones per quadrant) of domestic animals compared to that found in Early Iron Age levels. The amount of bones from domestic animals in the levels of the second part of the medieval period

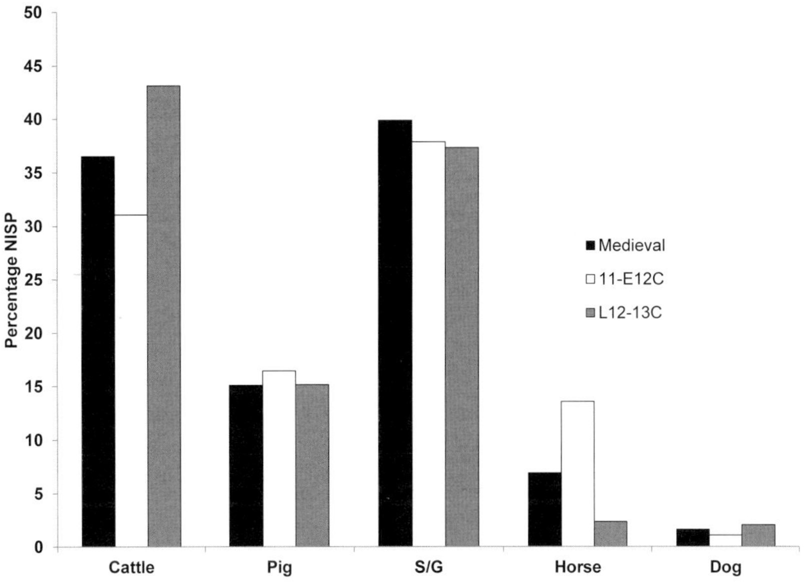

Figure 8.4: Percentages of domestic mammal species in medieval levels at Minino

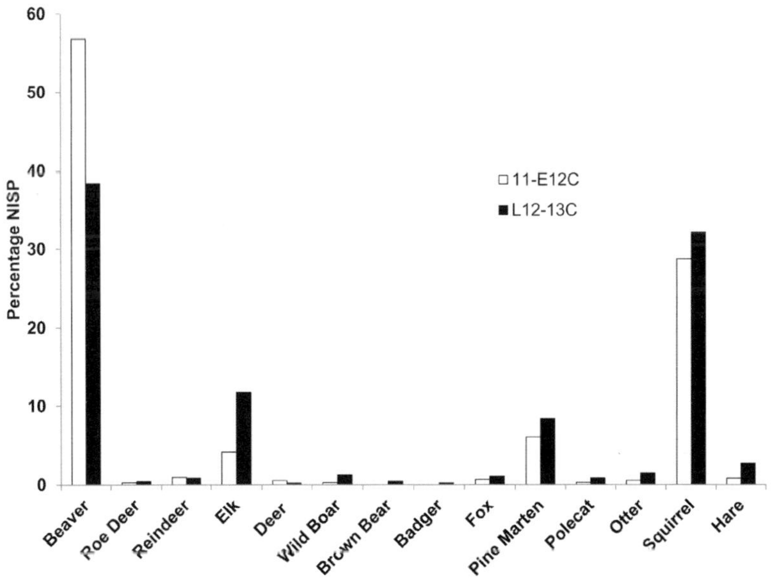

Figure 8.5: Percentages of wild mammal species in medieval phases at Minino

is approximately one and a half times greater than it had been in the first. This is particularly noticeable with regard to cattle and sheep/goat. The concentration of horse bones on the other hand decreases almost four times over.

It is in the Early Iron Age levels that we find the highest beaver-bone content. The concentration of these bones is almost twice as high compared with the medieval layers. In the levels relating to the second part of the medieval period (12th–13th century), the concentration of beaver bones was a little over half of what it had been in the earlier medieval period. In contrast, the concentration of squirrel bones is much higher in the medieval levels, almost five times greater than in the Early Iron Age levels. Changes were only minor variations in the concentrations of squirrel bones between the earlier and later parts of the medieval period.

In the medieval levels the concentration of elk bones is 1.6 times greater than in the Early Iron Age levels. The elk-bone content in levels from the latter part of the medieval period was more than twice as large as it had been in levels from the earlier part of that period. When we consider other well-represented species, there is a marked decline in the concentration of pine marten bones in the medieval levels in comparison with the Early Iron Age. The concentration of pine marten bones did not vary greatly in the two medieval periods.

Beaver ageing evidence

All of the beaver remains could be divided into four age groups: under one year; individuals aged one to two; individuals aged three to four and individuals aged four and over.

In the Early Iron Age the age structure of the beavers was as follows: 7.5% aged under one year; 11.8% were in the second age group of one to two years; 64.5% fell into the third group, three to four years; and 16.1% were old animals (aged four or

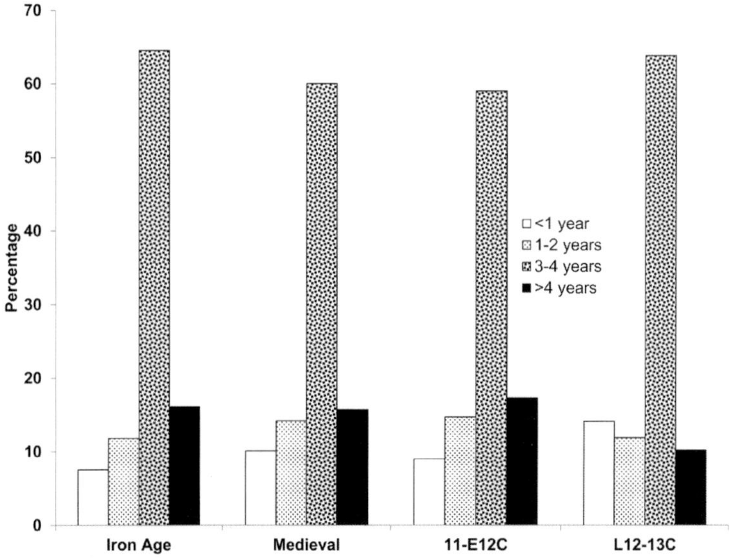

Figure 8.6: *Beaver ageing evidence from Minino*

over) (Figure 8.6). In the medieval period the percentage figures for these four age groups were very similar: 10.1% of the bones were from beavers under one year old; 14.2% from beavers aged one to two; 60% from individuals aged between three and four and 15.7% were old animals.

It is possible to note major changes in the age structure of the beaver caught, if we examine the two medieval periods (Figure 8.6): in the earlier medieval period we found 9% of beavers were under one year old, 14.7% of the bones came from the second age group, 59% from the third age group and 17.3% from the group of old animals. In the later medieval period the shares of the different age groups were as follows: 14.1% are animals under one year; 11.9% animals aged one to two; 63.8% aged three to four and 10.2% aged four and above. This means that in the two parts of the medieval period there were 1.5 times as many beavers aged under a year in the later period and a considerable decrease in the percentage of old animals.

GENERAL DISCUSSION

A characteristic feature of the material from the Minino archaeological complex is the predominance of wild mammals over domestic ones in the faunal remains for all periods during which the complex existed.

In the Mesolithic and Neolithic periods the only domestic animal in the region was the dog. The very small percentage of bones from domesticated animals in the Early Iron Age levels testifies to the substantial predominance of hunting over animal husbandry in that period. For the city sites situated further south in the Upper Volga region during the Early Iron Age, the reverse picture was typical: at most sites bones of domestic animals predominated (Alekseashina 1950; Bibikov 1950; Tsalkin 1956; Danilchenko 1984). It is clear that in the Kubenskoye Lake area animal husbandry was far less developed, but at the same time the species of domestic animals characteristic of more centrally located areas of Russia were already present.

In the medieval period domestic animals began to play a considerably more important part in the economic activity of the ancient inhabitants of Minino, but nevertheless remains of wild animals continued to predominate in the osteological material. It is generally accepted that not all bones from wild animals will be brought to a settlement. The preparation of carcasses of animals killed during hunting more often than not took place where the animals had been caught and only the most valued parts of the carcass would then be brought back to the settlement (those rich in meat or possessed of religious significance) together with the skin. This applied in particular to fur-bearing species, whose skins survive better if they are removed at once where the animal has been caught, and the only bones of such animals which later make their way into the cultural layer are the skulls, the mandibles and the bones from the distal parts of the limbs. The preparation of the carcasses of domestic animals takes place within a settlement and for that reason it is more likely that bones from more complete carcasses subsequently make their way into the cultural layer. All this testifies to the clear predominance of hunting over animal husbandry in the medieval period in the region in question.

In the medieval period, as in the Early Iron Age, all the species typical for the forest zone were already present. The predominance of cattle bones over those of sheep/goat also points to the predominance of cattle in the herds as well (Antipina 1997; 2003). Yet in the medieval period the picture changed with the passage of time. In the first part of the medieval period (11th and early 12th century) the quantity of sheep/goat bones was larger than that of cattle bones. This state of affairs can be interpreted in two different ways. Either there were more sheep/goat than cattle, or numbers of cattle and sheep/goat were equal, but cattle were not principally used for meat production, but for milk products (Antipina 1997). Later on in the medieval period (late 12th and 13th century) the number of cattle bones was greater than that of sheep/goat bones, which may reflect the actual situation pertaining at the time.

The third most prominent species of domestic animal among the identifiable bones in the osteological material from Minino was that of the pig. During both the Early Iron Age and the medieval period the share of pig bones was more or less the same (15%–16%). It is interesting to note that the amount of pig bones found in pits was little over half the amount found in the occupation levels.

It is interesting to note the small quantity of horse bones in the period lasting from the late 12th into the 13th century, in comparison with numbers found in the preceding period. This may indicate that horses had begun being used less frequently, if at all, for food and that they were only being used as working animals, which would have led to a reduction in the herds of those animals (Tsalkin 1956) or to a reduced possibility of horse bones turning up in the cultural layer of the settlement.

As regards wild animals, the main species upon which the local inhabitants appear to have concentrated their hunting efforts throughout the existence of the settlement were beaver, squirrel, pine marten and elk.

Beaver bones predominate among the wild mammals in all levels of the Minino archaeological complex. Even among the very small assemblage from the Stone Age levels, beaver bones have been found. The importance of beaver in the economic activities of the local population is difficult to exaggerate. However, the numbers of beaver declined both between the Early Iron Age and the medieval period and also during the course of the medieval period. Analysis of changes in the age structure of the animals caught by hunters enabled us to attempt to evaluate the degree to which hunting had affected the state of the beaver population in the vicinity of the Minino settlement.

When we compare the age structure of the beavers caught in the Early Iron Age and in the medieval period generally, no major differences stand out. All that emerges is that the percentage of young animals grows and that of old animals decreases. If, however we examine separately the earlier and latter part of the medieval period, then some differences do come to light. In the latter stage (12th and 13th centuries) an increase in the share of animals aged up to one year increases by 1.5 times, from 9% to

14.1% and there is a drop in the percentage of old animals. However, the percentage of animals aged below one year remained quite small and the large share of older animals suggests that hunting had only a small impact on the local situation, possibly simply reducing the rate of increase in the beaver population as a whole (Safonov 1966). In this case it may have even been useful even to intensify hunting so as to 'rejuvenate' the beaver population. According to data from excavations at the city-site of Indakar, in the upper medieval levels the percentage of animals under a year old accounted for almost 75% of the beaver bones (Bogatkina 1995). Yet, even given this high level of immature slaughter of beavers, they continued to exist in the region of the Kama River (Udmurtia) until as late as the 18th century. Bearing this in mind, it can be said that the pressure from hunters and fur-dealers on the medieval population of beavers in the Minino area was not catastrophic at that period. It is possible that in subsequent periods the pressure intensified so much that it led to the disappearance of this species in the area of the Kubenskoye Lake and in the Vologda Region, but not as early as the 13th century. It can definitely be stated that in 1492 beavers were still to be found at the north-western end of the Kubenskoye Lake, at which stage they were being registered in historical documents (Anon. 1958). It is possible that in this particular region there was a whole range of reasons for the disappearance of the beaver: population growth leading to an intensification of uncontrolled hunting; cutting back of woods to gain more arable land, which led to changes in the rivers' hydrological regimes; reduction in vegetation along the banks of waterways and so on. At the same time it would be wrong to rule out completely the possibility that there had been a sharp rise in the numbers of beavers being killed locally. The fact is that the capacity for the material currently available to enable us to resolve such questions is not very great and we can only make possible interpretations applicable to periods spanning several decades. The destruction of beaver settlements, however, could occur within a few years, a fact which would not have been reflected in the age-structure of the beaver population.

The second most important species in our faunal assemblage is that of the squirrel. In the Early Iron Age levels, however, squirrel bones were far from numerous (only 4.1% of the identified wild mammal bones). This indicates that squirrel hunting was not intensive and perhaps did not yet enjoy the major economic significance it acquired throughout the medieval period, when there was a marked increase in squirrel-hunting. In the two medieval periods squirrel accounts for between 28.7% and 32.1% of the total of the wild mammal bones. From archaeobotanical data (Makarov *et al.* 2000) we know that in the 11th and early 12th centuries there was a reduction in the area occupied by indigenous coniferous forests and these were partially replaced by secondary growth of plantings of birch and aspen in the immediate vicinity of the settlement. This should have had a negative impact on the numbers of squirrels, since mixed forests containing coniferous trees would have been preferable habitats. However, we know that in the medieval period squirrels were hunted in large numbers throughout the areas they inhabited. In the 19th century squirrels were one of the most important animals for the fur-trade and at that time tens of millions of squirrels were being caught each year (Kirikov 1960). Thus an increase in the numbers of squirrels caught indicates not an increase in their numbers in the wild but their enhanced role in the medieval economy.

The pine marten is the species accounting for the third largest amount of bones in the osteological collection from Minino. The largest quantities of this species' bones were recorded in the Early Iron Age. There were virtually no changes found in numbers of pine marten during the two phases of the medieval period, so that hunting cannot be said to have had a negative influence on this species' population.

The next most important species with regard to total numbers of bones is the elk. The share of elk bones in the total medieval assemblage from Minino is over twice as large as that found in the Early Iron Age (an increase from 3.1% to 6.7% of the wild mammal bones). In the medieval period the share of elk bones increased from 4.2% in the earlier phase to 11.7% in the later phase, which is perhaps connected with the increase of slash-and-burn agriculture and the spread of secondary growth, habitats which would attract elk. It is interesting to note that there was a similar increase in numbers of elk noted during this period in the Baltic region (Vereshchagin and Rusakov 1979).

CONCLUSION

The main animal species hunted for food and fur by the ancient inhabitants of the Minino area near Lake Kubenskoye in the Early Iron Age and the medieval period up until the 13th century were beaver, squirrel, pine marten and elk. Major changes took place in relation to elk and beaver during the time that these settlements existed. Relatively speaking, numbers of elk that were exploited grew, while numbers of beaver decreased somewhat. Hunting did not have a major negative impact on populations of wild animals before the middle of the second millennium AD, but we cannot rule out the rapid elimination locally of specific beaver colonies. At the same time the main changes in the geographical deployment and numbers of wild animals within the territory of north-west Russia were more likely to have been bound up with changes in the conditions in which the animals lived resulting from human activity.

ACKNOWLEDGEMENTS

This work was carried out with the financial support of the Russian Foundation for Fundamental Research (Grant No. 03-04-49323, No 17-06-00543) and the following programmes: 'The Origin and Evolution of the Biosphere', 'The Scientific Foundation for the Protection of Bio-diversity in Russia' and 'Fundamental Principles for the Management of Biological Resources'. It was also supported by INTAS project number 2000-154.

– 9 –

BIRD BONES FROM NOVGOROD AND SITES IN ITS IMMEDIATE HINTERLAND

Sheila Hamilton-Dyer and Mark Maltby with Mikhail Sablin and Andrei V. Zinoviev

INTRODUCTION

The first part of this chapter will present the evidence of the presence and exploitation of domestic and wild birds at Gorodishche and other sites in the immediate hinterland of Novgorod. The second part of the chapter will be focused on evidence from Novgorod itself, mainly from the Troitsky sites. As in previous chapters the evidence will be based mainly on the animal bone evidence. Other evidence related to the exploitation of birds will be incorporated into Chapter 11.

The bird bones analysed in this chapter are based on an updated dataset from those used in some previous publications. Some of the earlier reports of the bird bones (*e.g.* Hamilton-Dyer 2002) did not include the bones from the 1998 season at Gorodishche, which were only examined in 2003 and the data from both Gorodishche and the Troitsky sites in Novgorod have been refined and updated from those previously published by Maltby (2012), as a result of some re-identifications and other archive corrections. The current (and final) dataset from the Troitsky IX–XI and Fedorovsky sites is almost identical to that incorporated in a briefer review of the exploitation of wild animals published in 2017 (Hamilton-Dyer *et al.* 2017). This dataset also indicates how many of the bones from Troitsky X and XI were retrieved from sieving. We have also added counts of the species identified in a sample from Troitsky VIII, in which only measurable bones were fully recorded.

The vast majority of the bird bones from all these sites were identified and recorded by Sheila Hamilton-Dyer. Identification presented a major challenge, particularly in the initial stages of the work, as there was no reference material readily available in Novgorod and we had to rely on previous experience and knowledge alongside various reference manuals. Fortunately the great majority of the species were ones with which we were familiar, and we were able over the years to identify most of the 'mysterious' challenging specimens. However, we erred on the side of caution and a few bones that would be identifiable with access to good comparative collections were not recorded as identified. In the case of ducks, it is not possible to distinguish all bones to species due to close morphological similarities and overlapping sizes (Poland 2018). They were divided into three size categories; large (mallard-sized); small (teal-sized) and medium-sized (including species such as pochard and wigeon).

In a few cases we were able to postulate the most likely species of duck represented. The taxa identified in the main project (English and Latin names) are listed below:

Great-crested grebe	*Podiceps cristatus*
Cormorant	*Phalacrocorax carbo*
Heron	*Ardea cinerea*
Bittern	*Botaurus stellaris*
Stork	*Ciconia* sp.
Swan	*Cygnus* sp.
Domestic goose or cf. greylag	*Anser anser*
Goose, cf. white-fronted	*Anser albifrons*
Goose, cf. lesser white-fronted	*Anser erythropus*
Domestic duck or mallard	*Anas platyrhynchos*
Teal	*Anas crecca*
Other ducks including:	
cf. Garganey	*Anas querquedula*
cf. Wigeon	*Anas penelope*
cf. Shoveller	*Anas clypeata*
cf. Goldeneye	*Bucephala clangula*
cf. Pochard	*Aythya ferina*
cf. Tufted	*Aythya fuligula*
Eagle, cf. white-tailed	*Haliaeetus albicilla*
Sparrowhawk	*Accipiter nisus*
Goshawk	*Accipiter gentilis*
Buzzard	*Buteo buteo*
Hen harrier	*Circus cyaneus*
Kestrel	*Falco tinnunculus*
Hobby	*Falco subbuteo*
Merlin	*Falco columbarius*
Capercaillie	*Tetrao urogallus*
Black grouse	*Tetrao tetrix*
Hazel hen	*Bonasa bonasia*
Partridge	*Perdix perdix*
Domestic fowl (chicken)	*Gallus gallus*
Crane	*Grus grus*
Coot	*Fulica atra*
Wader, cf. Woodcock	*Scolopax rusticola*
Wader, cf. Snipe	*Gallinago gallinago*
Other waders	eg *Tringa* sp., *Limosa* sp.
Gull	*Larus* sp.
Woodpigeon	*Columba palumbus*
Stock dove or domestic pigeon	*Columba oenas/livia*
Owl, cf. Tawny	*Strix aluco*
Owl, cf. Tengmalm's	*Aegolius funereus*
Raven	*Corvus corax*
Rook/Crow	*Corvus frugilegus/corone*
Jackdaw	*Corvus monedula*
Small corvid, cf. Jay	*Garrulus glandarius*
Small passerine	

In addition, further bird bones were recovered during the 2000–2004 excavation seasons at Gorodishche and in the excavations of the Desyatinny-1 site in Novgorod. The species represented are described in separate sections below.

BIRD BONES FROM GORODISHCHE 1979 AND 1993–1998

All bird bones from the 1979 and 1993–1998 excavations at Gorodishche and the other hinterland sites were fully recorded individually. Therefore detailed records were made of all elements identified, parts of the bones represented, butchery marks, gnawing damage, ageing data, pathology evidence and the presence of medullary bone. Measurements of bird bones followed von den Driesch (1976) and were recorded to the nearest 0.1 millimetre.

Species represented

Although bird bones formed less than 4% of the total assemblage, they were much more frequently recovered than wild mammal bones, totalling 643 fragments, of which 470 (73%) were identified. At least 15 species were represented (Table 9.1). Ducks contributed most (293) of the bird bones, forming 63% of the identified elements. The second most frequent remains (118 bones) were of domestic fowl (chicken), providing 24% of the identified bones. Bones of geese were less common (8%) but were consistently present across the site. Bones of other birds were rare forming less than 5% of the identified elements (Figure 9.1). The most frequent of these was capercaillie (6 bones). Another large galliform bone could have belonged to capercaillie or possibly black grouse. Other taxa identified were white-tailed eagle, a medium-sized accipiter (probably sparrowhawk), crane, swan, cormorant, wader (cf woodcock), raven and smaller corvids (Table 9.1).

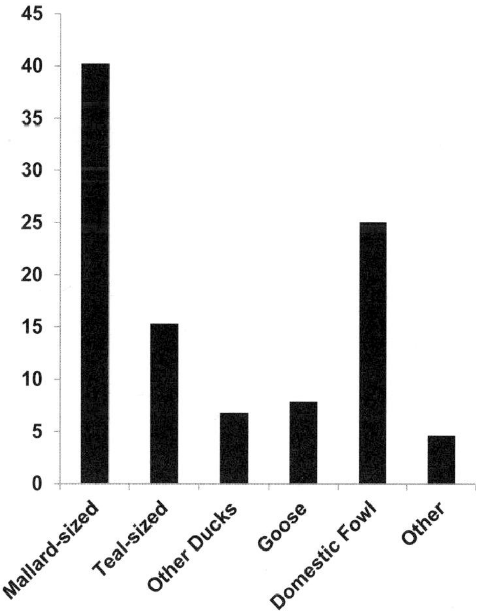

Figure 9.1: Gorodishche 1979–1998: NISP percentages of identified bird bones all seasons combined

Bird bones, particularly those of ducks, were better represented from deposits recovered between 1993 and 1997 from the North bank, where they formed 9% of the total fragments of mammals and bird (including unidentified fragments) compared with 6% from the South bank. There was a particularly high level of bird bones in the pits from the North bank (20%). In the

South bank pits the bird bones accounted for only 2% of the very small assemblage. The black levels in both North and South bank produced between 7% and 9% bird bones respectively, whereas the earlier brown levels contained very few bird bones, only 2% and 4% from the North and South banks respectively, although again assemblage sizes were very small. Only 19 bird bones were recorded in the 1979 hand-collected assemblage, representing 2% of the mammal and bird fragments. The 1998 excavations produced a further 76 bird bones, representing 2% of the total mammal and bird fragments.

Ducks

Ducks provided 62% of the identified bird bones in the Gorodishche assemblage. As noted above, positive identification to species level of individual duck bones in the assemblage was difficult; the duck family contains a large number of species, several of which can be found in the Novgorod region (Cramp and Simmons 1977), and many of the bones are very similar in size and morphology. The sizes of ducks will be considered further in the discussion of the metrical data. As indicated above, duck bones were initially assigned to one of three size categories. Most of the bones were from large ducks, most of which were probably mallard, perhaps including tamed or domestic birds. Large ducks provided over 40% of the identified bird elements from Gorodishche (Table 9.1; Figure 9.1). Their abundance did vary in the different assemblages, however (Table 9.2; Figure 9.2). They were the most common species represented in the 1993–1997 assemblages from both the North (41%) and South banks (50%). In contrast they provided only one of the 19 bird bones recovered from the

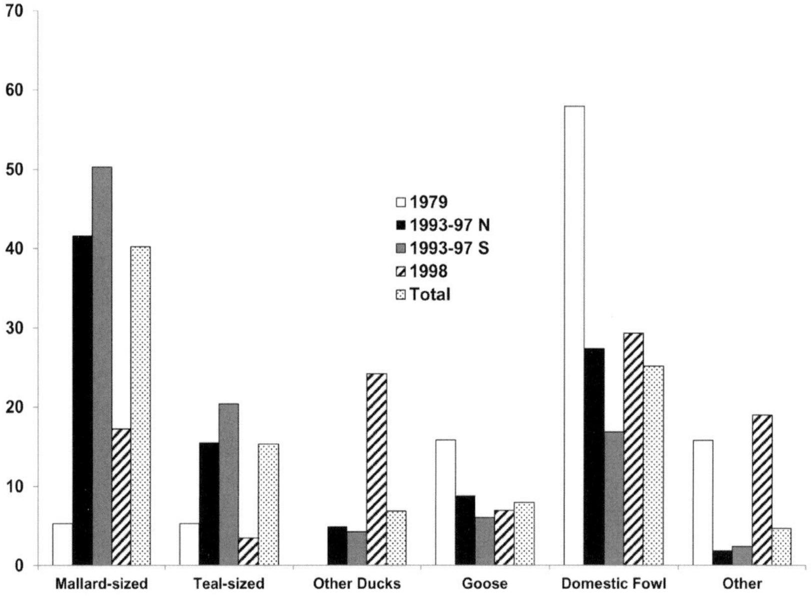

Figure 9.2: *Gorodishche 1979–1998 NISP percentages of identified bird bones by season*

1979 excavations and only 17% of the identified bird bones in the 1998 assemblage, in which they were outnumbered by domestic fowl and medium-sized ducks.

Bones of small, teal-sized, ducks were also present in good numbers forming 15% of the total identified avian bones (Table 9.1; Figure 9.1). Their relative abundance also varied in the sub-assemblages (Table 9.2; Figure 9.2), ranging from just over 3% in the 1998 assemblage to over 20% in the 1993–1997 South bank assemblage.

Bones of medium-sized ducks were consistently present but less frequently, providing 7% of the identified birds in the total assemblage (Table 1; Figure 9.1). However, they outnumbered both teal-sized and mallard-sized ducks in the 1998 assemblage, in which they provided 24% of the identified bird bones (Table 9.2; Figure 9.2). Their percentage in this relatively small assemblage was inflated by the inclusion of five shoulder and wing bones of one bird, probably a goldeneye. Other medium-sized duck bones were similar in size and morphology to those of tufted duck, garganey, wigeon and pochard. Other possible species which currently occur in the area include gadwall (*Anas strepera*), pintail (*Anas acuta*) and shoveller (Cramp and Simmons 2018). Pochard, tufted duck and goldeneye are diving ducks whereas all the others are surface feeders. Apart from some mallards, which are now resident all year round in the Novgorod hinterland, all the duck species are summer visitors to the region. Large numbers of ducks and geese are hunted today around Lake Ilmen. Many species of water birds, including mallards, migrate to this area from the west and south (Dobrynina and Kharitonov 2006).

Elements represented for all sizes of duck showed that wing bones were prominent, forming over 40% of the bones recovered (Table 9.3; Figure 9.3). In the case

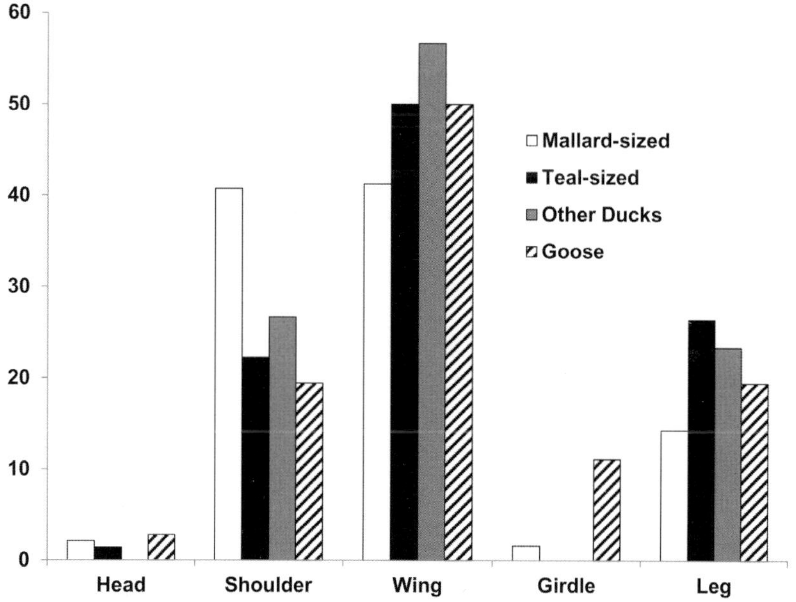

Figure 9.3: *Gorodishche 1979–1998 NISP percentages of ducks and geese body areas all seasons combined*

of large ducks, bones of the breast and shoulder also formed over 40% of the bones with the coracoid and furcula being particularly common (Table 9.3). Shoulder bones also provided over 20% of the bones of smaller ducks. Leg bones were recovered less frequently than wing bones in all three size groups (Figure 9.3). Biases towards wing bones are common in duck assemblages. It is often suggested that this is because the bones in the wings of ducks are generally larger than most of their leg bones, indicating retrieval bias. This is supported by the fact that the tibiotarsus, the largest leg bone, was recovered far more commonly than the smaller femur and tarsometatarsus (Table 9.3). However, this argument is less convincing with regards to the shoulder and wing bones. The smaller coracoid and carpometacarpus both outnumbered the longer humerus, radius and ulna in the large-sized duck assemblage. The coracoid and carpometacarpus are relatively sturdy and more easily identifiable bones than shaft fragments of humerus, radius and ulna. However, it is possible that some of the upper wings were transported off-site (Bovy 2012). Butchery marks in the form of fine knife cuts were observed on two mallard-sized coracoids indicating that the upper wings were separated from the shoulder. There was, however, no significant variation in their relative frequencies from the North and South banks.

Only three mallard-sized duck bones were recorded as porous, indicating that a few of the duck bones were from fledglings. High percentages of young birds can indicate that captive birds were being kept but it is probably more likely that a few fledglings were caught during hunting expeditions. None of the teal-sized and medium-sized duck bones were from immature birds.

Geese

As in the case of mallard, the bones of domestic geese can be very difficult to separate from those of its wild greylag ancestor, and captured and tamed birds are impossible to distinguish with confidence using standard identification methods unless they have increased in size (Poland 2018). All of the geese bones from Gorodishche were of a similar size to greylag goose. Although their distribution has diminished significantly because of hunting and drainage, wild greylag geese still breed in some parts of western Russia, many of them migrating to other parts of Europe in the winter (Kumari 1972; Cramp and Simmons 1978). However, genetic analysis has suggested that domestic geese were established in some parts of Russia during the early medieval period (Honka *et al.* 2018). The sizes of the geese will be considered further in the discussion of the metrical data. Geese provided 8% of the total number of identified bird bones (Table 9.1; Figure 9.1). There was a fairly consistent presence of geese in the different assemblages (Table 9.2; Figure 9.2). Their large size would have favoured recovery of their bones compared with those of smaller species in hand-collected assemblages.

As in the case of ducks, wing bones were the goose bones most commonly recorded (Table 9.3; Figure 9.3). This again probably mainly reflects their large size and relative robustness. None of the goose bones were recorded as porous indicating that they all belonged to adult birds. Cut marks were observed on the shaft of one of the goose bones. Unfortunately the element involved was not recorded in the archive.

Domestic fowl (chicken)

Domestic fowl formed about a quarter of the bones identified from Gorodishche (Table 9.1; Figure 9.1). Their bones were found in all of the assemblages with the exception of the very small sample from the brown layers on the South bank (Table 9.2). They were best represented (58%) in the small 1979 assemblage, where recovery methods favoured their collection compared with the bones of the medium-sized and small ducks in particular (Figure 9.2). In the 1993–1997 assemblages domestic fowl was better represented in the North bank deposits, contributing 27% of the identified bird bones, compared with only 17% in the South bank assemblage (Table 9.2; Figure 9.2). The high percentage of domestic fowl bones (52%) from the North bank pits was inflated by the inclusion of 14 bones from one cockerel. Overall, domestic fowl provided 23% of the identified bird bones from the 1993–1997 excavations (Table 9.2). They were recovered more commonly in the 1998 assemblage, in which they were the most frequently identified, providing 29% of the bird bones (Table 9.2; Figure 9.2).

There was a more balanced representation of wing and leg bones amongst the domestic fowl assemblage than in the case of ducks and geese. Shoulder bones were also quite well represented (Figure 9.4). Variations in element representation closely mirrored the size and robustness of the bones. Large bones such as the tibiotarsus and sturdy elements such as the humerus were the most frequently recovered whereas small bones like the carpometacarpus and fragile elements such as the furcula were poorly represented (Table 9.4).

Thirteen (11%) of the domestic fowl bones were recorded as porous and belonged to immature birds. This is a much higher percentage than for other species and indicates that chickens were bred at Gorodishche and some were culled at quite a young age for meat.

Three of the five tarsometatarsi had no spurs and belonged to adult hens. One had a spur scar and probably belonged to a young male. The only tarsometatarsus with a fully developed spur belonged to the cockerel skeleton. Males were more likely to have been culled immature, whereas many of the hens could have been keep as breeding birds and/or for egg production. One

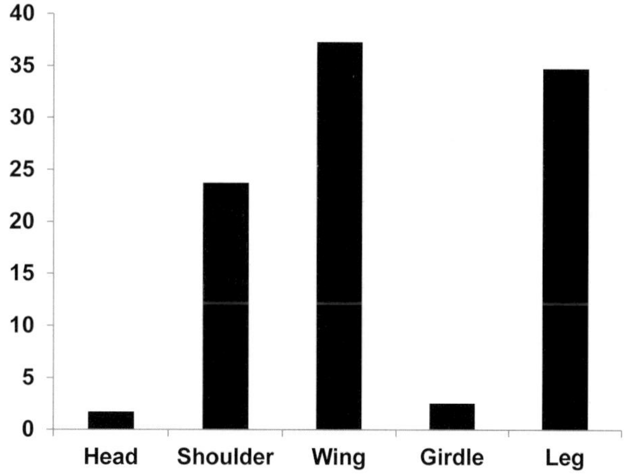

Figure 9.4: *Gorodishche 1979–1998 NISP percentages of domestic fowl body areas all seasons combined*

femur from the 1998 excavations had evidence of medullary bone within its shaft indicating that the hen had recently been in lay. One coracoid had a healed fracture indicating that the bird had survived the injury.

Other birds

The other bird species at Gorodishche were present in very small numbers, forming less than 5% of the identified bird bones (Table 9.1; Figure 9.1). Because of small sample sizes not much significance can be attached to their higher percentage frequencies in the 1979 and 1998 assemblages. They only formed around 2% of the 1993–1997 assemblages from both the North and South banks (Table 9.2; Figure 9.2).

At least six bones (coracoid; 2 furculae; ulna; carpometacarpus; vertebra) of capercaillie were recovered. These birds would probably have been captured in coniferous woodland. These large game birds were definitely exploited for meat, and perhaps imported for special occasions. A knife cut was observed on one of the furculae and the proximal end of the ulna had been severed with a heavy blade.

Other large birds such as grouse, swans and cranes may also have been acquired for special occasions. A tibiotarsus of a large galliform was recorded from the South bank. This may, based on its size and morphology, have belonged to a black grouse, possibly a male. A humerus of a swan was also found in the South bank excavations and a crane ulna was recovered in 1998.

Smaller species of waders were represented by a single humerus from the South bank. This was closely comparable in size and morphology to a woodcock. Woodcock are also present in large numbers around Lake Ilmen. A humerus of a cormorant was found in 1998. Cormorants nest in large colonies in the Gulf of Finland (Gaginskaya *et al.* 2012). It is therefore possible that it was imported. However, although cormorants are often regarded as a seabird, they can also be found in freshwater locations inland. There were cormorants breeding on Lake Ladoga in the early 20th century, for example Pihlström (2013). They can also migrate over long distances and therefore this bone may have belonged to a bird caught during migration.

Three carpometacarpi and two fragments of ulnae (possibly from the same bone) of eagle were identified. These were all good matches for white-tailed eagle. It is possible that eagles were attracted to the settlement as scavengers. However, it is more likely that these birds were killed elsewhere and parts of their carcasses imported to Gorodishche. It is significant that the three carpometacarpi were found in the same context from the 1998 excavations. It is probable that the lower wings of these birds were valued and processed for their feathers. A humerus of a raptor, probably a sparrowhawk, was discovered in the same layer as the eagle carpometacarpi. Sparrowhawks can be kept for falconry (Prummel 1997; 2018) but they could also have been valued for their plumage.

Two bones of raven (humerus; mandible) and three of smaller corvids were identified. A tibiotarsus was the size of a rook or crow; two ulnae were a good match for jackdaw. The corvids may have been natural mortalities, or were perhaps killed as scavengers.

Small passerines were absent amongst the bird bones. These are not usually common in early medieval assemblages and the surrounding countryside is better suited to waterbirds, but one might have expected to find a few of their bones.

BIRD BONES FROM GORODISHCHE 2004–2004 (Mikhail Sablin with Mark Maltby)

Species represented

The bird bones from the 2000–2004 excavations were identified by A.N. Panteleyeva of the Zoology Institute of the Russian Academy of Sciences. A total of 678 bones were recorded, of which 617 were identified from at least 25 species. The following species were identified:

Heron	*Ardea cinerea*
Whooper swan	*Cygnus cygnus*
Swan sp.	*Cygnus* sp.
Domestic goose or greylag	*Anser anser*
White-fronted goose	*Anser albifrons*
Domestic duck or mallard	*Anas platyrhynchos*
Teal	*Anas crecca*
Other ducks including:	
cf. Wigeon	*Anas penelope*
cf. Pintail	*Anas acuta*
cf. Shoveller	*Anas clypeata*
cf. Garganey	*Anas querquedula*
cf. Tufted duck	*Aythya fuligula*
cf. Ferruginous duck	*Aythya nyroca*
cf. Long-tailed duck	*Clangula hyemalis*
cf. Goldeneye	*Bucephala clangula*
White-tailed eagle	*Haliaeetus albicilla*
Sparrowhawk	*Accipiter nisus*
Goshawk	*Accipiter gentilis*
Buzzard	*Buteo buteo*
Peregrine	*Falco peregrinus*
Kestrel	*Falco tinnunculus*
Domestic fowl (chicken)	*Gallus gallus*
Willow ptarmigan	*Lagopus lagopus*
Eurasian eagle-owl	*Bubo bubo*
Raven	*Corvus corax*
Rook	*Corvus frugilegus*

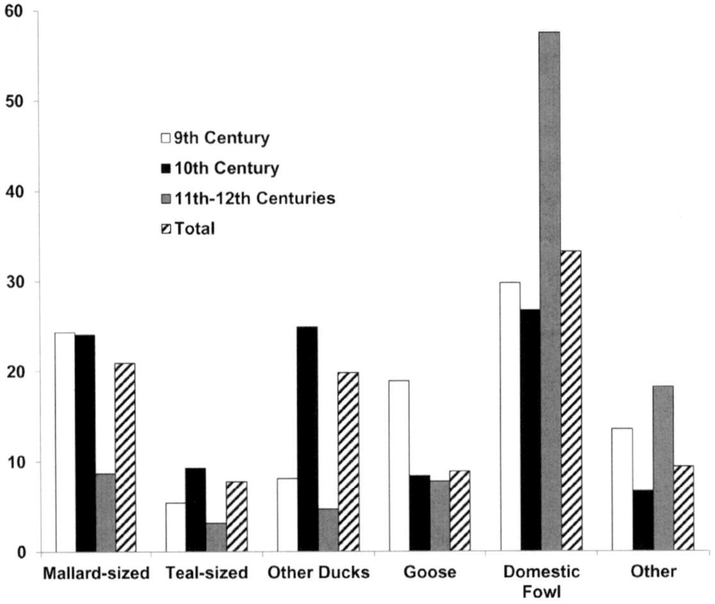

Figure 9.5: *Gorodishche 2000–2004 NISP percentages of identified bird bones by period*

Detailed NISP and MNI counts for the 9th, 10th and 11th–12th centuries are provided in Table 9.5 and the percentages of bones of the main categories in each period are summarised in Figure 9.5. The range of bird species was extremely varied, although in general the quantity of bird bones was quite small forming on average 8% of the identified assemblage (Table 9.5). There was an increase in the percentage of bird bones between the 9th and the 12th centuries, although the 10th-century assemblage was much larger than the others. The 9th-century assemblage was very small and should be treated with caution. The 11th–12th-century sample was also of modest proportions but does provide evidence for bird exploitation in a later period than represented in the previous excavations and is contemporary with the early deposits from Novgorod itself.

Ducks and geese

As in the previous excavations, ducks contributed most (299) of the bird bones from the 2000–2004 excavations. They provided 49% of the identified bird elements overall and 58% of the 10th-century assemblage (Table 9.5). However, the relative abundance of different species of ducks varied between periods and also diverged from the assemblages obtained from earlier excavations. Only 21% of the identified bird bones belonged to mallard-sized ducks compared with over 40% in the 1979–1998 assemblage (Tables 9.1 and 9.5; Figures 9.1–9.2 and 9.5). They were outnumbered by domestic fowl in all periods. They provided 24% of the assemblages from the 9th and 10th centuries but their contribution decreased to 9% in the 11th–12th-century sample.

Bones of teal formed 8% of the total identified bird elements (Table 9.5; Figure 9.5). Their percentage also decreased from 9% in the 10th-century assemblage to 3% in the 11th–12th-century assemblage.

Medium-sized ducks provided 20% of the identified bird bones outnumbering both teal and mallard in the 10th-century assemblage, in which they provided 25% of the bones (Table 9.5; Figure 9.5). As in the case of other ducks, their percentage decreased markedly in the 11th–12th-century assemblage to 5%. As indicated above, eight different species were recorded, confirming that the inhabitants of Gorodishche had a rich diversity of waterbirds available to be exploited (Table 9.5).

Geese provided 9% of the total identified bird bones. Most were identified as greylag/domestic goose but seven bones of white-fronted goose were also recorded. Geese were best represented (18%) in the small 9th-century assemblage (Table 9.5; Figure 9.5).

Domestic fowl (chicken)

Domestic fowl were the most common bird species identified, forming about a third of the bones identified in the 2000–2004 assemblage (Table 9.5; Figure 9.5). They were best represented (58%) in the 11th–12th-century assemblage compared with 27%–30% in assemblages of earlier date. This could imply that there was a greater reliance on domesticated birds in the later period.

Other birds

Bones of other birds were more common than in the assemblage from the earlier excavations at Gorodishche forming 9% of the identified bird elements overall and 18% of the 11th–12th-century sample (Figure 9.5). 7% belonged to seven species of raptor (Table 9.5). Amongst these, white-tailed eagle and sparrowhawk were also present in the assemblages from earlier excavations. In addition, bones of goshawk were recorded in all three periods and a bone of a peregrine was found in a 9th-century deposit. Some of these birds may have been kept in captivity for falconry. Four bones of kestrel and one each of buzzard and eagle owl were also recorded. Some of these may have been attracted to the site as scavengers as also could have been the corvids represented by seven bones of raven and one of rook. However, Zinoviev (2017) has suggested that some ravens and eagle owls may have been kept in captivity in Novgorod.

Although no further bones of capercaillie were recovered from the 2000–2004 excavations, a bone belonging to a willow ptarmigan was probably from a bird hunted for its meat.

Other large water birds probably exploited for food were represented by five bones of swan and one of heron, all from 10th-century deposits (Table 9.5). However, bones of smaller species of wader were not recovered.

BIRD BONES FROM OTHER HINTERLAND SITES

Small numbers of bird bones were recovered from the other three hinterland sites (Table 9.1). In all cases they were relatively less well represented than at Gorodishche, probably mainly reflecting poorer retrieval standards and preservation. All of the identified species were also found at Gorodishche and these were dominated by waterfowl.

Bird bones from Georgii

Twelve bird bones were recovered contributing 1.4% of the total assemblage. The nine identified specimens all belonged to geese and ducks (Table 9.1). The six bones of goose consisted of four humeri, a carpometacarpus and a wing phalanx, the last two from the same bird. A coracoid and carpometacarpus of mallard-sized ducks and a humerus of a medium-sized duck were also recovered. The dominance of wing bones mirrors the pattern found at Gorodishche. The higher incidence of goose may be a consequence of greater retrieval bias.

Bird bones from Prost

A tibiotarsus of a teal-sized duck was the only bird bone recovered from the Prost excavations. This represents only 0.03% of the total assemblage.

Bird bones from Vasilievskoye

A humerus and tibiotarsus of domestic fowl were the only bird bones recovered from the site. These bones formed 1.3% of the total animal bone assemblage.

BIRD BONES FROM NOVGOROD

The bird bones from the Troitsky and Fedorovsky sites

All bird bones from Troitsky IX–XI and Fedorovsky V were fully recorded individually using the same protocols used by Sheila Hamilton-Dyer on the hinterland sites.

Species represented

The bird assemblages from Troitsky IX–XI were subdivided into the same general assemblage groups as mammals. In addition, species counts were obtained for Spits 19–17 from Troitsky VIII, dating to the later 11th century, and from deposits from Fedorovsky V dating to the 13th and 14th centuries. The overall proportion of birds from Troitsky IX–XI formed 6.7% of the total mammal, bird and fish fragments. Although this is a higher figure than in the 1979–1998 assemblages from Gorodishche

(3.8%), the increase was largely the result of the poor recovery of fish bones in these hand-collected assemblages (Hamilton-Dyer *et al.* 2017). It is certain that the percentage of bird bones in relation to the larger mammals was also compromised by differential recovery.

Including bones from sieved samples, 4,497 bird bones were recorded from Troitsky IX–XI, of which 3,874 were identified. In addition, 305 and 43 bones were identified from Troitsky VIII and Fedorovsky respectively. At least 35 species were represented (Table 9.6). As at Gorodishche, ducks contributed most of the bird bones from Troitsky IX–XI, forming 52% of the identified elements. However, their percentages fluctuated within and between these sites (Figure 9.6). Ducks contributed 39% of the identified bird bones from Troitsky VIII but only 9% of the small assemblage from the Fedorovsky site. Domestic fowl provided 33% of the identified bones from Troitsky IX–XI but contributed over half the bones in some assemblages from those sites and in the samples from Troitsky VIII and Fedorovsky. Bones of geese were less common, providing 5% of the identified bird bones from Troitsky IX–XI (Table 9.6; Figure 9.6). Other species formed around 10% of the identified bird bones from Troitsky IX–XI, with percentages ranging between 5% and 25% in the major sub-samples. They were poorly represented in the Troitsky VIII assemblage (2%) but provided 21% of the identified bones from Fedorovsky (Figure 9.6). The other avian species identified varied between assemblages. They included birds hunted for food

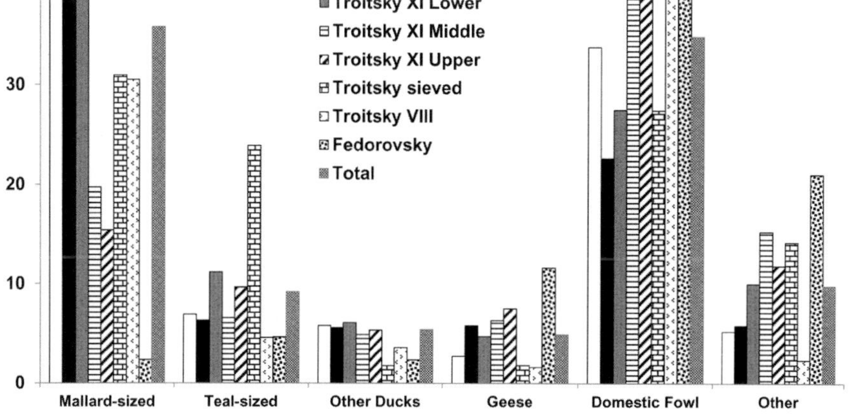

Figure 9.6: *Novgorod: NISP percentages of identified bird bones from Troitsky and Fedorovsky sites*

and feathers, birds kept in captivity for falconry and commensal species living within the town. 12% of the bird bones retrieved by standard collection methods in Troitsky IX–XI were unidentified. This figure rose markedly in the sieved samples, in which unidentified elements provided 53% of the total bird bones, reflecting the recovery of much greater numbers of small bone fragments, which were usually overlooked during normal excavation (Table 9.6).

Large (mallard-sized) ducks

Duck bones were initially assigned to one of three size categories. Most bones belonged to large ducks, many of which were a good match for wild mallard, which would have continued to have been in plentiful supply in and around Lake Ilmen and the River Volkhov. Although, it is possible that some of the bones belonged to birds kept in captivity, it is more likely that the vast majority were captured during hunting expeditions.

Mallard-sized ducks provided 37% of all the identified bird elements from Troitsky IX–XI (Table 9.6). They provided over 40% of the assemblages from Troitsky IX–X and the lower spits in Troitsky XI. In each case they were by some distance the most common category. Indeed, they formed over 53% of the identified bird bones obtained from Troitsky X (Figure 9.6). Large ducks, although still the second most common avian category, decreased significantly in relative abundance in the middle and upper spits of Troitsky XI, falling to below 20% of the identified bird bones in each case. This suggests that their contribution to the diet of the people who lived there decreased from the middle of the 12th century onwards, as domestic fowl became more important. This decline may have been widespread across the town. Large ducks formed only 30% of the Troitsky VIII assemblage, which came from spits dated to the later 11th century. They provided only 2% of the Fedorovsky assemblage, which was dated to the 13th–14th century. They were heavily outnumbered by domestic fowl in assemblages from two other Novgorod sites of 11th- to 14th-century date (Gorobets and Kovalchuk 2017), although more precise information about the sites and their date is not provided in the published article. They also ranked second behind domestic fowl in the assemblage from the Rogatitsky-3 site in deposits spanning the 11th to 16th centuries (Zinoviev 2016). Further comparisons are needed to establish whether variations in duck frequencies were principally due to chronological change or spatial variations.

As at Gorodishche, wing bones were prominent in the large duck assemblage from Troitsky IX–XI, forming 47% of the overall assemblage (Table 9.7). They provided over half the bones from both Troitsky IX and X and over 40% of the three assemblages from Troitsky XI (Figure 9.7). As discussed above, it is normal for wing bones to be prominent in duck assemblages, with differential recovery often a major factor. This is generally supported by the relative abundance of the different wing bones in the Troitsky assemblages. The largest wing bone, the humerus, was comfortably the most common of the mallard-sized duck bones in all the hand-collected assemblages, providing nearly a quarter of all their bones. The ulna (10%) was the next best represented, consistently outnumbering the adjacent but more slender radius (6%). The smallest of the wing bones, the carpometacarpus, was slightly less common than the radius (Table 9.7), in contrast to Gorodishche where it was the best represented of the wing bones (Table 9.3).

Again, as at Gorodishche, bones of the breast and shoulder formed a substantial portion of the mallard-sized duck assemblage, providing over 20% of the bones from Troitsky IX and X and over 30% of the three Troitsky XI assemblages (Figure 9.7). The sternum was particularly well represented, forming 12% of the total Troitsky IX–XI assemblage, mainly due to their unusually high prominence (16%) in the large and very well preserved assemblage from the lower spits of Troitsky XI (Table 9.7). Their abundance contrasts with Gorodishche, where fragments of sternum provided only 3% of the large duck assemblage (Table 9.3). The coracoid was also well represented, forming 10% of the assemblage despite its relatively small size. The slender furcula and scapula each provided around 6% of the assemblage from Troitsky IX–XI.

Leg bones were recovered less frequently than wing and shoulder/breast bones in all of the large duck assemblages from Troitsky IX–XI (Figure 9.7). As expected, the largest leg bone, the tibiotarsus, was recovered far more commonly (8%) than the smaller femur (3%) and tarsometatarsus (1%). No fibulae were recorded (Table 9.7). Relatively small numbers of bones from the head and girdle completed the assemblage (Table 9.7; Figure 9.7). The limited sieving programme produced only 35 bones of large ducks. Smaller elements, including mandibles and premaxillae were more prominent than in the hand-collected samples and there was a fairly even representation of

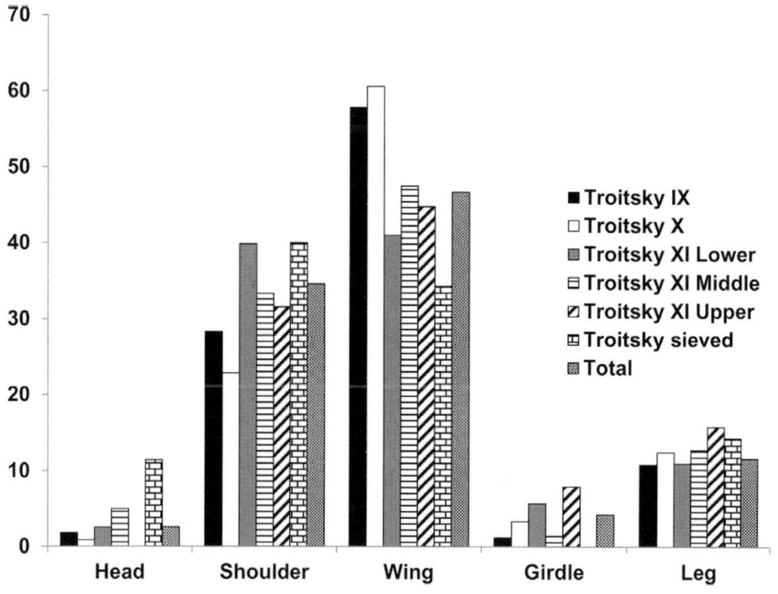

Figure 9.7: Novgorod: NISP percentages of mallard-sized ducks body areas from Troitsky sites

most of the major bones, although tarsometatarsi and the pelvis/synsacrum were absent (Table 9.7). The relatively poor representation of bones from the lower wing (carpometacarpus) and foot (tarsometatarsus) could indicate that some of the ducks had been brought to the properties after initial butchery elsewhere. However, more sieving is required to establish whether retrieval and preservation biases were the main factors that created variations in element abundance.

Twenty-three large duck bones were recorded as porous, indicating that only 1.6% of the bones were from young birds (Table 9.8). High percentages of immature bones can indicate that birds were being bred in captivity but only four of these bones were recorded as very porous and from young ducklings. It is more likely that most, if not all, of the immature ducks were caught during hunting expeditions.

Butchery marks were observed on 88 (6%) of the large duck bones (Table 9.9). Most of these were observed on bones of the wings and shoulder, reflecting the high numbers of these elements in the collection. They were particularly common on the coracoid (14%), scapula (9%), ulna (9%), humerus (8%) and furcula (8%). Despite being one of the most common bones recovered, no butchery marks were observed on the sternum. No marks were observed on the pelvis or synsacrum and only five leg bones had evidence of butchery (Table 9.10).

Most of the marks consisted of fine knife cuts but there were also some specimens which had been completely severed. Most marks were associated with the segmentation of the wings. Cut marks on the coracoid occurred in all areas of the bone as a result of separation from the breast and humerus. Cuts on the humerus were fairly evenly distributed between the proximal and distal ends, where the upper wing had been separated from the shoulder and lower wing respectively. In only one case were there cuts on both ends of a humerus. Ten of the 13 butchery marks on the ulna were located on or near the proximal end and made during separation from the humerus. All four cut marks on the tibiotarsi were on the distal end when they were separated from the tarsometatasus. Seventy-four (5%) of the surviving mallard-sized duck bones had been damaged by gnawing, in all but two cases by dogs or cats (Table 9.10). Two ulnae had rodent tooth marks. It is likely that many more bones would have been completely destroyed by these scavengers.

Small (teal-sized) ducks

Bones of small ducks were also commonly found forming 9% of the identified bird bones from all the sites investigated (Table 9.6). They provided less than 7% of the avian bones in most assemblages but formed over 10% in two of the three hand-collected Troitsky XI assemblages. Significantly, they were best represented in the sieved samples, forming 24% of the identified bird bones, suggesting that they were under-represented in the hand-collected assemblages (Table 9.6; Figure 9.6). Sieving may also account for their higher representation at Gorodishche (15%). Most of the bones of small ducks were a good match for teal but those of other small ducks, such as garganey, could be included. Like many of the bird species recovered from

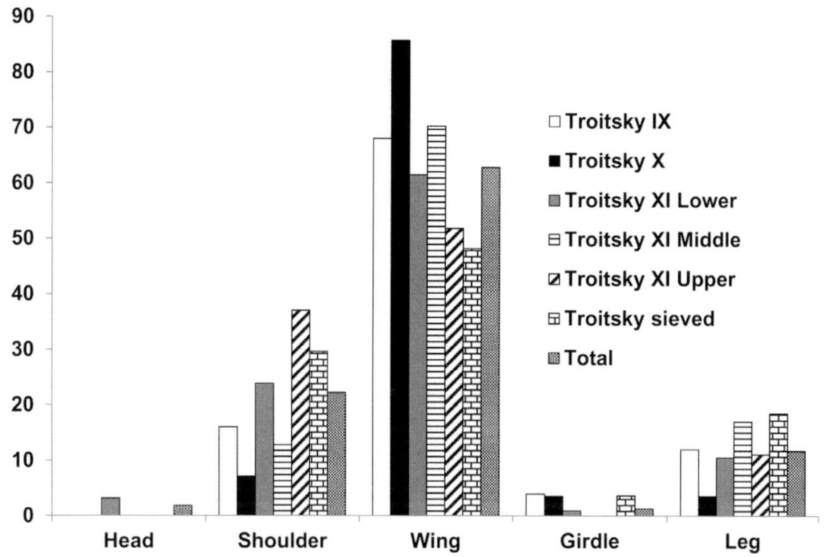

Figure 9.8: *Novgorod: NISP percentages of teal-sized ducks body areas from Troitsky sites*

Novgorod and Gorodishche, teal breed in the hinterland and migrate to the west and south in the autumn returning in the spring. Teal from northern Russia are commonly found in France (Guillemain *et al.* 2014). Birds identified as teal formed 11% of the birds from the Rogatitsky-3 site in Novgorod, on the opposite side of the Volkhov (Zinoviev 2016).

As at Gorodishche, the teal-sized duck assemblage was dominated by wing bones (Table 9.11; Figure 9.8). These formed the majority of bones in each of the hand-collected assemblages and nearly half the bones in the sieved samples. Overall, they provided 63% of the recorded bones. The humerus was comfortably the most commonly recovered, contributing 48% of the small duck assemblage (Table 9.11). The ulna was the second most common (15%). The dominance of the humerus is a consequence of its relatively large size and robustness. The only assemblage in which the humerus did not dominate was the one collected by sieving. The slender radius (3%) and carpometacarpus (2%) were both poorly represented.

Bones from the shoulder and breast were also well represented, providing 22% of the small duck assemblage (Table 9.11; Figure 9.8). The furcula, despite its small size, was particularly well represented, forming 8% of the assemblage, even outnumbering coracoids (7%). The sternum contributed 5% to the total assemblage. Again, the sternum was better represented in the lower spits of Troitsky XI than in other assemblages. This was probably due to the exceptional preservation conditions and possibly better retrieval rates. This was also the only assemblage to include the fragile skulls of small ducks. Leg bones provided only 12% of the overall assemblage with the larger tibiotarsus (7%) being much better represented than the smaller femur (5%) and particularly the tarsometatarsus (<1%).

Only three (0.8%) small duck bones were recorded as porous, and these presumably were from young birds caught alongside adults during hunting expeditions (Table 9.8). Cut marks were observed on 14 (4%) of the bones (Table 9.9). Ten of these were recorded on humeri and were inflicted during separation from both the shoulder and the lower wing. Cuts were also found on two ulnae and a coracoid. The back of a skull had cuts associated with beheading. Five bones had been damaged by dog or cat gnawing.

Other duck bones

This category covers bones that appeared smaller than those classified as large (mallard-sized) ducks but larger than those designated as small (teal-sized) ducks. Several species are likely to be represented in this category. Many are quite closely comparable with wigeon and pochard, but there are also a number of bones of other ducks, including several confidently identified as goldeneye and tufted duck. Pintail and shoveller may also be represented. These species were all likely to have been resident around Lake Ilmen in the summer months or during migrations (Cramp and Simmons 1978). Bones identified as wigeon, pintail and shoveller were recorded in the Rogatitsky-3 assemblage (Zinoviev 2016).

Duck bones assigned to this category provided over 5% of the identified bones from the sites investigated (Table 9.6; Figure 9.6). The assemblages, like the other duck categories, were dominated by wing bones which provided 54% of the total assemblage (Table 9.12; Figure 9.9). The humerus was the best represented element (32%), followed by the ulna (12%) but carpometacarpi (2%) were poorly represented. Leg bones (23%), particularly the tibiotarsus (12%) and femur (9%), although still comfortably outnumbered by wing bones, were better represented than in the other duck assemblages, outnumbering shoulder and breast bones (20%), with the coracoid (8%) again the best represented of these. No cranial elements were assigned to this category.

Nearly all the bones were from adult birds (Table 9.8). Fifteen (7%) bones were observed to have butchery marks (Table 9.9). These all consisted of knife cuts. They were most commonly recorded on humeri (7), including bones tentatively identified as wigeon, goldeneye, garganey and tufted duck. Cuts were also recorded on a scapula, radius, two ulnae, a carpometacarpus,

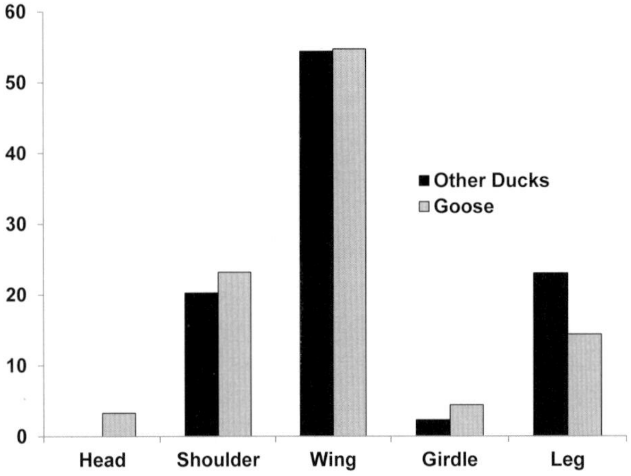

Figure 9.9: *Novgorod: NISP percentages of other ducks and geese body areas from Troitsky sites: totals from all sites combined*

a tibiotarsus and two femora, the latter identified as goldeneye. Carnivore gnawing damage was recorded on six bones.

Metrical analysis of duck bones

The greatest lengths of duck bones were routinely measured from both Novgorod and the hinterland sites. Breadth and depth measurements were also recorded for some bones. Full details of all measurements are stored in the archive. Summaries of the most common measurements are provided in Tables 9.13–9.20. In each case the data are presented in the three categories discussed above (large; small; other). Generally, there was little overlap in size between the three categories but we remain cautious about ascribing all the bones of large ducks to mallard/domestic duck and all the small bones to teal, although we are confident that these were the main two species involved. Amongst the large duck bones, there were very few that were larger than those we compared with modern wild mallard skeletons.

Measurements of the humerus were the most abundant (Table 9.15). There were large clusters of measurements of small and large ducks, with a more dispersed range of ducks of intermediate size (Figures 9.10–9.11). The very tight cluster of humeri with lengths averaging 58.4 mm (Table 9.15; Figure 9.11) suggests that most belonged to the same species and they were a good match for modern teal. The measurements of large ducks (nearly all over 90 mm) were mainly in the range of modern wild mallards. A few specimens over 100 mm were at the very top end of the modern wild mallard range and it is feasible that they were from domestic ducks. However,

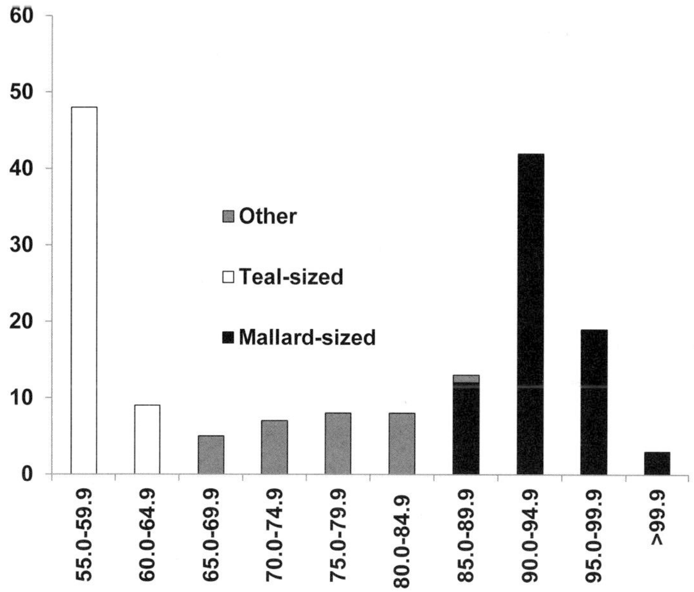

Figure 9.10: Novgorod. duck humerus greatest length measurements (mm)

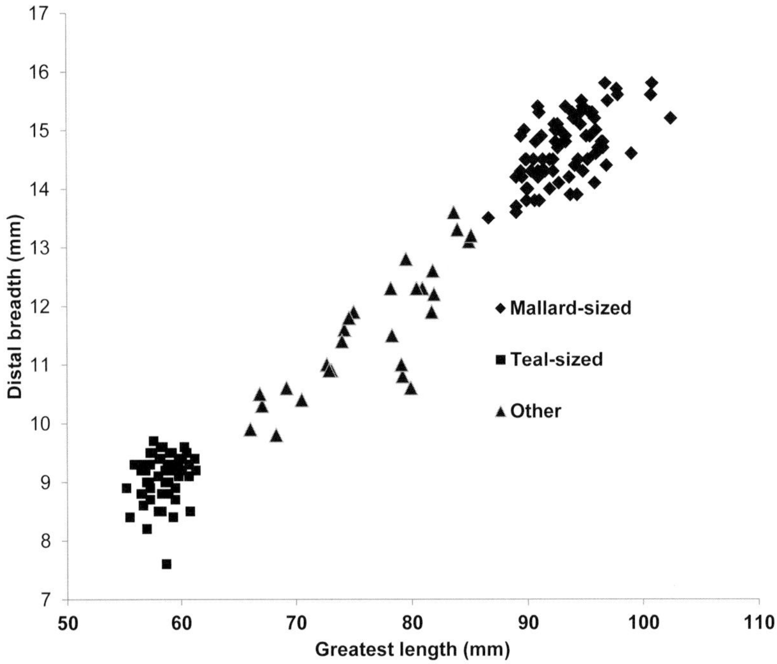

Figure 9.11: *Novgorod: duck humerus greatest length and distal breadth measurements (mm)*

it seems unlikely that domestic ducks were commonly kept, given the abundance of wildfowl available from the Lake Ilmen region. A range of other duck species is represented by the bones of intermediate size and these include some which were provisionally identified as wigeon, goldeneye, garganey, pintail and tufted duck. However, more confident identifications would require a greater range of measurements to be taken (Poland 2018). Measurements of small bones such as the coracoid (Figure 9.12) showed a greater predominance of large ducks, which indicates that more of the small duck bones were overlooked during excavation.

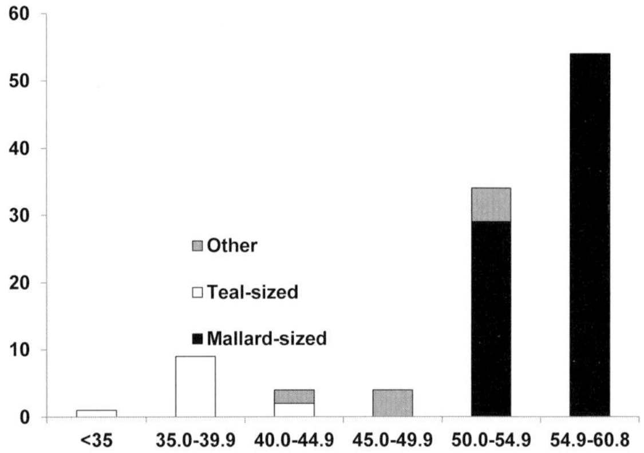

Figure 9.12: *Novgorod: duck coracoid greatest length measurements (mm)*

Geese

Geese provided 5% of the identified bird bones from the Novgorod assemblages, the majority coming from large geese (Table 9.6; Figure 9.6). The measured specimens fell within the range of modern wild greylag geese (Table 9.21). However, a few of the smaller specimens in this group overlapped in size with other large geese species such as the taiga bean goose (*Anser fabalis*) and therefore the presence of other species is possible (Poland 2018). Genetic analysis has shown that taiga bean geese were present on some early medieval sites in Russia including 9th–10th-century deposits at Staraya Ladoga (Honka *et al.* 2018). It should also be noted that none of the measured goose bones fell beyond the upper range of wild greylag, although many domestic geese are no larger than their wild counterparts.

Large geese were best represented in the upper spits of Troitsky XI (8%) and in the small Fedorovsky assemblage (12%). This may imply that relatively more wild greylag and possibly other large species of geese were being captured in the later medieval period, There is also the likelihood that more domestic geese were being kept. However, this needs verification from analysis of further assemblages. Bones recorded as greylag/domestic goose bones formed between 8% and 9% of other broadly dated medieval assemblages (11th–14th centuries) recently examined from Novgorod (Gorobets and Kovalchuk 2017) and 5% of the identified bird bones from the Rogatitsky-3 site (Zinoviev 2016).

Twenty goose bones were from geese smaller than wild greylag. Fourteen of these came from Troitsky XI, five from Troitsky VIII and one from Troitsky X (Table 9.6). None of these were confidently identified to species given the lack of an available reference collection and because of close morphological similarities between goose species. Most were, however, morphologically similar to whited-fronted or lesser white-fronted geese. In all cases these species are likely to have been birds caught during their migration passage, as none of them are currently resident in the Novgorod region during the breeding season (Cramp and Simmons 1978).

Body parts of geese represented followed the familiar pattern for Anatidae with the largest bones of the wings predominant. They formed 55% of the large goose assemblage, with the humerus (21%) being the best represented followed by the ulna (14%) and radius (10%) (Table 9.22; Figure 9.9). The carpometacarpus (5%) was much less well represented, either because of its smaller size or to the prior removal of wing extremities of some geese before being brought to the properties. Shoulder and breast bones contributed 23% of the large goose assemblage, with the coracoid (8%) and sternum (7%) the most common of these. Leg bones provided 14% of the assemblage, with the large tibiotarsus (8%), as usual, better represented than the smaller femur (4%) and tarsometatarsus (3%). Of the 15 bones assigned to smaller geese, eight came from the wings (3 humeri; 2 ulnae; 2 carpometacarpi; 1 radius). There were also four coracoids, two tarsometatarsi and one femur.

None of the goose bones were recorded as porous (Table 9.8). The absence of goslings could imply that few geese were bred in captivity, although immature domestic geese bones have not been found in great numbers elsewhere in medieval Europe, for example in England (Serjeantson 2002).

Butchery marks were observed more commonly on goose bones than on those of ducks, being recorded on 27 (14%) of the large goose bones (Table 9.9). This is a reflection of the larger bodies of geese, which required more partitioning than ducks. Most of the butchery marks were observed on the bones of the wings and shoulder but were also observed on several leg bones (Table 9.23). Cuts on the humerus were found on shafts as well as the proximal and distal ends. Fine knife cuts were also noted on five bones of smaller geese (2 humeri; coracoid; carpometacarpus; tarsometatarsus) from Troitsky XI (Table 9.9). Gnawing damage was recorded on 25 (13%) of the large goose bones (Table 9.23) and three of those of smaller geese. Most of the damage was inflicted by dogs and cats but three large goose bones had evidence of rodent gnawing.

Swans

Twelve bones were identified as swan, ten of which came from the lower spits of Troitsky XI. They formed only 0.3% of the total of identified bird bones from the Novgorod sites (Table 9.6). Five of these bones belonged to the lower right wing of the same bird (the distal halves of the radius and ulna; the carpometacarpus; and two wing phalanges). Cut marks were observed on the shaft of the radius and near the proximal end of the carpometacarpus. This could be interpreted as the disposal of bones of little meat value after butchery of a bird on one of the properties. Alternatively, the lower wing may have been detached prior to being brought to the property. If so, this associated bone group may indicate that swan's wing feathers were acquired for processing. The remaining bones identified as swan consisted of a coracoid, humerus, a radius, two ulnae, tibiotarsus and sternum. Cuts were found on an ulna shaft.

None of the swan bones were identified to species during initial recording. Whooper swans (*Cygnus cygnus*) currently have extensive breeding grounds in northern Russia (Cramp and Simmons 1978) and might previously have had more extensive breeding areas (Hokhlova and Artemjev 2002). Alternatively, some whooper swans may have landed in the region during migration. The presence of mute swan (*Cygnus olor*) cannot be entirely ruled out but they are not currently listed as breeding birds in this part of Russia (Cramp and Simmons 1978; Johnsgard 2016). Another possibility is that some of the bones belonged to migratory Bewick's swans (*Cygnus columbianus bewickii*) on their passageway to and from their breeding grounds around the White Sea (Cramp and Simmons 1978; Shchadilov *et al.* 2002; Kouzov and Losev 2016). However, the sizes of the measured bones are larger than those of Bewick's swan but fall within the range of whooper and mute swans (Table 9.21). Whooper swans therefore remain the most likely candidate. One bone from another site in Novgorod has been identified as whooper swan (Gorobets and Kovalchuk 2017) and three of the five swan bones from the 2000–2004 excavations from Gorodishche were also assigned to whooper swan (Table 9.5).

Other waterfowl (including waders and seabirds)

Bones of grebe were identified on all the Troitsky sites investigated but only provided a total of six bones (Table 9.6). All the bones were a good match for the great-crested grebe. They all came from the legs (femur; 3 tibiotarsi; 2 tarsometatarsi). The only porous bone from a young bird was a tibiotarsus that had cuts marks on the distal articular surface made when it was separated from the tarsometatarsus (Hamilton-Dyer 2002). Like many of the other species, the great-crested grebe is a migratory bird that frequently nests on the edges of freshwater lakes (Cramp and Simmons 1978). The area around Lake Ilmen is an ideal summer habitat.

Fourteen bones of crane were recovered from Troitsky X and XI including two from sieved samples (Table 9.6). Four of these were wing bones (2 humeri; radius; carpometacarpus). There was also a coracoid and scapula, both of which had gnawing damage. Leg elements were represented by two tarsometatarsi, a tibiotarsus, which was associated with one of these, and four ossified tendons from the tarsometatarsus region. Cranes inhabit a wide range of wetlands including river banks, lake margins and marshes. Despite losses of habitat, they are still abundant in north-west Russia in areas of lakes and swamps and congregate in large numbers prior to autumn migrations (Ilyashenko and Markin 2012). No butchery marks were observed on any of the crane bones but the carcasses of captured birds would have provided a substantial amount of meat and thus they may have been regarded as a luxury food. Cranes could also have been valued for their plumage. A tarsometatarsus of an immature crane was recovered from the Nerevsky excavations in association with a leather thong (jess), indicating that the bird was kept in captivity and perhaps used as a decoy bird or in falconry training (Hamilton-Dyer 2002; Hamilton-Dyer *et al.* 2017). Measurements of crane bones are included in Table 9.21.

The proximal half of a humerus (proximal breadth = 37.5 mm) of a stork was identified from Troitsky IX. This bore knife cuts around its shaft indicating processing. It was not possible to identify the bone to species and it could have been either from a white (*Ciconia ciconia*) or black stork (*Ciconia nigra*) (Hamilton-Dyer 2002). Both species currently breed in north-west Russia (Cramp and Simmons 1978; Konovalov *et al.* 2015).

Five bones (coracoid; ulna; femur; 2 tibiotarsi) of cormorant were recovered from Troitsky XI (Table 9.6). Six bones from Troitsky XI and one from Fedorovsky V were assigned to gulls. There is substantial overlap in the sizes of gulls and it was impossible to assign their bones to species. Most of them were similar in size to herring gull (*Larus argentatus*). The bones represented were a sternum, two humeri, a radius, ulna, tibiotarsus and tarsometatarsus. Although cormorants and gulls are often regarded principally as seabirds, they also commonly occur inland. Cormorants would have been able to feed on the abundance of fish available locally in Lake Ilmen and the River Volkhov. Modern studies have shown that cyprinids, perch and zander, the most common fish exploited in Novgorod (see Chapter 10), are sometimes an important component of the cormorant's diet in north-eastern Europe, even in coastal areas (Pūtys and Zarankaitė 2010; Salmi *et al.* 2015). In addition, gulls may have been attracted to Novgorod as scavengers. The discovery of two porous bones could indicate that some gulls were nesting in the town. Although no butchery marks were observed, cormorants and gulls may have provided an occasional addition to the diet.

Seven bones of heron were recovered from the lower spits of Troitsky XI. These consisted of two coracoids, a radius, two tibiotarsi and two tarsometatarsi. A tarsometatarus and a coracoid came from immature birds (Table 9.8). A radius and ulna probably from the same bittern were also found in a lower spit from Troitsky XI. No butchery marks were observed on the bones of either of these species but it is likely that these birds were captured along with other waterbirds in Novgorod's hinterland and brought to the town for consumption. Both species currently breed in many areas of north-west Russia (Cramp and Simmons 1978). Another common lakeside and riverside dweller is the coot, which was represented by a single humerus, also from the lower spits of Troitsky XI.

Nineteen bones from the Troitsky sites were classified as waders (Table 9.6). The land around Novgorod has many areas of raised bogs and water meadows that are ideal habitats for waders and they are currently abundant in the region around Lake Ilmen (Mischenko and Sukhanova 1998). Unfortunately, bones of waders are notoriously difficult to speciate due to the close morphological similarities of many species and the limitations of modern reference collections. Species mentioned in the following discussion represent potential rather than confirmed identifications.

Two humeri and two ulnae were quite closely comparable in size and morphology to woodcock. Woodcock are found in large numbers in north-west Russia and many are still shot in the Novgorod area particularly in the spring (Blokin *et al.* 2006; 2015). Many woodcock migrate from this region to winter as far afield as France and Spain (Cramp and Simmons 1983; Bauthian *et al.* 2006; Guzmán *et al.* 2011).

Six bones (2 humeri; radius; carpometacarpus; femur; tibiotarsus) were quite similar in size and morphology to the common snipe. This species is currently commonly found in meadows around Lake Ilmen (Mischenko and Sukhanova 1998). Apart from woodcock, snipe are nowadays the wader species most commonly shot in north-west Russia (Blokin *et al.* 2015).

A larger humerus was tentatively ascribed to curlew (*Numenius arquata*). Curlews currently breed on local bogs and are also found on ungrazed flooded meadows in the vicinity of Lake Ilmen (Mischenko and Sukhanova 1998). A tibiotarsus and two tarometatarsi from the middle spits of Troitsky XI belonged to a wader with long slender legs, comparable in size to whimbrel (*Numenius phaeopus*) or black-tailed godwit (*Limosa limosa*). Whimbrel are currently less common than curlews in the Novgorod region but do inhabit some of the bogs. Godwits also currently breed in the area (Mischenko and Sukhanova 1998). A radius from Troitsky IX was similar in size and morphology to redshank (*Tringa tetanus*), which is currently commonly found in flooded meadows in the region. A tarsometatarus from Troitsky VIII was similar to greenshank (*Tringa nebularia*), which currently inhabits similar environments but is less common (Mischenko and Sukhanova 1998).

Raptors and owls

Nine species of raptor provided a total of 3% of the identified bird bones from the assemblages examined (Table 9.6). Twenty-one bones were identified as eagle, most, if not all, of which are likely to have belonged to white-tailed eagle. Its geographical range extends over the northern Palaearctic including north-west Russia. White-tailed eagles inhabit a wide range of different habitats, ranging from terrestrial, freshwater, brackish and marine. Many of their common prey species (obtained from both hunting and scavenging) are those which were also represented in the archaeological assemblages in Novgorod and Gorodishche. These include pike, perch, cyprinids, ducks, geese, cormorants, gulls and wild boar (Nadjafzadeh *et al.* 2016). Bones of white-tailed eagle have been recorded from other Novgorod sites (Gorobets and Kovalchuk 2017). However, other eagle species (golden; spotted; lesser spotted) also breed in north-west Russia (Pchelintsev 2006) and it is feasible that some of the bones belonged to these species. Some species of eagle can be used in falconry but the very large white-tailed eagle is not generally regarded as a suitable bird for training (Prummel 1997; 2018).

Eleven of the eagle bones came from the wings, four from the shoulder, five from the legs and one from the pelvis (Table 9.24). The bias towards wing bones was accentuated by the recovery of an associated group of five bones from Troitsky XI spit 13. The group consisted of the left humerus, radius, ulna, carpometacarpus and third wing phalanx of an adult eagle. The proximal end of the humerus had been severed when the wing was detached from the shoulder. Fine cut marks were also recorded on the distal end of the carpometacarpus. The distal end of the ulna was slightly charred. The cuts on the carpometacarpus may indicate the removal of feathers for the manufacture of arrows (Hamilton-Dyer 2002; Orten Lie 2018; Prummel 2018). Another eagle humerus from the lower spits of Troitsky XI had been severed in the middle of the shaft above the surviving distal half of the bone, again indicating that the wing was separated for processing. The tubular shaft of an ulna from the upper spits of Troitsky XI had evidence of working. Three eagle bones had been slightly damaged by carnivore gnawing. All the bones came from adult birds (Table 9.8).

Sparrowhawk was the most common raptor recovered from the excavations, providing over 1% of the identified bird bones overall and nearly 3% of those from the middle spits of Troitsky XI (Table 9.6). The sparrowhawk is the most common European raptor. Sparrowhawks that breed in Finland and north-west Russia often migrate to western Europe in the winter. The ideal habitat for Eurasian sparrowhawks lies in areas of dense woodland adjacent to open hunting grounds, often near a stream or river, but they can adapt to more open areas as well (Cramp and Simmons 1980; Lehikoinen *et al.* 2010). However, it is likely that most, if not all, of the sparrowhawks found in Novgorod were captive birds trained for falconry. Of the ten bones where comments about the possible sex of the birds were recorded in the archive, eight were ascribed to females and only two to males. Greatest length measurements were obtained on 30 bones (Table 9.26). These were subsequently compared to those of a modern female sparrowhawk skeleton (coracoid 34.0 mm; humerus 59.8 mm; radius 63.1 mm; ulna 68.7 mm; carpometacarpus 38.8 mm; femur 52.1 mm; tibiotarsus 70.1 mm; tarsometatarsus 56.4 mm). All measurements were converted into base-10 logarithms with those of the modern skeleton used as the standard (0.00) for log-ratio comparisons. Twenty-five (83%) of the bones from Novgorod were of a similar size to those of the female sparrowhawk (–0.01 to 0.05), whereas only five (17%) much smaller bones (–0.09 to –0.06) could be assigned to males (Figure 9.15).

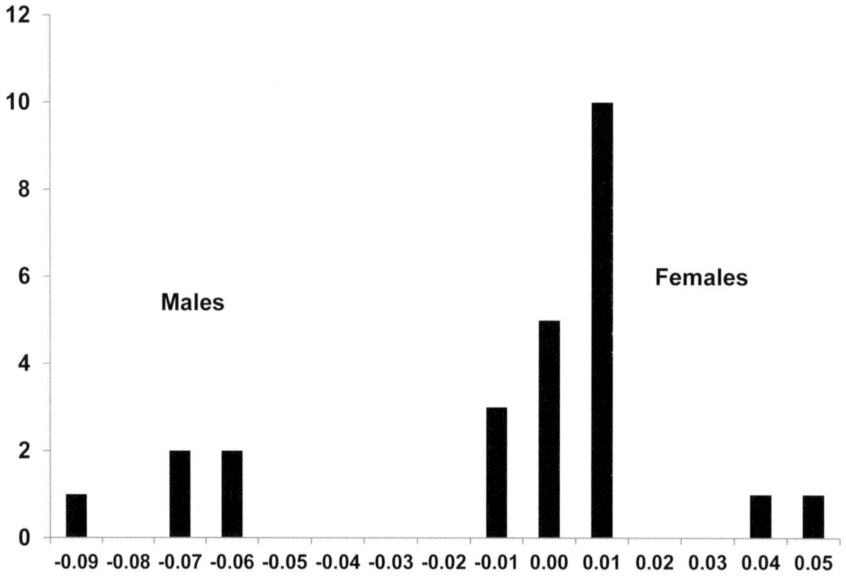

Figure 9.13: *Novgorod: log ratios of sparrowhawk length measurements*

Two sparrowhawk bones from the Rogatitsky-3 site also came from females (Zinoviev 2016). Falconers preferred to train the larger females as they could capture larger quarry than males (Prummel 2018). Typical quarry would have been small birds such as thrushes and smaller corvids but larger prey including woodcock and partridge could also have been hunted.

All the sparrowhawk bones were from adult birds, 45% of the 51 bones coming from the wings, 10% from the shoulder, 6% from the pelvis and synsacrum and 40% from the legs (Table 9.24). Although it was not possible to determine whether any of the bones were still articulated, the even distribution of wing and leg bones indicates that all parts of the body were represented, which supports the contention that the birds were kept in the town. Only one humerus showed evidence of carcass processing in the form of knife cuts on the distal part of the shaft.

A total of 42 bones of goshawk were recovered from the Troitsky and Fedorovsky excavations, forming 1% of the identified bird bones (Table 9.6). These included 11 associated bones (coracoid; 2 humeri, radius, 2 ulnae, pelvis, synsacrum, femur, tibiotarsus and tarsometatarsus) from the skeleton of an adult female found in a lower spit from Troitsky XI. In total, 20 (48%) of the goshawk bones came from the wings, eight (19%) from the breast and shoulder, five (12%) from the os coxae and synsacrum, eight (19%) from the legs and one from the cranium (Table 9.24). Although all parts of the body were represented, there was a bias towards wing bones. Cut marks were observed on a furcula and the proximal joint surface of a humerus indicating that some of the wings had been separated for processing for meat and/or feathers. However, these and all the other goshawk bones are likely to have belonged to birds kept in captivity for hunting.

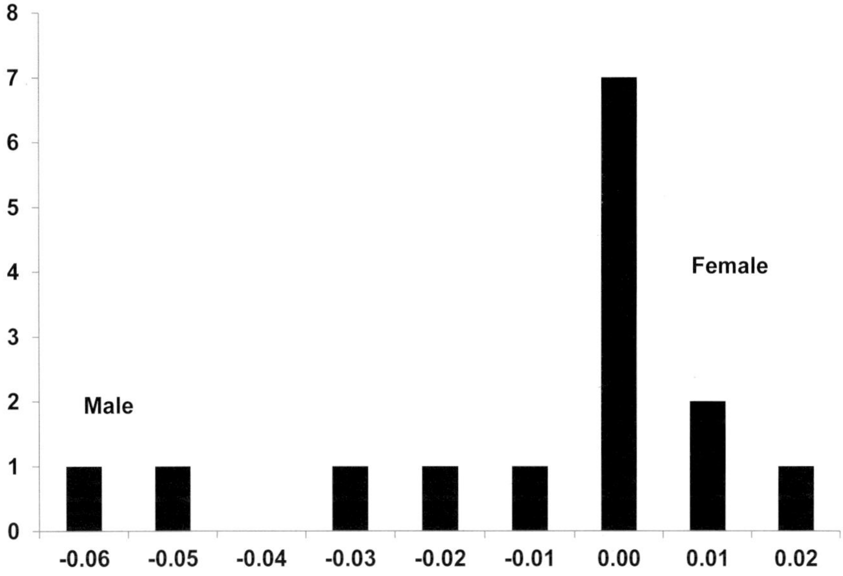

Figure 9.14: *Novgorod: log ratios of goshawk length measurements*

Greatest length measurements were obtained on 15 bones including six from the associated bone group (Table 9.25). These were compared to the equivalent measurements of a modern female goshawk skeleton (coracoid 52.3 mm; humerus 102.7 mm; radius 105.6 mm; ulna 115.5 mm; carpometacarpus 63.5 mm; femur 88.0 mm; tibiotarsus 116.7 mm; tarsometatarsus 75.7 mm), which was used as the standard (0.00) for log-ratio comparisons. Thirteen (87%) of the bones from Novgorod, including six of the associated bones, were of a similar size to those of the female goshawk (–0.03 to 0.02), whereas only two smaller bones (–0.06 and –0.05) could be assigned to males (Figure 9.14).

Four goshawk bones recovered from the Rogatitsky-3 site belonged to at least two females (Zinoviev 2016). As in the case of sparrowhawks, female goshawks were often preferred for falconry because of their ability to capture larger prey (Prummel 2018).

The goshawk was another species commonly trained by medieval falconers across Europe (see papers in Gersmann and Grimm 2018). Wild goshawks are resident throughout the year in large parts of Eurasia. They are forest-dwelling raptors whose preferred habitats lie in mature forests but they can also reside and hunt in more open areas and indeed in towns (Cramp and Simmons 1980; Clark 1999; Conenna *et al.* 2017). Their diet varies greatly in different regions dependent on the bird and mammals available (Rutz *et al.* 2006). In northern latitudes, wild goshawks, commonly hunt various species of grouse, including all the species represented in the Novgorod assemblage, but they will diversify their diet in the summer when more species, including waterfowl, are available (Tornberg *et al.* 2009). The prey of trained goshawks included hares, squirrels, corvids, pigeons, partridges, larger galliforms and waterfowl including ducks, crane and heron (Prummel 2018)

Only one bone, a humerus from the Fedorovsky excavations, was positively identified as buzzard, although a furcula and ulna from the lower spits of Troitsky XI were assigned to either buzzard or goshawk (Table 9.6). Buzzards, which are the most common of the European raptors (Cramp and Simmons 1980) are usually regarded as scavengers and they would have been attracted to the urban refuse that would have been created in Novgorod. However, a buzzard tarsometatarsus found in excavations of the Nerevsky site was associated with a leather thong, probably a jess. This suggests that some buzzards may have been captive birds, and possibly used as decoys (Hamilton-Dyer 2002; Hamilton-Dyer *et al.* 2017; Zinoviev 2017).

A humerus (distal breadth 14.3 mm) and ulna (greatest length 94.8 mm) of a hen harrier, possibly from the same bird, was recovered from a middle spit (13) of Troitsky XI. The measurements indicated that it was from a larger female. Hen harriers are commonly recorded throughout Russia and often hunt in pastures (Cramp and Simmons 1980; Men'shikova 2006).

Seven bones of falcons were recovered. A sternum, femur and tibiotarsus, all from lower spits of Troitsky XI, were a good match for kestrel. Troitsky XI also produced an ulna and femur, which were identified as hobby, and two humeri, which were recorded as merlin (Table 9.6). All of these birds could potentially have been kept for falconry, although these smaller species captured smaller prey (Prummel 2018). The hobby ulna (greatest length 68.1 mm) was from a large female and the femur was from a sub-adult, both of which would support the contention that these birds were kept in captivity.

Ten bones from the lower and middle spits of the Troitsky sites were recorded as owl (Table 9.6). Seven bones (radius; 2 ulnae; femur; 2 tibiotarsi; tarsometatarsus) were a good match for tawny owl. One of the tibiotarsi had cut marks on the distal articulation, indicating that the carcass of at least one of the owls had been processed, presumably for meat. Tawny owls have a wide distribution in Eurasia including north-west Russia (Cramp and Simmons 1980). They can feed on a large range of small mammals, birds and amphibians (Balčiauskienė *et al.* 2005). Three ulnae were a better match for the Tengmalm's owl. In boreal forest regions Tengmalm's owl's main prey are voles (Jacobsen and Sonerud 1993). Owls can be a quarry in falconry and can also be used as decoys for capturing potential birds of prey to be trained (Zeiler 2010; Orten Lie 2016; Zinoviev 2017).

Domestic fowl and other galliforms

These were separated where possible into species. The majority of the bones were a close match for domestic fowl (chicken) and have been assigned to that species. Technically not all of the bones had surviving features that were diagnostic specifically of domestic fowl but many positive identifications of domestic fowl were made and none had attributes diagnostic of peafowl or pheasants. A few bones were assigned to large wild galliforms such as capercaillie and black grouse. It is possible that a few bones recorded as domestic fowl in this archive have been misidentified, as it was not possible to check the provisional identifications of them all. However, we are confident that the vast majority of the identifications of domestic fowl are correct.

Domestic fowl formed 35% of the identified bones from all the sites investigated at Novgorod (Table 9.6; Figure 9.6). As indicated above, percentages of domestic fowl did vary significantly between sites and assemblages. In those from Troitsky IX and X and the lower spits of Troitsky XI, domestic fowl bones were significantly less common than those of large (mallard-sized) ducks, forming between 22% and 34% of the identified bird bones. They were also slightly less common in the sieved samples (27%). In contrast, domestic fowl bones were the most common species in all the other assemblages, forming 48% of the assemblage from the middle spits of Troitsky XI and over 50% of the bones from the upper spits. They provided 57% of the Troitsky VIII sample and 58% of the assemblage from Fedorovsky V. The sequence from Troitsky sites suggests that domestic fowl became more important to the residents of that area from the 12th century onwards and perhaps from the later 11th century, based on the Troitsky VIII evidence. Domestic fowl also outnumbered other bird species in the Desyatinny-1 (see below) and several other assemblages from Novgorod of 11th- to 14th-century date (Gorobets and Kovalchuk 2017) including the assemblage from the Rogatitsky-3 site (Zinoviev 2016). To what extent there were variations in domestic fowl abundance in different parts of the town and in different periods remains to be investigated further.

Leg bones were the most commonly recorded domestic fowl elements from all of the Novgorod sites, forming 40% of their overall assemblage (Table 9.26; Figure 9.15). This was a slightly higher percentage than at Gorodishche (35%), where they were narrowly outnumbered by wing bones (Figure 9.4). Leg bones formed 54% of the domestic fowl assemblage from Troitsky IX. The dominance of leg bones may principally be a reflection of differential recovery with the three main bones (femur; tibiotarsus; tarsometatarsus) all being larger than most of the major bones of the wing (humerus; radius; ulna; carpometacarpus). However, leg bones were also the most commonly recovered from sieved samples. As expected, the largest bone, the tibiotarsus, was the best represented (18%), compared with the femur (14%) and tarsometatarsus (8%). The comparatively poor representation of the tarsometatarsus (it is roughly the same length as the femur) may indicate that the feet of chickens were sometimes removed before they were brought to the properties. This may have become more common in the later period as the ratio of tarsometatarsus:femur fell from 1:1.86 in the lower spits of Troitsky XI to 1:3.5 in the upper spits (Table 9.26).

Wing bones provided 31% of the overall domestic fowl assemblage with only modest variations in this figure in the larger assemblages (Table 9.26; Figure 9.15). The largest wing bone, the humerus, provided 18% of the overall assemblage, followed by the second largest, the ulna (9%). The slender radius and the small carpometacarpus provided only 3% and 1% of the domestic fowl assemblage respectively (Table 9.26).

Shoulder and breast bones provided 17% of the overall domestic fowl assemblage (Table 9.26; Figure 9.15). Apart from the sternum, these are all quite small bones, which

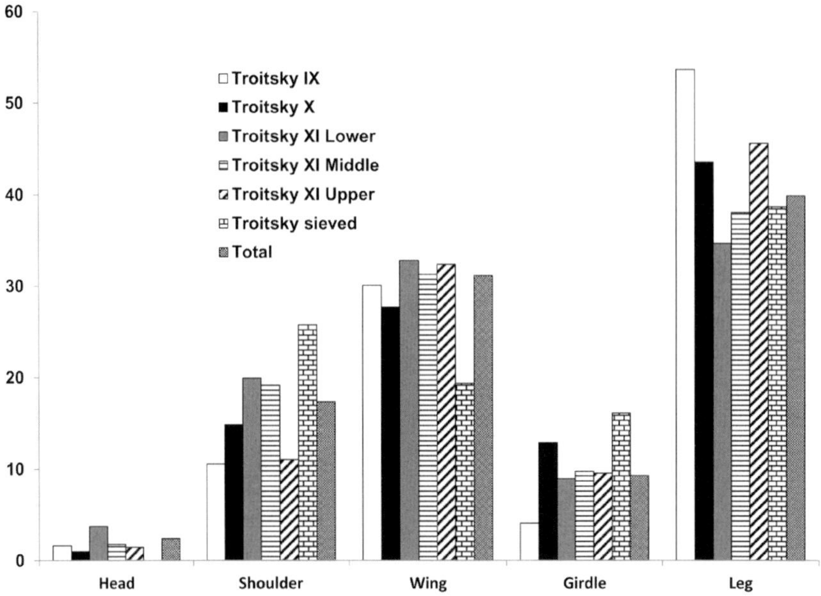

Figure 9.15: *Novgorod: NISP percentages of domestic fowl body areas from Troitsky sites*

accounts for their better representation in the sieved samples (25%). The sternum was the most commonly identified (8%), followed by the coracoid (4%) and scapula (4%). The pelvis and synsacrum provided 9% of the total domestic fowl assemblage, and were unusually well represented in the Troitsky X collection (13%). Bird cranial elements generally preserve poorly because of their fragility, which largely explains their low representation in the domestic fowl assemblage (2%). Unsurprisingly, cranial elements were most abundant (4%) in the exceptionally well-preserved assemblage from the lower spits of Troitsky XI (Table 9.26). A further factor for their infrequency could have been that sometimes their heads had been removed before they were brought to the properties.

A major contrast to the duck and goose assemblages was the presence of many more porous bones, which constituted 16% of the domestic fowl assemblage (Table 9.8). These bones came from both juvenile chicks (2%) and immature birds (14%). The presence of the chicks, most of which were found in the lower spits of Troitsky XI, indicates that some chickens were being kept in the properties. The high incidence of bones from immature fowl also reflects the raising of chickens in captivity, which enabled many young birds to be culled as they approached full size. In the Troitsky XI sequence, the percentage of porous bones decreased from 20% in the lower spits, to 17% in the middle spits to 7% in the upper spits. This may reflect a decrease in intensity of exploitation for meat, although factors such as differential preservation (which was exceptional in the lower spits) and recovery must also be considered. It is noteworthy that 35% of the domestic fowl bones in

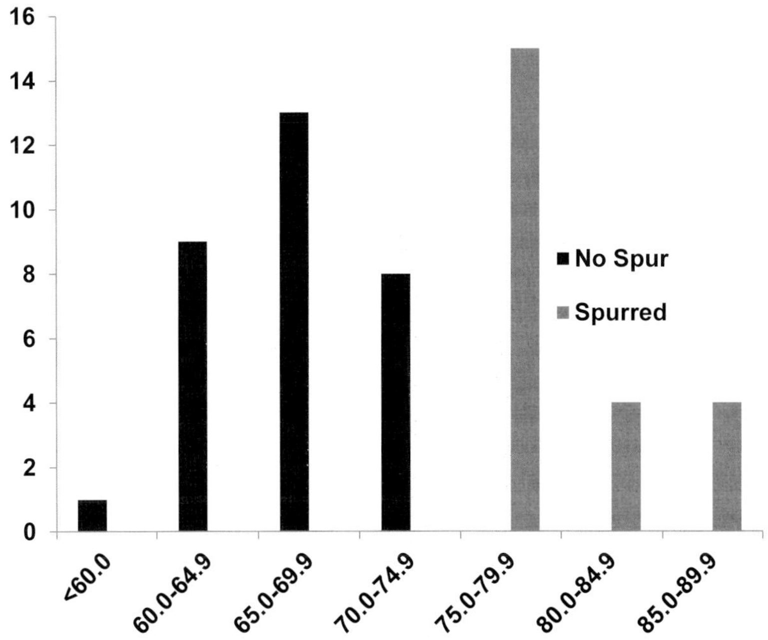

Figure 9.16: *Novgorod: domestic fowl tarsometatarsus greatest length measurements (mm)*

the sieved samples were porous, which suggests that retrieval bias may have been a significant factor.

The presence or absence of spurs on tarsometatarsi was noted in all the assemblages but recorded most consistently on the bones from the middle and lower spits of Troitsky XI (Table 9.27). The complete absence of a spur indicated that the bone was from a hen. The presence of a substantial spur indicated that it belonged to a cockerel, although smaller spurs can develop in some adult hens. Tarsometatarsi with no spurs provided 71% of the recorded specimens from the lower spits of Troitsky XI. Males are more likely to have been killed immature whilst adult hens would have provided eggs. The percentage of tarsometatarsi with no spurs decreased in the middle spits (36%), although this was a fairly small sample. Comparisons of the greatest length measurements of spurred and unspurred tarsometatarsi showed no overlap in size. All of the unspurred tarsometatarsi of hens had lengths of less than 75 mm, whereas the spurred tarsometatarsi were all over 75m long. Several of the largest tarsometatarsi had spur scars. These were sometime recorded as slightly porous and came from young males culled for meat (Figure 9.16).

Medullary bone was recorded on broken femora in all of the assemblages (Table 9.27). Medullary bone forms in the shafts of most bird bones when a female is preparing to lay and gradually diminishes during the laying season. It tends to be most prominent in the leg bones, particularly the femur (Van Neer *et al.* 2002). Although the presence of medullary bone does not prove that the eggs laid were destined for human consumption, a high incidence of bones with medullary deposition is usually taken to indicate that hens were providing eggs to be eaten. Thus, the fact that in the lower spits of Troitsky XI 62% of the femora were recorded as having medullary bone is a strong indication that egg production was an important factor in the exploitation of domestic fowl. Hens not required further at the end of the laying season could be culled for meat. The incidence of medullary bone was also recorded for tibiotarsi on some of the sites. In this case 65% of the specimens from the middle spits of Troitsky XI had medullary bone compared to 39% in the lower spits. This may indicate an increase in the importance of egg production but the absence of medullary bone in tibiotarsi may not have been recorded as systematically in the middle spits, biasing the results. Measurements of the femora and tibiotarsi showed that bones with medullary bone were all amongst the smallest specimens confirming than most hens were smaller than cockerels (Figures 9.17–9.19).

Butchery marks were observed on 53 (4%) of the domestic fowl bones (Table 9.28). Most of these marks were observed on the tibiotarsus and humerus reflecting the abundance of these elements in the assemblage. All the marks on the tibiotarsi were made with fine knives. Nearly all of these were found on the distal end. A corresponding cut was found on the proximal end of a tarsometatarsus. This confirms that the two bones were commonly separated and supports the contention that the relatively poor representation of tarsometatarsi could be the consequence of the removal of the feet of some of the birds prior to their carcasses being brought to the properties. Cuts on the humerus and femur were located on both the proximal and distal ends of the bones and occasionally on the shafts. Several of the femora and tibiotarsi with medullary bone had cut marks, confirming that hens were commonly butchered for meat after they were no longer required for egg production.

Metrical data from the major domestic fowl bones are summarised in Tables 9.29–9.35. As noted above, a major factor that creates size variation in domestic fowl

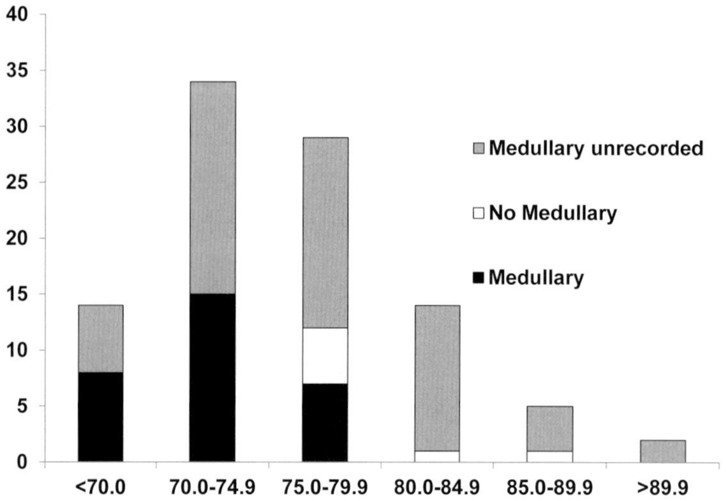

Figure 9.17: *Novgorod: domestic fowl femur greatest length measurements (mm)*

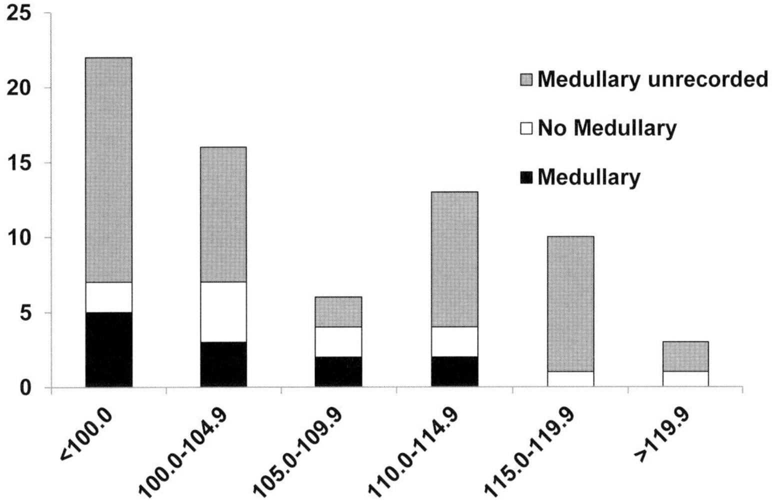

Figure 9.18: *Novgorod: domestic fowl tibiotarsus greatest length measurements (mm)*

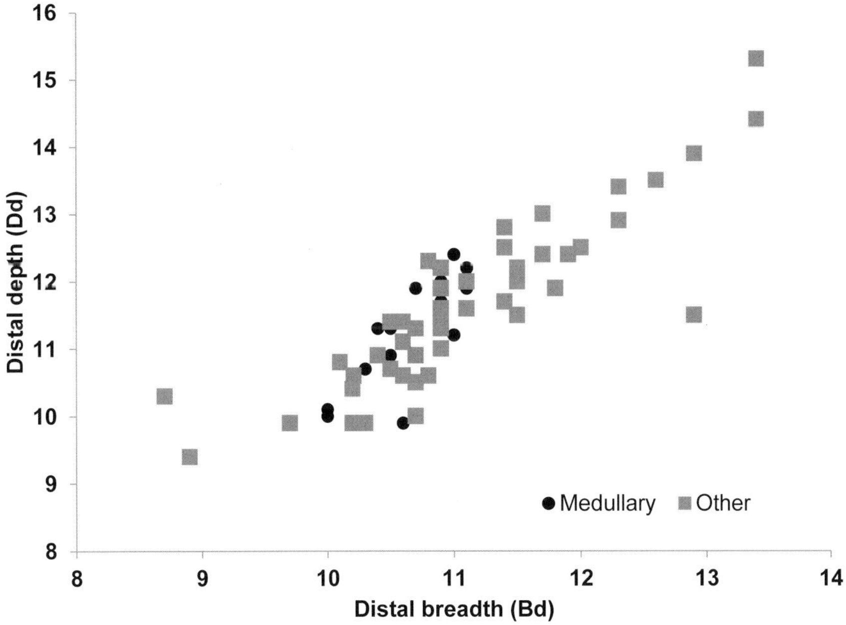

Figure 9.19: *Novgorod: domestic fowl tibiotarsus distal breadth and depth measurements (mm)*

is sexual dimorphism, which makes the investigation of possible changes in the overall stature of domestic fowl challenging. It was possible to separate probable males from females with regard to the spurred and unspurred tarsometatarsi respectively. Unfortunately the samples of measured bones from the middle and upper spits of Troitsky XI were too small to establish whether there were significant chronological variations in the sizes of hens or cockerels, although most of the largest tarsometatarsi with spur scars came from the middle and upper spits of Troitsky XI and there was also a large spurred specimen from Troitsky VIII. This provided tentative evidence that some larger fowl were bred from the late 11th century onwards (Table 9.35). However, there were no consistent trends in the range of sizes of other domestic fowl bones from different sites and periods (Tables 9.29–9.34; Figure 9.20). All the length measurements produced bimodal distributions, with the smaller mode mainly comprised of bones from hens and the larger one consisting mainly of bones of cockerels.

Measurements of the greatest lengths (GL) and distal breadth of humeri produced two main clusters (Figure 9.21). It is likely that all the 29 bones in the smaller cluster (GL <68 mm) came from hens. Although it is tempting to ascribe all the 39 bones in the other cluster (GL>70 mm) as cockerels, such a bias towards males would not match the results from the tarsometatarsus and femur that indicated that adult hens outnumbered cockerels (Figures 9.16–9.17). It is feasible that most of the humeri with lengths of 70–74 mm also belonged to hens. If bones of over 74 mm long all belonged to cockerels, a minimum of 22% of the measured humeri were from males, which is more in line with the percentage of spurred tarsometatarsi (27%). The three largest specimens (GL 81–82 mm) all came from the lower spits of Troitsky XI. This raises the possibility that hens and cockerels of two different types (one larger than the other) were exploited in Novgorod. However, much more analysis is required to test this

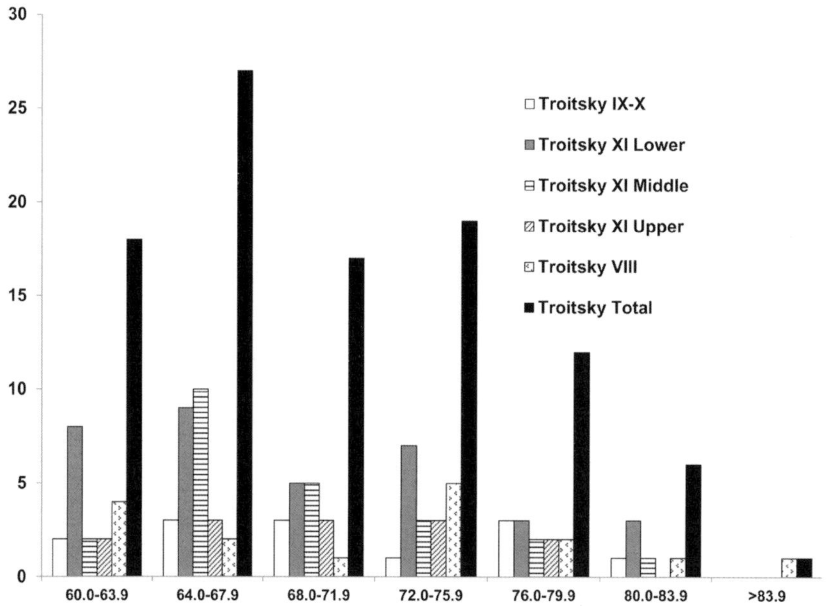

Figure 9.20: *Novgorod: domestic fowl humerus greatest length measurements (mm)*

observation. There were too few measured specimens from Gorodishche to assess whether there were significant differences in the stature of domestic fowl in the 9th and 10th centuries (Tables 9.29–9.35).

Bones of large galliforms formed 3% of the identified bird assemblage (Table 9.6). Their large size favoured their recovery in comparison with those of many avian species. Bones that were a good match for capercaillie were recorded the most frequently (2%) and examples were found in all of the hand-collected assemblages from Troitsky IX–XI. Wing bones (41%) outnumbered leg bones (33%) with the humerus (19%), tibiotarsus (17%) and femur (16%) the best represented elements (Table 9.36). The absence of tarsometatarsi may indicate that the feet were frequently removed before the birds were introduced to the properties. Only one of the bones (from Troitsky IX) was from an immature bird. Fourteen (18%) of the bones were recorded as butchered (Table 9.9). Butchery marks were observed on two skulls, a sternum, a scapula, three humeri, three ulnae, a carpometacarpus, two femora and a tibiotarsus. Most of the marks were fine cuts on the surface of the bone but the back of one of the skulls and the proximal end of the carpometacarpus had been completely severed. The high incidence of butchery marks is consistent with the frequency observed on other large birds, such as geese and swans that were exploited for food. Large male capercaillie can weigh as much as 5.5 kg. These birds would probably have been hunted in areas of coniferous woodland and brought to the town. In pine forests capercaillie rely mainly on eating pine needles in the winter but have a more varied diet during other seasons (Borchtchevski 2017).

Ten bones were assigned to the smaller black grouse, all but one of which came from Troitsky XI (Table 9.6). Femora, coracoids, carpometacarpi and tarsometatarsi were represented (Table 9.36). None of the bones bore butchery marks but it is very

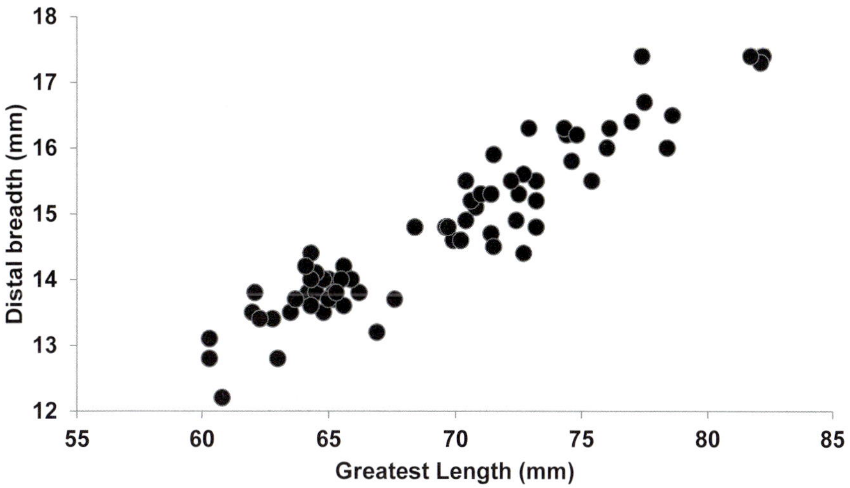

Figure 9.21: Novgorod: domestic fowl greatest length and distal breadth measurements (mm)

likely that they belonged to birds that were eaten. Capercaillie and grouse would have been regarded as prestige birds to hunt. They are still popular quarries for hunters in the Novgorod hinterland. A reference to black grouse was found in one of the birch-bark documents, which is a good indicator of its high status (BBD 42). Black grouse, like capercaillie, can be found in abundance in boreal forests, particularly in clearings and on the forest edge (Angelstam 2004). They are found across large tracts of northern Europe including Russia (Corrales *et al.* 2014).

Twenty-eight bones from Troitsky IX–XI were assigned to a 'large galliform' category (Table 9.6). There is a small overlap in size between male grouse and female capercaillie and there is also overlap between large male domestic fowl and black grouse. Additionally, nine of the bones were porous specimens from immature birds, which made them harder to identify (Table 9.8). All three species could therefore be represented in this category. Sixteen of the bones were from the leg, nine from the wing, plus two fragments of pelvis and part of a sternum (Table 9.36). A porous tibiotarsus from Troitsky IX had knife cuts on the distal articular surface. Its large size (length 132 mm) makes it feasible that this specimen was from an immature capercaillie.

Two species of smaller galliforms were represented by three bones. A femur and a tibiotarsus of hazel grouse, and a humerus of a partridge were recovered from the upper spits of Troitsky XI (Table 9.6). Hazel grouse mainly inhabit dense mixed deciduous-coniferous forests (Mathys *et al.* 2006; Ludwig and Klaus 2017). The grey partridge has a wide distribution in Europe occupying more open habitats than the grouse species discussed above.

Pigeons and doves

Only four bones were identified (Table 9.6). A sternum and tibiotarsus from Troitsky XI spit 20 were recorded as woodpigeon. The tibiotarsus was from a sub-adult bird. Two tibiotarsi from immature stock doves or domestic/feral pigeons were found in Troitsky XI spit 10. Although domestic pigeon were kept in the medieval period throughout Europe, wild pigeons are also a commensal species and these birds could have been resident in the town, a contention supported by the presence of the immature birds. They are also birds commonly caught by raptors and some of the birds could have been caught during falconry (Prummel 2018).

Corvids

A total of 72 bones from members of the crow family were recovered forming 1.7% of the identified bird bones. They were represented in all of the hand-collected assemblages (Table 9.6). The largest of the corvids, the raven, was identified in four assemblages and provided 15 bones (Table 9.20). Five of these (coracoid; humerus; radius; ulna; pelvis) were found in close association in Troitsky XI Property G, spit 16 and probably belonged to the same adult bird. A femur from the Fedorovsky site was from a sub-adult, the rest were from adults. No butchery marks were observed on any of the raven bones. Eleven bones were recorded as rook/crow, eight of which were recovered from the lower spits of Troitsky XI (Table 9.6). The humerus (5) and, unusually for bird bones, the skull (3) were the best represented rook/crow elements (Table 9.37). The skulls all came from the lower spits of Troitsky XI. Two bones were recorded as porous and belonged to immature birds (Table 9.8). Three of the bones were a close match for crow.

The most common corvid identified was the jackdaw, which was represented by 37 elements, all but two of which came from the lower spits of Troitsky XI (Table 9.6). Six bones (sternum; humerus; 2 ulnae; pelvis; tarsometatarsus) from spit 21 probably belonged to the same bird. A second associated group of five bones (sternum; humerus; radius; pelvis; synsacrum) was found in spit 16 (Property G). Both groups belonged to adult jackdaws. The radius had been cut though the distal end when separated from the lower wing. Wing bones provided over half of the jackdaw assemblage with the two largest elements, the humerus and ulna, being the best represented (Table 9.37). One humerus belonged to an immature bird. Nine corvid bones could not be assigned to species with confidence (Tables 9.6 and 9.37). They included bones of larger (*e.g.* rook, crow or raven) and smaller corvids (jackdaw; jay; magpie). A femur of one of the larger corvids had been chopped through the proximal end.

All the corvids could have been resident within, or in close proximity to, Novgorod and would have found ample food through scavenging. The presence of cut marks on at least two of the bones indicates that their carcasses were sometimes processed for food and/or feathers.

Passerines

All nine passerine bones (coracoid; humerus; radius; ulna; carpometacarpus; pelvis; synsacrum; tibiotarsus; foot phalanx) were recovered from sieved samples suggesting that many were overlooked during normal retrieval (Table 9.6). The tibiotarsus of a fledgling was large enough to be from a small corvid or a bird of similar size. Most of the bones, however, were from small passerines similar in size to sparrows. It is likely that many passerines were resident in Novgorod and would be vulnerable to hunting by domestic cats but it is also feasible that some of the bones belonged to birds caught by smaller raptors during falconry.

BIRD BONES FROM DESYATINNY-1 (Andrei Zinoviev)

As with the Troitsky assemblages, bird bones from Desyatinny-1 were significantly outnumbered by mammal bones and form only 2.3% of the total assemblage.

Thus, birds can be considered as only a supplementary addition to the diet. In contrast to the Troitsky assemblages where bones of mallard-sized ducks often formed the highest percentage of bird bones, bones of domesic fowl dominate in Desyatinny-1, forming 89.3% of 191 bird bones (Table 6.38). The trend for the increase in the percentages of domestic fowl bones from the 12th century onward is also supported by Desyatinny-1.

Bones of geese of the size of greylag goose (most of these almost certainly were bones of domesticated birds) are the second most commonly represented. Various species of waterfowl were also identified. Bones of mallard-sized ducks were much less numerous than in the Troitsky assemblages. Several bones were identified as swan, which is commonly found on Lake Ilmen during spring and winter migration and is known to have been regularly consumed in medieval Russia. Great-crested grebe, of which only a single tibiotarsus was uncovered, might have also been captured for food (Zhilin and Karhu, 2002; Guminski, 2005). Occasionally caught in fish nets, these birds can provide tasty meat if the fat is removed before cooking.

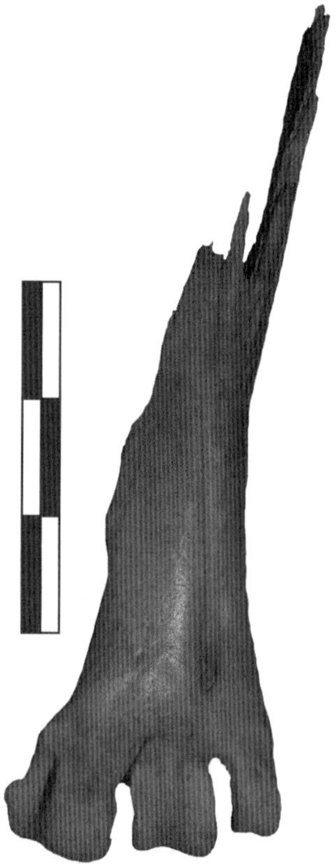

Figure 9.22: *Novgorod: white-tailed eagle from Desyatinny site. Photograph by Andrei Zinoviev*

A single tarsometatarsus of white-tailed eagle has been uncovered (Figure 9.22). It is highly unlikely that the white-tailed eagle and other birds of prey represented in Novgorod were exploited for meat. Whereas species such as the peregrine falcon, goshawk, sparrowhawk, hobby and kestrel can be used in falconry, the larger and differently specialised white-tailed eagle must have been used for other purposes. Reichstein and Pieper (1986), in their analysis of a large number of wing bones of white-tailed eagles from medieval sites in Germany, suggested their use as a source of high-quality feathers for arrows. This also appears to be true for medieval Novgorod.

A single wing bone of a starling indicates its presence in Novgorod in the 13th–14th century. It is not surprising to find this species in urban sediments, as starlings are still common inhabitants of contemporary Novgorod.

– 10 –

FISH BONES FROM NOVGOROD, SITES IN ITS HINTERLAND AND MININO

Sheila Hamilton-Dyer with Mark Maltby, Mikhail Sablin and Andrei V. Zinoviev

METHODOLOGY (Sheila Hamilton-Dyer)

Similar identification criteria were employed in relation to the fish bones as was carried out on the mammal and bird bones. Where possible these were identified to species but in some cases limitations of reference material and close similarities of the skeleton meant that some bones could, at best, only be identified to genus or family.

Identification aids included, among others; a limited reference collection made from locally caught fish, images of specimens in the reference collection of S. Hamilton-Dyer, the manual by Lepiksaar (1983), Libois *et al.* (1987), Libois and Hallet-Libois (1988) and Wheeler (1978). A few bones were retained for further checking of identification in England.

Unlike the situation in mammals the individual elements of the fish head are not fused but remain separate and number well over 100 individual pieces in most cases. The recording was restricted to the following; the larger and most recognisable of the paired elements of the face and jaws, the major elements of the neurocranium, pectoral girdle, pelvic girdle and the vertebrae. The elements vary in comparative size and robustness in different families and in some cases special elements were also recorded. In the sturgeons, for example, the skeleton is rarely fully ossified but the modified scales or scutes are robust and easily recognisable. In the cyprinids the inferior pharyngeals are large, toothed and species specific but they were not recorded for the other fish. No attempt was made to identify the vertebral sequence; instead the vertebrae were divided into caudal (tail end) and precaudal (anterior body). In some cases the first precaudal was also recognised and recorded separately. Fragmentary vertebrae that could not be identified to body area are listed as unassigned vertebra in tables. A high proportion of the remains from sieved samples consisted of fin rays, spines and other undiagnostic fragments, which could not be identified. All fin rays, branchiostegal rays, ribs and fragments thereof were grouped together. In a similar manner all the bones not included in the above list, together with cranial fragments, were recorded simply as fish fragments. Scales were usually not present in the hand-collected material but were sometimes recovered from sieving (and in large quantities at Minino). Where largely intact they can be divided into type but are not often identifiable to species. The sturdy and distinctive ctenoid (toothed) scales of perch and zander have been listed separately and the least damaged of the larger cycloid scales are assumed to be cyprinid, the remainder were recorded as indeterminate.

Sizes of fish were estimated mainly by comparison with recent material. Some elements of certain fish species were measured, mainly following the manual of Morales and Rosenlund (1979).

THE EXPLOITATION OF FISH ON THE HINTERLAND SITES (Sheila Hamilton-Dyer with Mark Maltby)

Gorodishche (1979 and 1993–1998)

Species representation

Except from the 1979 material, fish bones were very numerous, comprising 5,549 bones which constitute 41.3% of the total number of bones recorded. It is notable that the material collected in 1979, before the sieving programme was instigated, contained no fish bones at all (Table 10.1).

As expected, a high proportion of the remains were of fin rays, spines and other undiagnostic fragments. Over 58% of the fish bone total could be identified to at least family level and over 35% to species. Bones of cyprinids (members of the bream family) were the most frequent of the identified taxa at 22% of all fish bone and 39% of the identified fish bone. Without an extensive reference collection to hand, many of the cyprinids bones could not be exactly identified to species but most were comparable with bream (*Abramis brama*). Other cyprinid species positively identified were roach (*Rutilus rutilus*), chub (*Leuciscus cephalus*), ide (*Leusiscus idus*), blue bream (*Abramis ballerus*), silver bream (*Blicca bjoerkna*) and rudd (*Scardinius erythrophthalmus*). Cyprinids were closely followed by pike (*Esox luscus*), and by zander (*Sander lucioperca*), each providing 30% of the identified fish bones. A small number of perch (*Perca fluviatilis*), wels catfish (*Siluris glanis*) and whitefish (*Coregonus lavaretus*) remains and a single bone of a sturgeon (*Acipenser sp.*) were also present (Figure 10.1; Table 10.1).

There were notable differences in the proportion of identified fish bones found in different deposits. From the North bank, 53% of the fish bones could not be identified even to family level whereas from the South bank only 36% were unidentified. The amount of unidentified material from the black layer stood at 40% from both banks. The unidentified portion from the pits amounted to only 29% (Table 10.2). These results parallel those for the large mammal bones and this is almost certainly related to preservation conditions, which were poorer generally from the North bank and best in the pits where material was quickly buried and protected.

Although pike and zander did vary in relative proportions, it was the amount of cyprinids which fluctuated the most, ranging from 6% of the identified fish bone in the northern pits to 66% in the South bank 'other contexts' (Table 10.2). As the bone from the pits was well preserved, this must indicate a genuine difference in disposal patterns. The actual numbers were, however, quite variable too and the sample from the North bank pits consists of only 148 bones including the unidentified fragments.

Four of the five bones of catfish and two of the three whitefish bones came from the South bank, but with such small numbers it is not possible to be certain that this is significant. A single fragment of a dermal scute of a sturgeon was recovered from the black layer of the 1998 excavation.

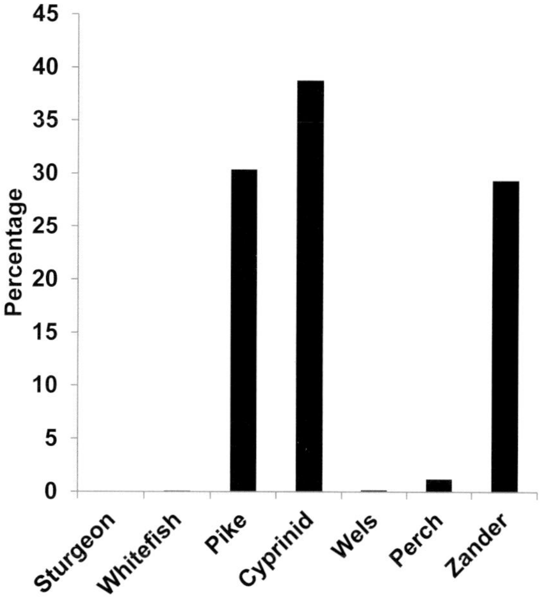

Figure 10.1: *Gorodishche 1993–1998: NISP percentages of identified fish taxa*

Anatomical distribution

Bones of the head, pectoral girdle and vertebrae were all represented for the most frequent species but with considerable variation (Table 10.3). As expected, the smallest and the most fragile elements within each taxon were less well represented than the more sturdy bones and those that are easily recognised even when incomplete. Vertebrae were common, being more frequent in the skeleton. For pike and zander the dentary was the best represented of the head bones with the articular and cleithrum also frequent. The ceratohyal, quadrate and preoperculum were also well represented in the zander bones. These are all elements that are sturdy and easily recognisable in this species. The pattern was different for the cyprinids, where the mouth bones are relatively small and without teeth. Dentaries were less common than the pharyngeals. The latter lie in the throat of the fish and, in cyprinids, are large highly specialised toothed structures that are species specific. The preoperculum and operculum are the largest and sturdiest of the other cyprinid head bones and these were the best represented elements. The cleithra, in the pectoral girdle, are also large in cyprinids but can be fragile, perhaps explaining why none were recognised from the North bank and relatively few from elsewhere. Comparing the North bank assemblage (1993–1997) with the better preserved assemblage from the South bank, it can be seen that vertebrae are better represented in the latter (Table 10.4). This is almost certainly due to better preservation because there were fewer vertebrae that were not identified to taxon than from the North bank. The spread of other elements was also better balanced for the cyprinids. The balance of head bones to vertebrae is about the expected number for cyprinids (about 45 vertebrae in each fish) and zander (about 62 vertebrae) but there were insufficient pike vertebrae in comparison with the numbers of their head bones, especially dentaries. All the pike dentaries consisted of the most cranial part or the bone, sometimes with the middle section. None were complete and no caudal portions were recovered. Although cut marks were not observed on the bones, it seems likely that this unusual distribution resulted from butchery and unequal disposal practices.

Butchery

Butchery traces are generally not common in fish bone assemblages as small fish are frequently consumed whole and larger ones may be prepared with minimal butchery.

However, there are definite cut marks on a few of the bones. Most of these probably relate to removing the head and gutting/gill removal. The distribution of observed butchery marks by species and element is given in Table 10.5.

Georgii

Only 71 fish bones were recovered from Georgii, where sieving was not carried out. Rays, spines and other indeterminate elements accounted for over half of the bones (NISP=40). The 31 bones identified to taxon were mostly of cyprinids (17); pike, zander and perch were also present (Table 10.6). The identified bones consisted of head or pectoral elements, except for one vertebral fragment from a large pike; the other seven vertebral fragments were probably of cyprinids. The five pike bones include three comparable in size with specimens of about one metre in length. At least five of the six zander bones came from fishes of about 60 cm long and were probably from one individual. The cyprinid bones were not identified to species but several are comparable with bream and would have been about 30 cm in length. The only measurable perch bone is from a large fish of about 40 cm in length, the other two bones coming from smaller fish. These are fish of a good size for eating but, as no sieving was carried out, it is not certain whether smaller fish were also exploited.

Prost

Excavation at Prost included the same sieving and sampling strategy as at Gorodishche and a good sized group of 1,181 bones was recovered, although nearly 70% were indeterminate rays, spines and other fragments (Table 10.7). The 395 bones identified to taxon were dominated by those of cyprinids at (52.9%). These remains include a few inferior pharyngeals, the distinctive toothed elements that are often identifiable to species. The three most complete pharyngeals could be identified as ide of two different sizes and one of tench (*Tinca tinca*). Several dentaries and basioccipitals were a good match for bream. The other bones were not distinguished to species. Vertebrae were well represented and the other elements present indicated that whole fish are present. Bones of pike and zander were almost equally represented (18.7% and 20.5% respectively), perch contributed 7.8%. The pike bones were biased in favour of vertebrae and cleithra but most of the larger head elements were also present. The bones include some from large individuals but smaller fish of around 50 cm total length were also present. At least five, probably more, different individuals were represented. One vertebral centrum has an etched appearance typical of canid digestion. At least four individuals were represented in the zander bones, including several from large fish of 60 cm or more in length.

Vasilievskoye

A total of 102 fish bones were recovered from this site, over half of which could be identified to taxon. The identified elements belonged to zander, cyprinids and pike (Table 10.8). Zander (29 specimens) were slightly more frequent than the cyprinids (22). Several of the cyprinid bones are a good match for bream, whereas one is comparable with chub. There were only four pike bones in this assemblage. Most of the bones came from the head; vertebrae numbered just three, one of pike and two of zander. A cleithrum of zander was the only bone with a visible butchery mark.

Gorodishche 2000–2004 (Mikhail Sablin)

Fish provided a high proportion of the bones recovered from these excavations. Overall they provided 41% of the total mammals, birds and fish dated between the 9th and 12th centuries. Attention should be drawn to the large number of fish bones (51%) in the 10th-century levels. However, these bones have not as yet been identified.

THE EXPLOITATION OF FISH IN NOVGOROD (Sheila Hamilton-Dyer with Mark Maltby)

Troitsky hand excavation

The hand collection of bones on the Troitsky sites produced 1,292 fish remains of which 820 were identified to taxon (Table 10.9). At least nine species were represented with bones of zander the most frequent at 316 specimens (38.5% of the identified bones). Pike and cyprinids contributed about a quarter each with wels catfish, perch and sturgeon making minor contributions (Figure 10.2). Within the cyprinid bones most of those identified to species were bream with at least three bones of ide and one each of silver bream and chekhon (*Pelecus cultratus*). In addition to these species, three bones of whitefish were identified in material from mixed levels at site IX (not included in the tables).

The distribution of anatomical elements varied considerably across the main taxa (Table 10.10). Both head bones and vertebrae were present but their relative proportions differ from their expected numbers. For all taxa there are insufficient vertebrae for the number of head bones and, while this could indicate the presence of discarded heads, this is at least partly because the vertebrae are in most cases smaller and may have been overlooked. Proportionally, the large vertebrae of the wels catfish were better represented. The main head and pectoral bones of zander were relatively

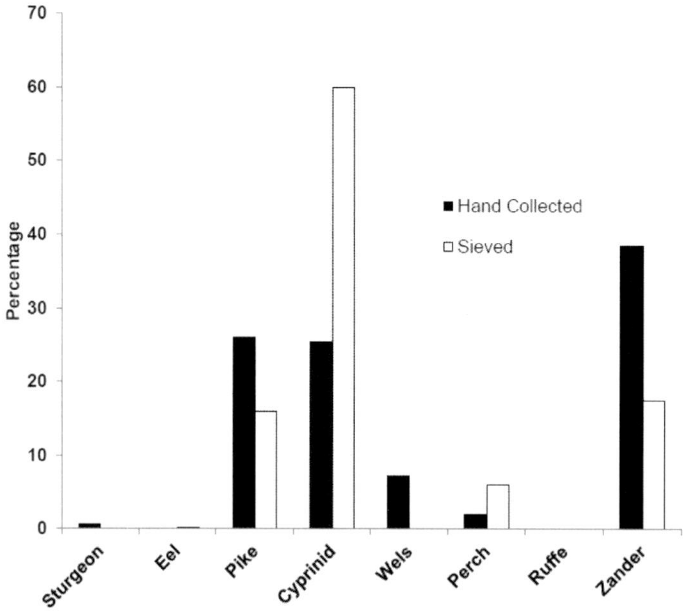

Figure 10.2: *Novgorod: NISP percentages of identified fish taxa from hand-collected and sieved samples from Troitsky IX–XI*

evenly represented, mainly varying with size and ease of recognition when fragmented. Cyprinids were best represented by the large opercula and cleithra. The bones of pike were the most unequally distributed, with a significantly high number of dentaries and cleithra. Again this is likely to be at least in part due to their relatively large size and sturdy nature. The five bones of sturgeon were all from site XI property G and were from at least two individuals of different size. The remains included dorsal scutes, a pinna caudialis from the tail and a supracleithrum.

Troitsky sieved samples

The small-scale sieving programme at Troitsky recovered an impressively large number of fish remains, a total of 4,621 specimens, of which 1,499 were identified to taxon (Table 10.11). In contrast with the hand-collected material a much larger proportion (60%) of the identified bones belonged to cyprinids (Figures 10.2 and 10.3). In addition to bream, there were a few bones of ide, roach, rudd, chub and at least one probably of silver bream. Zander (Figure 10.4) and pike were at a correspondingly lower percentages, though they are still common at 17.5% and 15.9% respectively. Perch, though still a minor species, was more frequent at 6.1%, increasing from 17 hand-collected specimens across the sites to 91 from the limited sieving. The higher number of cyprinids and perch is clearly related to bone size. The other three species found, eel (*Anguilla anguilla*), wels catfish and ruffe (*Gymnocephalus cernuus*), occurred as a few bones only but eel and ruffe have very small bones and were not found in the hand-collected material (Figure 10.2; Table 10.11). The significant bias in the anatomical distribution that was seen in the hand-collected bones is much less noticeable in the sieved bones, again almost certainly related to size (Table 10.12).

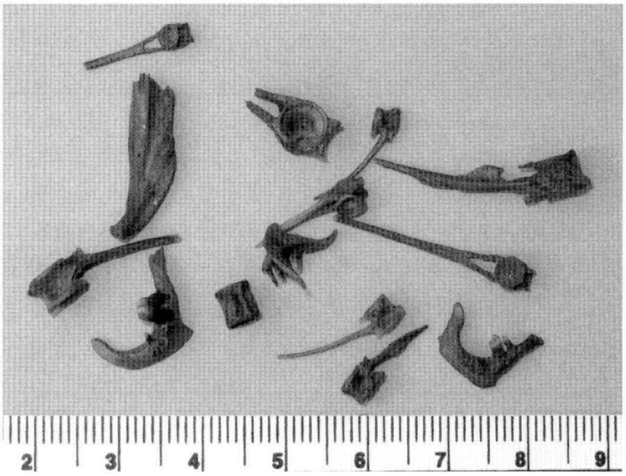

Figure 10.3: *Selection of cyprinid bones from Troitsky IX–XI. Photograph by Sheila Hamilton-Dyer*

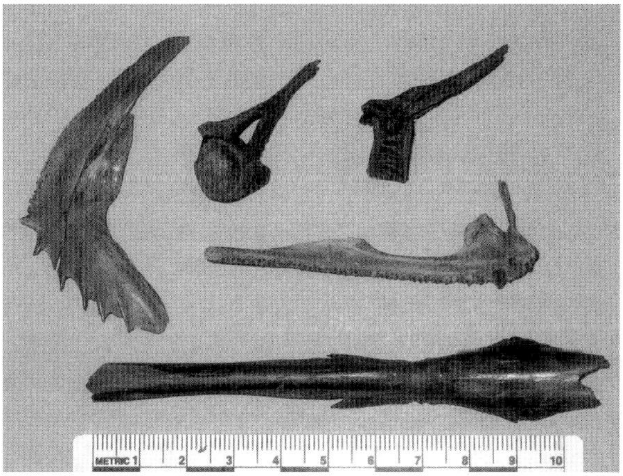

Figure 10.4: *Selection of zander bones from Troitsky IX–XI. Photograph by Sheila Hamilton-Dyer*

Comparison of species representation with the Nerevsky site and Gorodishche

An earlier report on archaeological fish remains, from excavations at the Nerevsky site to the north of the Kremlin (Sychevskaya 1965), listed 15 species amongst the bones with another three represented by scales only. All are freshwater species. The main species identified in the bones were zander followed by bream and pike, very similar to the hand-collected data from Troitsky. Eel was not found but there are four species not identified at Troitsky, all of which were cyprinids represented by single bones (asp, vimba) or by a few scales (bleak, Crucian/Prussian carp). The bone counts taken from the individual bone descriptions in the text are summarised in Table 10.13 alongside the Troitsky data.

The species list from the combined hand-collected and sieved fish remains from the Troitsky sites is very similar to that produced from Gorodishche. The only species not found at Gorodishche were eel, ruffe and chekhon. These were infrequent finds at Troitsky and their absence from the smaller assemblage at Gorodishche may not be significant. With the exception of the whitefish and sturgeon, all the species are still widely available from the River Volkhov and Lake Ilmen. The highly valued whitefish and sturgeon are migratory and have not been caught since the completion of the Volkhov hydro-electric station in 1926 (Sychevskaya 1965). Eel, so common in most north-western European sites of this period, is also a migratory species and Lake

Ilmen would have been at the limit of its natural distribution. It is in the lake now (Savenskova *et al.* 2010) but whether deliberately maintained is not indicated. The eel bones found at Troitsky were of good size and could represent traded preserved fish.

Butchery

Traces of butchery were observed on a total of 63 bones from the Troitsky sites, 16 of these were sieved bones and the others were hand-collected. This should be regarded as a minimum figure due to the poor lighting conditions and the rapidity of analysis. The distribution of butchery by species and element is given in Table 10.14. Proportionally, butchery traces were most frequently observed on the wels catfish bones, perhaps in part because these are large fish that require division. The nine bones include three vertebrae chopped across, four head bones with cut marks, and two pectoral spines with chop marks near the base, probably made when removing this large barbed spine. Butchery marks on pike were found mainly on the dentaries, which were chopped in a consistent manner: the dentary was chopped right through, mainly at right angles to the bone rather than from the ventral or dorsal, and in some cases it can be seen that the chop is from the inner side. This implies that the head had already been split open and the gill and tongue area removed. It has been suggested that cut marks on pike dentaries can result from the removal of the gills in order to dry the heads (Jonsson 1986 and pers. comm.), in that case the cuts are mainly from the ventral direction. Some of the butchery from Novgorod could represent this type of preparation but not all; some bones represent quite small fish and others show evidence of a double 'trial' mark from heavy chopping straight through the already opened head. Several of the cleithra were also chopped; this bone lies just behind the gills and this is a convenient place to chop off the head. Pike heads were observed for sale in the modern local market, for making fish soups and stock. These were of mixed sizes and roughly chopped into quarters; the archaeological remains may show a similar food use. Some of the zander bones had been chopped and/or cut but there were fewer butchered bones in total and a lower proportion with marks. Several were chopped vertebrae. Cyprinids were by far the most frequent bones but very few bore butchery marks; the majority of these fish are smaller than the other taxa and require less division in preparation. Today, many are gutted and then smoke-dried whole for winter use, a technique that was probably also employed in the past.

The final butchery mark was found on the supracleithrum of a large sturgeon from Troitsky XI spit 2; this was chopped at both ends, possibly while separating the body of the fish from the head. The few other sturgeon remains are of the dermal scutes that run down the back and sides of the fish.

The sizes of fish from Novgorod and the hinterland

A selection of measurements was taken on the main elements, mainly following Morales and Rosenlund (1978). Additional measurements include one taken on the cleithrum between the most caudal point of the spina dorsalis and the fossa muscu-interna; named the 'point to bend'. This was employed as many cleithra were either damaged or chopped and this portion of the bone was more often preserved and the measurement easy to take. For similar reasons there are many measurements listed in Morales and Rosenlund that were rarely or never available and were not taken. In the case of some taxa and elements that meant only a few measurements were possible.

Selections of the most frequent measurements for pike, zander and cyprinids from the Troitsky and hinterland sites are given in Tables 10.15–10.20. Although the cyprinid metrics include more than one species, the morphology of many bones is very similar and, for site comparative purposes, the species composition is likely to be similar with the majority probably of bream.

At Gorodishche the cyprinids were mostly comparable with fish of 30–40 cm length, while the zander were generally slightly larger, usually 30–50 cm and some larger than 55 cm. These are typical sizes of fish caught today. Small fish are absent, presumably indicating that net sizes were large enough to allow them through and that hooked ones were discarded or used as bait. Taphonomic bias may also play a part. Pike, as the fish with the largest potential size in the common species, were the most variable in size with a few specimens from small fish of around 20–30 cm, several around 50 cm and some approaching one metre. The three whitefish bones were from fish of about 30–50 cm long. A wels catfish pectoral spine represents a fish of 100–150 cm and two conjoined vertebrae were from a large fish but there was also a precaudal vertebrae from a much smaller individual.

The sieved bones from Troitsky were from fish that ranged from quite small specimens to large ones, depending on the taxon. The hand-collected bones, on the other hand, were on average larger with none of the small specimens present. Some of the pike bones were from fish of at least a metre long, as were several of the wels catfish bones. This species can reach an impressive size, fish of at least two metres and 65 kg are known and potentially the wels can grow to 500 cm long and around 300 kg. Sturgeon can also reach several metres in length but the few remains are difficult to size, given a lack of comparative material. Such large fish may be desirable but will always be comparatively rare, being at the top of the food pyramid.

Although some of the data is limited, the sizes of the fish from Novgorod do not seem to be substantially different from those from Gorodishche and the other hinterland sites, apart from some of the pike. This implies that the fish sources, fishing techniques and supply demands were similar. Sychevskaya (1965) suggested that, while the maximum sizes were similar, the medieval fish were on average larger and grew more slowly in comparison with recent specimens. If so, then the fishing did not overexploit the fish stocks, at least not to the extent that can be seen today. Demand for fish and changes in habitat due to pollution among other factors can also change the relative proportions of species, often with an increase in those that are smaller and less useful as food sources (Savenskova *et al*. 2010)

FISH FROM THE DESYATINNY-1 EXCAVATIONS (Andrei Zinoviev with Mark Maltby)

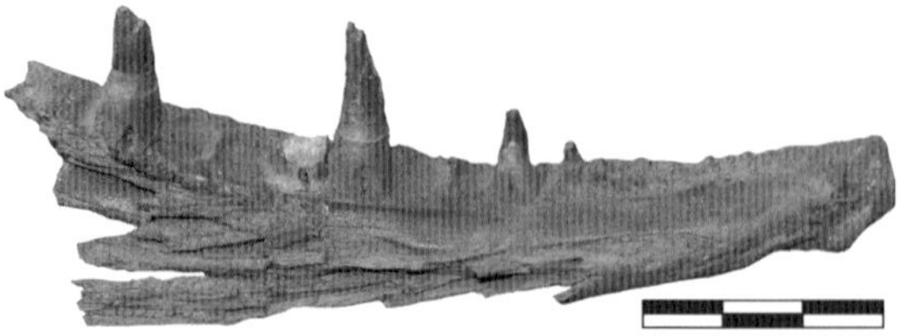

Figure 10.5: *Pike jaw from Desyatinny-1. Photograph by Andrei Zinoviev*

Keeping in mind an abundance of lakes and rivers in the vicinity of Novgorod, it is not surprising to expect a significant role of fish in the diet of citizens. In hand-collected samples from Desyatinny-1 only 26 fish bones were recovered comprising only a diminutive 0.3% of all bones. As discussed above, hand-collected samples do not provide a full picture, since only the bones of large specimens attract the attention of the collector. While fish species identified in the sieved samples from the Troitsky sites were dominated by cyprinids, hand-collected samples from Desyatinny-1 produced more bones of pike (16 specimens). Dental bones of the largest specimens furnished by sharp teeth are difficult to overlook (Figure 10.5). Most of them belonged to fish weighing 7–15 kg, which would have been caught in lacustrine environment. Cyprinids hold the second place in the Desyatinny-1 assemblage (six specimens), whereas only two bones of zander were identified.

FISH REMAINS FROM MININO SITES (Sheila Hamilton-Dyer)

Introduction

A large quantity of fish remains from recent excavations from several sites at Minino were made available for study. The remains had been collected by sieve and the larger contaminants, such as stones, removed. All the bones were identified and recorded at the Novgorod Archaeological Expedition headquarters. The bones were initially recorded on paper forms and transferred to computer in the UK.

The greatest volume of fish remains, especially from Pit 3, consisted of scales. These were briefly examined and noted but not systematically recorded. The majority belonged to cyprinids, which have large, fragile, cycloid scales. Some were of pike, which has smaller ovoid scales, and others were the distinctive toothed ctenoid scales of perch. Fishery biologists frequently report research involving scales, and sometimes provide detailed analyses of age, size, spawning etc. This has, however, often proved to be difficult to replicate with archaeological material, as analysis usually requires undamaged scales from specific areas of the individual animal (Wheeler and Jones 1989). Cyprinid scales, for example, usually have damaged edges and are rarely able to be distinguished to species. For these reasons this rapid appraisal of the material did not include the scales, apart from a brief note of species presence and comparative amounts.

Results

A grand total of 3,848 individual bone specimens were recorded, but not all of the 58 packets were analysed in detail and a great deal more material is still available for additional recording. All of the packets were scanned to check for species and to get an impression of proportions and anatomical representation. The overall summary is shown in Table 10.21 and the detail for each packet is provided in 10.24.

Species representation

Bones of perch were the most frequent of the taxa with 822 specimens (51% of the identified total). Cyprinids follow closely at 627 specimens (38.9%). Many of these bones could not be identified to exact species. The dentaries and inferior pharyngeals are, however, usually species specific and where sufficiently complete the species could be distinguished; bream (*Abramis brama*), roach (*Rutilus rutilus*) and ide (*Leuciscus idus*) were definitely present. Remains of roach were identified more frequently (42) than those of bream (27), while just four specimens were positively identified as ide. The remains of pike (*Esox lucius*) formed the third most frequent group at 159 specimens. Other fish species were very rare, during the scan of all the material just four remains were noted. These were a preopercular of ruffe (*Gymnocephalus cernuus*), two dorsal scutes of sturgeon, (*Acipenser sturio*) and a zander dentary (*Sander lucioperca*) (Figure 10.6; Table 10.21).

Anatomical representation

The major elements were recorded for all taxa with additional elements for some taxa, such as inferior pharyngeals for the cyprinids. A comparison of the results is given in Table 10.22. It can be seen that elements of the head, pectoral and pelvic girdle, and vertebrae were all present. The unidentified bones included large amounts of ribs, fin rays and elements of the gill apparatus. It is clear, therefore, that all parts of the fish are represented. Where bulk processing was carried out one can expect a large number of head bones and, conversely, a greater proportion of vertebrae at a consumption site. The relative amounts of particular elements within a species and between species were not the same however. The differences are largely due to recording and survival; some elements are very small or fragile in

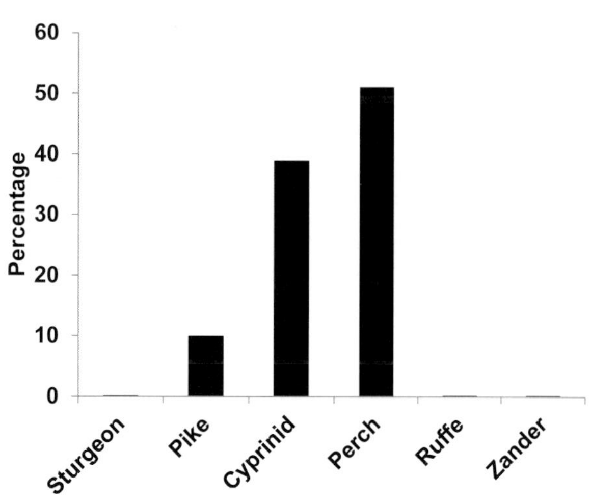

Figure 10.6: *Minino: NISP percentages of identified fish taxa*

one species yet large and robust in another. In this case the bones of perch were all relatively similar in size, robust and distinctive even when fragmented. The distribution of the elements was therefore, as expected, relatively even. For pike and the cyprinids, other than the vertebrae, it is the largest and most recognisable elements that were well represented. Several of the elements of the cyprinids are particularly thin and it is difficult to identify them when fragmented, and it is probably for this reason that the vertebrae were better represented.

Sizes

A selection of elements was measured as follows: the greatest length of the dentary of perch and cyprinids, the chord length of the cyprinid inferior pharyngeals (these as given in Morales and Rosenlund 1979), and for pike cleithra the p-b length (from the medial fossa to the caudal process as described above). The measurements are given in Table 10.23. The values for the pike cleithra indicate fish of between about 35–100 cm in length. These were not large pike; most would have been under half a metre and about a kilogram in weight. This is in contrast to Novgorod where several specimens were of fish as large as or larger than any available today, 1.5 m and perhaps as heavy as 30 kg (Table 10.16). The other fish from Minino also appeared to be of 'average' size. Young, very small, individuals were not present either – some of the roach and perch were from smaller fish than the pike, but were themselves not particularly small for their species. They were also large enough to have been butchered; five perch cleithra were chopped through, indicating that the fish were beheaded. Four pike cleithra were cut in a similar manner and there were 18 butchered fish bones in total. The good state of preservation and the recovery by sieving shows that the lack of small fish is not a taphonomic factor but reflects a deliberate selection.

The deposits

The species composition of each deposit is noted in Table 10.24. Whilst there are some differences, the impression overall is of similar activities, modified by sample size and probably by individual catch. Almost every sample contains all three main taxa. Pit 3 and surrounding contexts provided a huge collection, much of which belonged to cyprinids and perch with slightly less pike than in some other samples. This exceptionally large collection was only sampled for detailed analysis, but a scan of the remainder showed the same content. There appears to be an over-abundance of scales in this pit (Figure 10.7) and in the surrounding deposits, which may indicate where the catch was processed, either for immediate consumption or for smoking for winter use. The smoking of cyprinids is commonly practised today throughout north-west Russia.

Of special interest were the fish remains found in four inhumation graves, detailed below. These graves were discovered at Minino II, where 80 individual graves (65 inhumations; 15 cremations) were excavated. These date from the 11th to early

13th centuries (Makarov 2007). In each case the fish remains were found in the region of the feet. Pre-Christian traditions of placing food for the journey of the deceased, or a share of the funeral feast, are common throughout Europe and may sometimes include fish. One is reminded of the use of pike, catfish and sturgeon bones for amulets and protective necklaces in Poland (Makowiecki 2003). This author is not able to check the literature for similar findings in Christian Russia but the fact that it is fish, a strong Christian symbol, and that the remains are placed at the feet is likely to be highly significant. The results indicate that these were not complete individual fish but were perhaps a portion from a graveside feast. This may have been a pre-Christian practice, adapted and continued into the medieval period.

Figure 10.7: *Fish scales from Minino Pit 3. Photograph by Sheila Hamilton-Dyer*

Burial 10

The 22 bones were identified as a pike right cleithrum and supracleithrum, three precaudal vertebrae and 17 unidentified rays, probably pectoral fin rays, from the same fish. The cleithrum measurement, at 26.4 mm, is slightly over the average but is not the largest recorded. One of the vertebrae had been cut laterally, damage that can occur when filleting.

Burial 17

In this burial a much larger number of bones was found, 135 individual specimens. Of these 59 could be identified as cyprinid, the remaining 76 are fin rays and other fragments probably also of cyprinids. The pair of inferior pharyngeals could be identified as bream. The other bones are the left cleithrum, the first precaudal vertebra, 19 other precaudal vertebrae and 34 caudal vertebrae. This is more than is found in one fish and therefore the remains are of two or more individuals.

Burial 19

A similar collection was found in this burial with 44 cyprinid bones and 50 fin rays and other fragments. In this case the remains are 16 precaudal vertebrae and 15 caudal vertebrae of one size and a further 13 caudal vertebrae of much smaller size. Two of the larger caudal vertebrae were pathologically fused.

Burial 36

Perch accounted for 11 of the remains identified in this burial, along with 10 fin rays, which could well be also of perch. The bones are a left cleithrum, a left supra-cleithrum, paired opercular, paired preopercular, a ceratohyal and four precaudal vertebrae. One of the caudal vertebrae was much larger than the others and would have come from a different individual.

Discussion

Minino is situated next to Kubenskoye Lake in north-west Russia. It is part of the Arctic White Sea drainage system and, in comparison with some others in the region, is a relatively small, narrow, body of water, which has been modified into a reservoir of 13 m depth and 417 km^2. In comparison the neighbouring Beloe (White) Lake is at the northern end of the Volga drainage basin and is a larger but more shallow reservoir of 6 m and 1284 km^2. Beloe Lake is frequently mentioned in medieval documents concerning fishing and taxation, whereas references to Kubenskoye are rare.

Zander is only found today in Kubenskoye Lake due to introduction of stock in 1936 (Bolotova *et al.* 2010). The fish were brought from Beloe Lake, where it dominates the biomass (Dumnich *et al.* 2003) and is almost certainly a native species there. In spite of much effort the zander has not proved very successful and comprises only 2% of the commercial catch. Similarly sturgeon is not listed in the fauna for Kubenskoye whereas it does occur in Beloe (although wild fish are now very rare). The fauna in Kubenskoye today comprises 19 species (two introduced) but only 10 of these (including the introduced zander) are fished, with bream and pike being the most important. The nine species recorded from the Minino excavations include seven of these species, plus sturgeon (Table 10.25). The three species fished for today but not found at the excavation are members of the arctic fish community; the two whitefish species and the burbot. These three are indeed desirable food fish but are rare in the lake.

The frequency of the excavated taxa from the Minino sites is not exactly the same as the current frequency of the taxa in the lake. The most frequent finds from the excavations were of perch; although this is common today, it is not as abundant as bream, roach and ruffe. It is difficult to assess how much the differences between the proportions of taxa are due to human choice and how much it reflects recent changes in the lake. The relationship between species is a complex one and can be affected by fishing pressure; the population structure in Beloe is being affected by current over-fishing (Dumnich *et al.* 2003). The balance in Kubenskoye Lake now may not be exactly the same as in the past, but the proportions of carnivore to prey are likely to be similar. Of the species found in the excavations, sturgeon is the largest and in any waters will always have been the most rarely found. Pike as a top-level predator will always be less common then the prey it consumes. In this species, each size-class will consume any fish (including other pike) smaller than itself, along with frogs and waterfowl chicks in the case of large fish. The cyprinids vary in size but all feed on invertebrates and plant material. Today bream and roach are amongst the most frequent fish in Kubenskoye Lake, and this was probably the case in the past. In the present-day list ide is moderately common but was rare from the excavations. This is curious as ide is a good-sized fish and of commercial value. It will readily take baited hooks and may even be caught in gill nets but, unlike the bream, does not feed along the bottom in shoals and therefore may not be so easily targeted if nets and traps are used. Perch also tend to feed in shoals, feeding on invertebrates and fish eggs and fry (Phillips and Rix 1988). The perch breeds quickly and can adapt to overpopulation by remaining stunted. Ruffe belongs to the same group of fish (Percidae) and also breeds quickly, but has a much smaller adult size and most will weigh only a few grams. It is benthivorous and an important source of food for larger species. The near absence of ruffe from the excavations is likely to be on account of its small size and spiky appearance; the inhabitants seem to have been concentrating on capturing the more common of the best food fish. The type of fishing gear may also be relevant; if nets were being used the mesh size may have been too large to retain such a small fish. This may also explain the lack of very small pike and cyprinids and is likely to have been a deliberate policy. Observation of the catches and nets currently used around Lake Ilmen would seem to bear this out. Fishing can of course be done with spears for large fish and with hook and line. In the latter case it is usual practice to use the smallest fish as bait for more desirable ones.

There are two alternative explanations to account for the presence of the zander and sturgeon in the excavations; they either represent very rare catches, or were special trade items from nearby Beloe and its river system. The high numbers of squirrel and beaver from Minino (see Chapter 8) suggest this was a settlement specialising in the acquisition and trade of skins and therefore the inhabitants may well have had contact with, or travelled to, settlements around Beloe.

Minino comparison with Novgorod and Gorodishche

At Minino the identified fish remains come from a restricted fauna with perch numerically dominating the assemblage. Cyprinids were a major component, while pike were less common and other species are rare. The percentage of cyprinids is very similar to Gorodishche (Table 10.26). Pike are less common than in the assemblages from Troitsky and Gorodishche but it is the amount of perch that is most striking. Although this carnivore is a smaller fish, here it replaces the (almost) absent zander. Much of this difference must relate to the ecological difference between the fishing grounds; zander are abundant in Lake Ilmen and the river Volkhov and would also no doubt be a preferred catch over the smaller perch. The same reason explains the lack of wels catfish, whereas the presence or absence of the rarer and also the smaller species is subject to the effects of sample sizes and, in the case of Troitsky, collection bias.

-11-

HUMANS AND ANIMALS IN NORTHERN MEDIEVAL RUSSIA

Mark Maltby

INTRODUCTION

This concluding chapter has two main aims. It will first draw together the evidence accumulated in Chapters 3–10 and will review how different species of domestic and wild animals were exploited within and around the medieval town of Novgorod. The evidence drawn from the zooarchaeological data will be supplemented by evidence derived from other sources of archaeological and documentary data. Comparisons will be made with evidence from other medieval towns and settlements in northern Europe. Although, as described in Chapter 2, some of the formative conclusions have been previewed in other published outputs, this will be the first time that the final sets of data have been considered together.

The second section will provide a critical review of the project and the results of the analysis, reiterating some of the problems regarding sampling raised in Chapter 2. It will also consider pathways for further research. The book will conclude with an assessment of the achievements of this project and how far we have developed our knowledge of human-animal relationships in Novgorod and its territory.

HUMAN-ANIMAL RELATIONS IN NOVGOROD AND ITS TERRITORY

The evidence base

The zooarchaeological evidence based on excavations in Novgorod, Minino, Gorodishche and other hinterland sites has been set out in detail in Chapters 3–10. In this chapter this evidence will form the basis of the interpretation of how different species were exploited in Novgorod and its territory. In order to better appreciate the results from the recent excavations, we need to compare these results with other zooarchaeological evidence drawn from European sites of the same period, particularly evidence from towns in northern Europe. In addition to reports from individual sites, a number of general surveys of animal exploitation in the medieval period in Europe have been published (*e.g.* Tsalkin 1956; Benecke 1986; O'Connor 2010; Iwaszczuk 2014; Maltby *et al.* 2019) that can assist in comparing the results from Novgorod.

Deliberately, there were only brief references in Chapters 3–10 to other types of evidence that can contribute to our understanding of human-animal relationships in medieval Novgorod and its territory. There is a substantial amount of information available from other sources, which can be used in conjunction with the zooarchaeological evidence. Various documentary sources have been used in previous publications in support of discussions about various aspects of the medieval economy in the region. Particular attention has been paid to evidence that has highlighted the importance of various traded items such as furs and salt in the Russian medieval economy (*e.g.* Smith and Christian 1984; Martin 1986; Makarov 2006; Etkind 2011). In addition to many works by Russian scholars, previous discussions in English of the history and role of Novgorod that are based largely on historical evidence include works by Birnbaum (1981; 1989; 1996) and Halperin (1999). Books and articles incorporating both historical and archaeological evidence can be found in Thomson (1965), Brisbane (1992), Price (2000) and Brisbane *et al.* (2012). In addition to these secondary sources, this chapter will also make reference to vernacular works pertaining to Novgorod itself. The first of these are the Novgorod Chronicles.

The contents of Chronicles I and IV spanning the period 1016–1471 were translated by Michell and Forbes (1914). The Chronicles provide, in various levels of detail, annual accounts of events that occurred in Novgorod and its territory. The Chronicles are ecclesiastical documents and there are accordingly frequent references to the foundation of churches and other events pertaining to the clergy. The Chronicles also provide information about major political events and military campaigns within and beyond Novgorod's borders. However, in addition to the ecclesiastical, political and military accounts, the Chronicles also provide some information about aspects of everyday life in the town and its territory, particularly with regards to bad news. For example, in some years there are references to severe weather conditions and crop failures that resulted in famine and many deaths. There are reports of major fires that destroyed buildings and sometimes laid waste to substantial areas of Novgorod. There are references to major outbreaks of disease amongst the people and sometimes their animals. In places, there are specific mentions of animals that provide some insights into their relationship with the human inhabitants of Novgorod and these will be noted in this chapter's narrative.

The exceptional anaerobic conditions in Novgorod have resulted in the survival of many organic materials not usually preserved in medieval towns. The most unique of these are the birch-bark texts. The first of these was discovered in 1951 and by 2019 over 1,100 have been recovered from excavations in the town. In addition, over 100 similar documents have now been found in other towns, including 48 from Staraya Russa (situated beyond Lake Ilmen c.100 km south of Novgorod) and 19 from Torzhok (near Tver). The Cyrillic texts are written in an early form of Slavic and many can be translated. They date from the mid-11th to late-15th centuries, although their frequencies fluctuated within that period (Schaeken 2012; 2017). Most of the texts are concerned with the business transactions of the merchants (boyars) but their subject matter is diverse. Topics include legal matters, wills, debts, tax lists, complaints, prayers and personal letters (Yanin 1992; Dekker 2018). Although many fragments contain only a few words, which provide very limited information, the texts have furnished many insights into the lives of Novgorodians, particularly those of the boyars. Most of the texts were probably brought by messengers, who would present them as proof of their mission (Dekker 2018, 10).

A database of the birch-bark documents created by the Institute of Slavonic Studies of the Russian Academy of Sciences can be found at http://gramoty.ru. Although English translations are not always available, references to animals and their products feature in well over 100 of the texts from Novgorod. These provide interesting snippets of information about the use of animals that can be compared with, and contrasted to, the zooarchaeological evidence. The evidence about Novgorod's domestic economy, including the use of animals, was reviewed by Rybina (2001). This was based on information extracted from the 915 texts available by the end of 1999. We can now add information from texts discovered in excavations of the last two decades.

Other archaeological data can also be incorporated into the discussion. These include antler and bone artefacts and offcuts, which have been studied separately from the rest of the zooarchaeological evidence (*e.g.* Smirnova 2001; 2005). Leather objects and offcuts also survive in abundance in Novgorod (Izyumova 1959; 1967; Kurbatov 2012) and it is possible to identify the species that provided the leather (Solovyov 2012; Sully 2012). Numerous objects have been found that were associated with transport, ploughing, fishing, hunting and the processing of meat, skins and other animal products. These include wooden artefacts associated with fishing, hunting, wool and leather production (Kublo 2007a; 2007b; Rybina 2007), which rarely survive in aerobic conditions. A suite of geographical and environmental data has also provided additional complementary information of relevance to the use of animals. For example, studies of the historical landscape and settlement patterns (Yeremeyev 2012), pollen analysis (Spiridonova and Aleshinskaya 2012) and analyses of wooden timbers (Tarabardina 2007; Helama *et al.* 2017) have all provided information about the landscape, climate and vegetation in the immediate hinterland of Novgorod and (in the case of pollen) the area around Minino. Plant macrofossil and insect data have shed light on living conditions within Novgorod, including evidence for the presence of live and dead animals and their products within some of the properties (Monk and Johnston 2012; Reilly 2012).

Using this more holistic approach, in which the zooarchaeological evidence will be complemented by these other types of data sources, we will consider the evidence for the relationships between Novgorodians and their animals on a species by species basis, beginning with the domesticated mammals.

Cattle

Abundance

Although preservation and excavation methods, particularly in Novgorod, have favoured the recovery of cattle bones, there can be little doubt that beef was the most important source of meat in Novgorodian lands throughout the medieval period. However, they were not always the most common species represented by NISP counts in all the 9th–10th-century assemblages from settlements in Novgorod's hinterland. They were outnumbered by pigs at Georgii and in the northern part of the Gorodishche excavations and only narrowly outnumbered pigs in the Prost assemblage (Figures 3.5 and 3.22). However, if one takes into account their larger carcass size, cattle would have provided far greater amounts of meat than any other species on these early hinterland settlements. Cattle were also slightly outnumbered by pigs in the medieval assemblage from Minino (Figure 8.4).

In Novgorod itself cattle bones formed the bulk of the identified assemblage throughout the deposits of the three Troitsky sites. They consistently provided over 60% of the domestic mammal bone assemblages on Troitsky IX and XI and over 70% on Troitsky X (Figures 4.3–4.10). The dominance of cattle bones was consistent throughout the spit sequences from all three sites. Comparisons between spits in Troitsky XI showed some fluctuations with cattle being particularly well represented in the 12th-century spits (Figure 4.9). There were also some variations between the percentages of cattle within different areas and properties of these sites but none which consistently showed that the deposition of cattle bones varied significantly spatially, at least at this scale of comparison (Figures 4.4–4.7). It was not possible to investigate spatial variations in more detail within properties or in the streets. Excavations on the Desyatinny and Kremlin sites also produced very high percentages of cattle (Figure 4.53; Tables 4.74 and 4.77).

The traditional way of collecting animal bones from the excavations in Novgorod has undoubtedly favoured the recovery of cattle. Their elements formed only 46% of the domestic mammal assemblage in the sieved samples from the Troitsky sites (Figure 4.3).

Comparisons of the relative abundance of the domestic mammals that supplied the bulk of the meat supply (cattle; pig; sheep and goat) can be found in Table 11.1. Excluding sieved samples, cattle provided between 69% and 75% of the NISP counts of these species on the three Troitsky sites. Elsewhere on the Kremlin side of Novgorod, cattle provided 80% of these species in the large assemblage from the Nerevsky site (Tsalkin 1956), 75% from the Desyatinny-1 site and 73% in the small late medieval sample obtained from the 1996 excavations within the Kremlin itself. Similarly high percentages of cattle have been obtained from sites on the market side of the town. The large Slavensky assemblage consisted of 80% cattle (Tsalkin 1956) and cattle also provided 80% of the bones of cattle, pig and sheep/goat from the Nutny site that were scanned in 2011.

By contrast, cattle provided only between 42% and 53% of the total cattle, pig and sheep/goat elements from the three hinterland sites and only 40% at Minino. As discussed above, the lower percentages of cattle may largely reflect the more efficient recovery of the smaller bones of pig and sheep/goat at Gorodishche and Minino in particular. However, it should also be noted that the percentage of cattle in the 11th–12th-century assemblage from the 2001–2004 excavations at Gorodishche increased to 70% compared with 43% in the 9th–10th-century deposits (see Sablin in

Table 11.1: *NISP percentages and ratios of cattle, pig and sheep/goat from Novgorod town and territory and selected other northern European Towns and Proto-Towns. Counts are of number of individual specimens (NISP); S/G = sheep/goat. C:P = NISP cattle/NISP Pig; S/G:P = NISP sheep/goat/NISP pig*

			% NISP						
Town/site	*Country*	*Period*	*Cattle*	*Pig*	*S/G*	*C:P*	*S/G:P*	*NISP*	*Source*
Gorodishche (1979–1998)	Russia	9–10	53	41	6	1.31	0.16	3906	Table 3.1
Gorodishche (2000–2004)	Russia	9–12	50	41	9	1.21	0.22	3434	Table 3.36
Georgii	Russia	9–10	42	52	6	0.82	0.12	483	Table 3.28
Prost	Russia	9–10	47	45	8	1.02	0.15	187	Table 3.32
Minino	Russia	L10–E13	40	16	44	2.41	2.64	794	Table 8.2
Novgorod, Troitsky IX	Russia	10–E12	71	22	8	3.19	0.34	3423	Table 4.1
Novgorod, Troitsky X	Russia	10–E12	74	20	6	3.47	0.29	6464	Table 4.1
Novgorod, Troitsky XI	Russia	10–E15	69	22	10	3.18	0.46	22065	Table 4.1
Novgorod, Troitsky Sieved	Russia	10–M13	49	37	14	1.33	0.39	280	Table 4.1
Novgorod, Desyatinny-1	Russia	10–16	75	21	4	3.57	0.20	5503	Table 4.74
Novgorod, Kremlin	Russia	Med	73	28	7	2.33	0.25	86	Table 4.77
Novgorod, Nerevsky	Russia	11–15	80	15	6	5.45	0.39	10416	Tsalkin 1956
Novgorod, Slavensky	Russia	13–17	80	13	7	5.98	0.51	10056	Tsalkin 1956
Novgorod, Nutny 2011	Russia	Med	80	17	3	4.80	0.18	1117	Unpublished scan
Staraya Russa	Russia	Med	68	30	3	2.27	0.08	708	Unpublished scan
Shekshovo 2, Suzdal	Russia	Med	19	68	13	0.27	0.16	123	Unpublished scan
Staraya Ladoga	Russia	8–10	56	36	8	1.55	0.22	12216	Tsalkin 1956; O'Connor 2010
Pskov	Russia	9–12	70	22	8	3.14	0.35	8001	Tsalkin 1956
Moscow, Zaryadrey	Russia	10–17	65	25	10	2.62	0.39	22468	Tsalkin 1956
Moscow, Romanov Dvor	Russia	12–19	44	26	30	1.49	1.13	11092	Glazunova *et al*. 2009
Staraya Ryazan	Russia	11–13	60	28	12	2.18	0.44	7234	Tsalkin 1956
Tver, Kremlin	Russia	13–E15	61	27	12	2.23	0.45	31485	Lantseva, Lapshin 2001
Turku	Finland	13–19	49	13	38	3.95	3.08	37667	Tourunen 2008
Torino	Finland	17	61	10	29	6.01	2.84	1939	Puputti 2008

Table 11.1: *NISP percentages and ratios of cattle, pig and sheep/goat from Novgorod town and territory and selected other northern European Towns and Proto-Towns. Counts are of number of individual specimens (NISP); S/G = sheep/goat. C:P = NISP cattle/NISP Pig; S/G:P = NISP sheep/goat/NISP pig*

Town/site	*Country*	*Period*	*Cattle*	% NISP *Pig*	*S/G*	*C:P*	*S/G:P*	*NISP*	*Source*
Pappila, winter market	Finland	17–18	15	0	85			74	Harlin *et al.* 2019
Markkina winter market	Finland	17–18	27	0	73			67	Harlin *et al.* 2019
Karski Castle, Period II	Estonia	1250–1350	31	42	27	0.75	0.66	2750	Rannamäe and Lõugas 2019
Karski Castle, Period III	Estonia	1250–1550	57	14	30	4.13	2.15	1208	Rannamäe and Lõugas 2019
Karski Castle, Period IV	Estonia	1550–1700	63	12	25	5.40	2.17	789	Rannamäe and Lõugas 2019
Viljandi Hills & Settlement	Estonia	10–13	39	23	38	1.64	1.62	3237	Rannamäe and Lõugas 2019
Viljandi suburb	Estonia	1250–1550	63	14	23	4.56	1.66	6861	Rannamäe and Lõugas 2019
Viljandi town Period II	Estonia	1250–1350	43	26	31	1.68	1.22	1773	Rannamäe and Lõugas 2016
Viljandi town Period III	Estonia	1250–1550	51	18	31	2.84	1.72	5458	Rannamäe and Lõugas 2016
Viljandi town Period IV	Estonia	1550–1700	71	10	20	7.26	2.02	483	Rannamäe and Lõugas 2016
Viljandi Castle Period II	Estonia	1250–1350	66	1	34	97.70	50.17	3424	Rannamäe and Lõugas 2016
Viljandi Castle Period III	Estonia	1250–1550	41	13	46	3.09	3.52	7950	Rannamäe and Lõugas 2016
Viljandi Castle Period IV	Estonia	1550–1700	44	14	42	3.14	2.94	2200	Rannamäe and Lõugas 2016
Riga town	Latvia	1300–1500	64	25	11	2.61	0.46	2727	Maltby *et al.* 2019
Riga Castle	Latvia	1200–1550	54	22	24	2.40	1.09	368	Maltby *et al.* 2019
Vecdole Castle	Latvia	1220–1290	47	37	16	1.29	0.43	953	Maltby *et al.* 2019
Āraiši Castle	Latvia	1300–1420	31	28	41	1.12	1.49	513	Maltby *et al.* 2019
Āraiši Castle (Polish phase)	Latvia	1573–1624	34	24	42	1.40	1.71	946	Maltby *et al.* 2019
Ventspils town	Latvia	1300–1450	55	16	29	3.46	1.81	439	Maltby *et al.* 2019
Cēsis Castle (1996)	Latvia	1350–1600	53	12	35	4.55	2.96	42385	Maltby *et al.* 2019
Klaipėda Castle	Lithuania	14–17	51	39	10	1.30	0.27	545	Zulkus and Daugnora 2009
Klaipėda	Lithuania	16–17	75	14	11	5.50	0.79	1269	Piličiauskienė and Masiulienė 2013

Table 11.1: *(Continued)*

				% NISP					
Town/site	*Country*	*Period*	*Cattle*	*Pig*	*S/G*	*C:P*	*S/G:P*	*NISP*	*Source*
Grodno	Belarus	11–15	40	38	22	1.06	0.57	6546	Tsalkin 1956
Wolin	Poland	9–12	17	71	12	0.24	0.17	28717	Filipowiak 1979; O'Connor 2010
Szczecin	Poland	9–11	12	78	9	0.16	0.12	22088	Benecke 1986; O'Connor 2010
Poznan	Poland	10–12	35	24	41	1.51	1.74	10081	Benecke 1986; O'Connor 2010
Legnica	Poland	10–12	43	50	6	0.86	0.12	4727	Benecke 1986; O'Connor 2010
Opole	Poland	10–12	30	56	14	0.52	0.25	11561	Benecke 1986; O'Connor 2010
Gdansk	Poland	13–16	74	15	12	5.06	0.82	28672	Makowiecki *et al.* 2019
Elblag – Old Town	Poland	13–16	57	33	11	1.73	0.35	634	Makowiecki *et al.* 2019
Gniew	Poland	13–15	43	35	22	1.24	0.62	7354	Makowiecki *et al.* 2019
Torun	Poland	13–16	93	3	3	29.30	1.10	911	Makowiecki *et al.* 2019
Kaldus, V–IX	Poland	10–13	42	42	17	1.00	0.40	34873	Makowiecki, Makowiwcka 2014
Eketorp III	Sweden	11	36	15	49	2.38	3.23	187427	Boessneck *et al.* 1979; O'Connor 2010
Birka, 7–8	Sweden	E–M10	41	43	16	0.93	0.37	6027	Wigh 2001; O'Connor 2010
Sigtuna, St Gertrud 1–2	Sweden	L10–E11	29	20	52	1.45	2.61	2471	Wigh 2001; O'Connor 2010
Sigtuna, Tradsgard.0–3	Sweden	L10–11	45	19	36	2.38	1.94	18644	Wigh 2001; O'Connor 2010
Paviken, Gotland	Sweden	9–M11	64	22	13	2.88	0.60	1398	Wigh 2001; O'Connor 2010
Lund	Sweden	E11	48	28	24	1.74	0.86	3681	Benecke 1986; O'Connor 2010
Oslo	Norway	E11–M12	62	22	16	2.75	0.72	11071	Lie 1988; O'Connor 2010

Chapter 3 and Table 3.36). This could imply that cattle became more important in the hinterland after the foundation of Novgorod, reflecting increased demand for beef to provision the town. Cattle were also less dominant in the lowest spits from Troitsky XI (Figure 4.5). After the spring floods have receded, the area around Novgorod has abundant pasture available for cattle grazing, particularly in the floodplain of the River Volkhov and along the shore of Lake Ilmen. Cattle still form an important part of the modern local agricultural economy. The forest clearance in the medieval period would have increased the amount of pasture available for local herds.

Looking further afield, cattle formed 68% of the cattle, pig and sheep/goat NISP counts in an assemblage that we scanned from excavations in the town of Staraya Russa, situated about 100 km to the south of Novgorod (Table 11.1). Cattle bones have consistently been very well represented in Russian medieval and post-medieval urban assemblages, forming, for example, over half of the total cattle, pig and sheep/goat NISP counts in assemblages from sites in Staraya Ladoga, Pskov, Staraya Ryazan, Moscow (Tsalkin 1956; 1971) and Tver (Lantseva and Lapshin 2001; Zinoviev 2008). There have been a few exceptions. For example, cattle were less well represented at the Romanov Dvor site in Moscow (Glazunova *et al.* 2009). Cattle elements are also dominant in some medieval and early post-medieval urban and castle assemblages elsewhere in north-east Europe, although only a few contained more than 70% cattle and in a number of assemblages cattle were outnumbered by pig or sheep/goat (Table 11.1). Variations in landscape and climate would have been major factors in accounting for these variations.

Even in assemblages where cattle elements were outnumbered by other species, there can be little doubt that beef was more commonly consumed than pork or mutton given that an adult cattle carcass would have provided around ten times more meat than a sheep or goat and over four times as much as a pig. It is not surprising that cattle feature in reports of warfare, raiding and enforced procurement throughout the region. The Novgorod Chronicles report a number of such cases. For example in AD 1191, 'the men of Novgorod went in sailing vessels with the Korel people against the Yem…and made war upon their land and burnt it and cut to pieces the cattle' (Michell and Forbes 1914, 35). In AD 1212, 'Mstislav went with the men of Novgorod against the Chud people called Torma and made many captives and brought back countless cattle' (Michell and Forbes 1914, 52). In AD 1240, Novgorod was on the receiving end of cattle raids when the Lithuanians, Nemtsky and Chud people invaded Novgorodian lands and seized all of the horses and cattle about Luga (a village c. 95 km west of Novgorod) (Michell and Forbes 1914, 86).

Cattle are mentioned as tribute in several of the birch-bark documents (Rybina 2001), although perhaps not as frequently as could be expected given their importance in provisioning the town.

Cattle element representation and carcass processing

With regards to variations in element representation of cattle, Gorodishche produced the most convincing evidence for spatial variations. The high percentage of cattle elements from the southern area of the excavations included a higher proportion of foot bones and cranial elements (including horn cores) than encountered in the assemblages from the northern part of the investigated area. This suggests that the southern area was where more primary carcass processing took place. Cranial and foot elements were not uncommon in the northern area but were not as dominant (Figure 3.9). These results demonstrate the value of studying intra-settlement variations in species abundance and body part representation.

Although there were also some variations in the relative abundance of different types of cattle elements in the assemblages from all three Troitsky sites, the presence of large amounts of both high and low quality meat bones found in all the properties investigated suggests that much of the butchery took place within these properties. This observation is supported by the presence of insects associated with carcass processing (Reilly 2012). There were no consistent trends in the variations encountered between areas, properties and spits (Figures 4.13–4.20). The discovery of a number of calf mandibles in the upper spits of Troitsky XI may indicate that calfskins were being brought to one of the properties for processing. The assemblage from Desyatinny-1 also contained a mixture of high and low quality meat bones, suggesting carcass processing within those properties as well.

Cattle carcass processing was carried out with a range of tools both within Novgorod and on the earlier hinterland sites. Marks made with heavy blades were encountered more frequently than those made with fine knives, although the relative frequency varied between different bones. As discussed in Chapter 5, analysis of the frequency of butchery marks on cattle bones could indicate that carcass processing became more intensive in the town. Marks associated with many aspects of carcass processing were recognised. These included marks associated with skinning, carcass segmentation, filleting, marrow extraction, pot-boiling, and the removal of brains, tongues and horns. This again suggests that many stages of cattle carcass processing including initial slaughter took place within the properties. This is supported by the analysis of the insect remains. Some samples produced species commonly associated with carrion, bones or skins and indicate that primary processing waste was discarded in some parts of the properties (Reilly 2012).

Fine incisions, indicative of initial skinning, were observed on several shafts of cattle phalanges and metapodials from Novgorod, Gorodishche and Georgii (Tables 5.24–5.28). Although some of these bones could have been imported whilst attached to imported skins, it is likely that many were detached from the hides of animals slaughtered in Novgorod and processed within the properties. Perhaps in support of this, it is interesting to note that reference to cattle hides as items of tribute are almost absent from the birch-bark documents. Five calfskins are mentioned alongside other goods in BBD 500, which was found on the Slavensky site and dated to the 14th century (Rybina 2001) but no other documents that refer specifically to the hides of adult cattle appear to have been found. The virtual absence of cattle hides from specific mention in the documents could imply that cattle hides were too common a commodity to receive special attention in such records and were supplied by other networks.

There is abundant evidence from artefacts and offcuts that cattle hides were frequently used in the creation of leather goods. Novgorod is famous for the superb preservation of footwear, purses and other leather objects and a sequence of advances in technology and scale of production have been described (Kurbatov 2012). The 11th–12th-century deposits at Gorodishche have produced evidence for the presence of craftspeople making high quality footwear on this high status site, whereas contemporary craftspeople in Novgorod were producing less elaborate products perhaps for a wider market (Kurbatov 2012, 406).

Workshops associated with the manufacture of shoes and other leather objects have been found on properties on the Nerevsky and Nutny sites (Izyumova 1967) as well as on several Troitsky sites (Rybina 1992, 183; Kublo 2007a). Some of the insects directly associated with leather offcuts from a workshop on Troitsky XIII are species commonly associated with hides and skins (Reilly 2012).

Analysis of 42 leather offcuts from the Troitsky and Kremlin sites revealed that half belonged to cattle, including several from calves (Sully 2012). The leather used to manufacture most types of shoes and boots found on Troitsky XI were also mainly made from cow hides (Solovyov 2012).

Horn-working was also carried out both at Gorodishche and within Novgorod. Discrete concentrations of butchered cattle horn cores were recorded at Gorodishche, indicating that small-scale horn-working was taking place there. Processing marks at the base of many cattle horn cores in the Troitsky and Desyatinny assemblages showed that horns were commonly separated from the skulls at their base (Figures 5.8–5.9). Even allowing for the possibility that some horns were imported with hides and not worked, the evidence suggests that horn-working was probably an industry that was more important than has previously been realised and can be added to the list of animal products acquired by specialist manufacturers.

A group of over 100 sawn cattle proximal and distal metatarsals were found in 13th–14th-century deposits on the Fedorovsky site (see also Chapter 6). This provides an example of the stockpiling of bones for working. In this case it was the shafts of the bones that were required for further manufacture whereas the proximal and distal ends were surplus to requirements and discarded.

Although there was a fair amount of variation in the location and nature of the butchery marks, there were sufficient consistencies encountered (for example in the severing of the scapula blades, fragmentation patterns of upper limb bones and the splitting of the vertebrae) to suggest that specialist trained butchers were involved (Figure 11.1; see also Figure 5.12). Perhaps they were hired by the owners of the properties to carry out processing when required. One assumes that butchers also operated in the town's marketplace and some carcasses or joints would have been acquired from there but the impression remains that a lot of processing took place within the properties themselves.

It was not possible to carry out chronological or spatial comparisons of butchery practices, as detailed records were only made for bones from Gorodishche and some of the lower spits of Troitsky XI. However, although there were many similarities in the marks encountered, there were some interesting variations. For example, some of the types of butchery marks found in Troitsky XI were not encountered at Gorodishche and there appears to have been a decline in the practice of splitting metapodials longitudinally to extract marrow within the Troitsky XI sequence (see Chapter 5 for details and further examples).

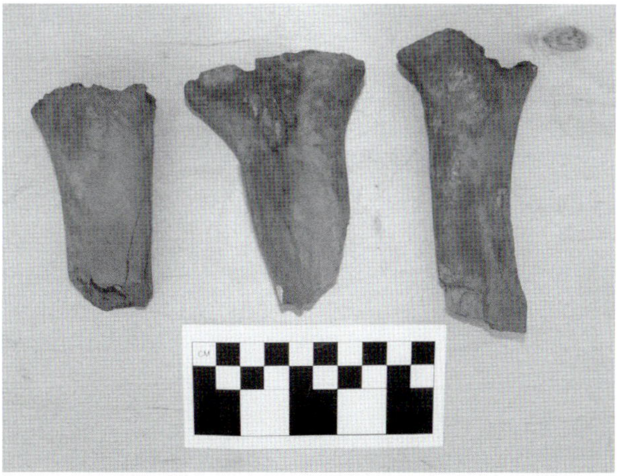

Figure 11.1: *Proximal portions of cattle radii from Novgorod showing consistency in breakage patterns. These would have been a suitable size for pot-boiling. Two of these bones have been damaged by gnawing. Photograph by Krish Seetah*

The birch-bark documents have so far provided no information about the status or operations of butchers and the only mention of butchers in the Chronicles of Novgorod dates to AD 1403 when Novgorod merchant butchers erected a church in the town of Russa (presumably Staraya Russa) (Michell and Forbes 1914, 176). This implies that by the early 15th century some butchers had acquired substantial wealth. However, their role and status in the earlier medieval period in Novgorod is unclear.

Cattle mortality patterns

Detailed ageing analysis of cattle mandibles was carried out on all assemblages examined during the project. These included 50 mandibles from Gorodishche, which showed that 60% of the mandibles belonged to adults over four years of age but few belonged to very old cattle (Figure 3.11). Analysis of the 782 cattle mandibles from Troitsky IX–XI showed that a minimum of 42% also belonged to adult cattle. However, again only a small percentage of these belonged to very old animals (Figures 4.21–4.22; Table 4.21). Although the 284 mandibles from Desyatinny-1 contained fewer adults (minimum of 33%) and greater chronological variations, very few of these were from old adults (Figure 4.54; Table 4.75). This indicates that although many cattle were old enough to have been exploited for milk production, traction and transport, most animals were slaughtered prior to very old age. Many cattle were slaughtered between around four to seven years of age. In addition, over half the cattle from the Troitsky sites were under four years old. There were a significant number of cattle culled between two and three years of age. These represent animals raised primarily for beef. It is interesting to note that only a single mandible from Gorodishche belonged to an animal of this age, yet they were present in fair numbers in the lower spits of all of the Troitsky and Desyatinny-1 sites. This could reflect a change in the exploitation of cattle during the 11th–12th centuries with a greater emphasis on meat production in response to the need to provision the expanding town.

At least 14% of the mandibles from the Troitsky and at least 20% from Desyatinny-1 belonged to cattle younger than 18 months of age. Most of these were from calves of around 2–6 months old. Given the likelihood that such mandibles are under-represented because of retrieval and survival biases, their frequency indicates that veal was fairly commonly consumed in the town. The presence of substantial numbers of calf mandibles in these assemblages indicates that milk would have been readily available for human consumption, as veal production is often closely linked with dairying. There were, however, some fluctuations in the percentages of mandibles of young calves. They are relatively uncommon in the 12th–13th-century assemblage from Troitsky XI and in the early 12th-century levels at Desyatinny-1 compared to earlier and later levels on both sites. They were also quite well represented in the 9th- and 10th-century levels at Gorodishche. Therefore, although veal appears to have been a common component of the meat diet during most of the medieval period, there may have been periods when dairy production declined resulting in fewer calves being slaughtered. As noted above, some calf mandibles may have been brought to the properties attached to hides that were destined for leather processing.

Metrical analysis of metacarpals revealed that the majority of specimens with fused distal epiphyses from Gorodishche (Figures 3.13–3.14, 6.11), the Troitsky sites (Figures 4.25–4.26, 6.11–6.12) and Desyatinny-1 (Figure 4.55) came from females, showing that most of the adult cattle were cows. There were some chronological and spatial variations in the relative abundance of metacarpals of male and female cattle. Metacarpals of adult males were always in a minority but they were more commonly recovered from Troitsky IX and X than from Troitsky XI, where the percentage of male cattle over three years old decreased to c. 10% in the upper spits. It is understandable that a higher proportion of bulls and steers would be slaughtered at a young age for meat. Adult cows would then act as both a source of milk and as breeding stock. However, as noted above, most of the cattle slaughtered at these settlements did not live to a very old age. A 13th-century birch-bark document (BBD 718) refers to a tribute of three dry cows (Rybina 2001, 129), indicating that those particular cattle at least were not slaughtered until after their value as dairy providers had diminished.

The manufacture and trade of cheese and butter is mentioned in several birch-bark documents (Rybina 2001, 130). For example BBD 406, dated to the late 14th century, refers to tributes of butter and cheese (Dekker 2018, 15). BBD 540, dated to the same period, also gives instruction about the sale of butter and cheeses. Cheese is also mentioned in a mid-12th-century document (BBD 831). Amongst other household objects found in Novgorod that could be associated with butter- and cheese-making are perforated wooden moulds (Khoroshev 2007: 276).

There is no specific reference to working oxen in any of the birch-bark documents. The entry for AD 1240 in the Chronicles did note that because all of the cattle and horses had been seized in the villages around Luga, it was impossible to plough (Michell and Forbes 2014, 86). However, although some of the adult oxen could have been used as plough animals for several years prior to slaughter, the importance of cattle seems to have lain mainly in their value as meat and milk producers.

In assemblages from several medieval settlements in Estonia and Latvia, the percentage of young calves was generally lower than at Novgorod and Gorodishche but many sub-adult and young adult cattle were also represented. Most of the adult cattle were females. There was, however, a lot of variability between assemblages (Maltby *et al.* 2019, 144).

Cattle epiphyseal fusion data collected from Gorodishche and Novgorod generally supported the mandibular ageing evidence but fusion points were subject to greater taphonomic decay, particularly from gnawing (Figure 11.1), which meant that the limb bones of young calves were less well represented than their mandibles and unfused late-fusing epiphyses suffered more destruction by carnivore gnawing than fused specimens (Figures 3.12, 4.22–4.23).

The sizes of cattle

Metrical analysis of cattle bones from Novgorod and Gorodishche supported previous observations that cattle in this region were of small stature (Tsalkin 1956). Twenty-nine complete limb bones from 9th–10th-century levels at Gorodishche excavated between 1979 and 1998 ranged between 100.5 cm and 123.2 cm, with a mean of 112.0 cm (Figure 6.23). The 37 complete specimens from the 2000–2004 excavations produced a mean withers height of 112.4 cm. The 155 complete limb bones from the Troitsky sites had withers heights estimates ranging between 95.6 cm and 127.5 cm with a slightly lower mean of 109.8 cm (Figures 6.24–6.25). Other measurements also indicated that the cattle in Novgorod decreased slightly in size during the medieval period. Mean values of length, breadth and depth measurements of limb bones were between 3% and 10% higher in the 9th–10th-century levels at Gorodishche than in the 13th–15th-century levels from Troitsky XI (Figures 6.1–6.31), although the low numbers of bones of bulls and oxen in the later medieval deposits could account for some of this decrease.

Cattle from Novgorod were on average slightly smaller than those from medieval Riga but larger than those from Ventspils in Latvia, which were of very small stature (Maltby *et al.* 2019, 145). The average height of cattle from later medieval levels (AD 1250–1550) at Karski and Viljandi in Estonia lay around 105 cm, slightly smaller than the average from contemporary sites in Novgorod (Rannamäe and Lõugas 2019, 71). Travellers to Russia in the 16th and 17th centuries noted the small size of cattle in the region (Kurbatov 2012, 393).

Pigs

Abundance

At Gorodishche, pig was the second best represented species overall forming 36% of the domestic mammal elements collected between 1979 and 1998 (Figure 3.4). Pig elements contributed 53% of the identified domestic mammal remains from the northern area of Gorodishche (Figure 3.5). In the assemblage collected between 2000–2004, examined by Mikhail Sablin, pigs provided 39% of the 9th–10th-century domestic mammal assemblage but only 18% of the 11th–12th-century assemblage, when cattle elements became more dominant (Table 3.36). Pigs were slightly better represented in 9th–10th-century sites of Prost and Georgii forming over 40% of the mammal elements recovered from both sites (Figure 3.22).

Pigs were the second most common species recorded from Troitsky IX–XI, forming around 20% of the identified domestic mammal counts from all three sites (Figure 4.3). They were better represented in the sieved samples (35%). This again demonstrates that traditional excavation methods have favoured the retrieval of cattle bones. Pigs were slightly better represented in the Desyatinny-1 assemblage, examined by Andrei Zinoviev, forming between 15% and 33% of the domestic mammals, with the percentage of pigs gradually increasing in the upper spits (Figure 4.54), in contrast to Troitsky XI where pig percentages tended to decrease in the upper spits (Figures 4.5–4.7).

At Minino, pig elements formed around 15% of the domestic mammal assemblage but they were outnumbered by sheep/goat as well as cattle (Figure 8.4).

Comparisons between cattle, pig and sheep/goat only showed that pigs formed 41% of the totals from Gorodishche, 45% from Prost and 53% from Georgii (Table 11.1). They provided around 20% of the overall totals amongst the hand-collected bones from the three Troitsky sites, their percentages decreasing slightly in the higher spits of Troitsky IX and X (Figures 4.8–4.9). Pig percentages from Troitsky XI also showed some fluctuations with pigs being better represented in some of the lower spits dated to the 10th and 11th century and again in some of the 13th-century upper spits (Figure 4.11).

The pig percentages at Troitsky were broadly similar to those from other sites in Novgorod (13%–28%), although, as noted above, pigs were heavily outnumbered by cattle on all these sites (Table 11.1). In Novgorod, it seems that although pork and bacon were regularly consumed, they were much less important than beef and veal.

Comparisons with other Russian and eastern European urban and castle assemblages display substantial variations in the abundance of pigs (Table 11.1). Pig percentages were generally slightly smaller than assemblages from Staraya Russa (30%), Pskov (22%), Moscow (25%–26%), Staraya Ryazan (28%) and Tver (27%). Pigs were better represented at Staraya Ladoga (36%), although still heavily outnumbered by cattle. Most assemblages from nucleated sites in Latvia and Estonia have produced similar ranges of pig percentages as those encountered in Novgorod (Maltby *et al.* 2019). None of the Russian urban assemblages were dominated by pigs in contrast to assemblages from some towns in northern Poland, for example at Wolin and Szczecin (O'Connor 2010).

Comparisons of species represented in all these assemblages were handicapped by the varied methods of retrieval, with cattle being favoured in hand-collected samples. If cattle bones are excluded, the relative frequencies of pig and sheep/goat can be compared using the sheep/goat:pig ratio (after O'Connor 2010). Retrieval bias between these medium-sized mammals should be less than between cattle and pig, for example. The ratio of sheep/goat ranged between 0.18 and 0.51 in the hand-collected assemblages from Novgorod, indicating that pigs were consistently better represented than sheep/goat (Table 11.1). The ratio of sheep/goat was generally lower in the earlier assemblages from Georgii, Prost and Gorodishche (0.12–0.22), probably indicating that there was more woodland available for the foraging of pigs prior to the main phases of agricultural clearances in the Novgorod hinterland during the later medieval period. In contrast, the sheep/goat:pig ratio in the Minino assemblage was 2.64, despite the presence of high percentages of wild mammal species that favour woodland habitats. It is possible that climatic and environmental conditions were less suited for pig keeping in that region. The area has lower mean winter temperatures and longer and deeper coverings of snow (Spiridonova and Aleshinskaya 2012) that would be less favourable for winter foraging by free-range pigs. Modern wild boar population densities decrease significantly in northern latitudes in relation to lower winter temperatures (Melis *et al.* 2006) and the need to supply their domestic pigs with supplementary winter fodder may have limited the numbers kept by the inhabitants of Minino.

Only in one (Romanov Dvor, Moscow) of the eight assemblages from other parts of Russia were pigs outnumbered by sheep/goat. None of the others had a sheep/goat:pig ratio of over 0.45 (Table 11.1). In contrast, pigs were very poorly represented in comparison with sheep on Finnish sites, again indicating unfavourable high latitude habitats. Pigs were also outnumbered by sheep/goat in all but one of the 11 Estonian assemblages compared here. The same was true in the case of five of the seven Latvian assemblages. In contrast, pigs outnumbered sheep/goat in eight of the ten Polish assemblages. Therefore, although there were some variations in the percentages of pigs represented in Novgorod and other Russian sites, they appear to have been exploited more heavily than sheep and goats than in areas to the north and west of Russia.

Pig element representation and carcass processing

Although there were some spatial variations, there was quite an even representation of body parts in the pig assemblage from the 1979–1998 excavations at Gorodishche. Forty-three per cent of the fragments consisted of skull, mandibles and loose teeth. Upper limb bones provided 29% and 18% of the elements came from bones of the limb extremities (Table 3.13). It seems likely that most pigs were slaughtered at the settlement, (Figure 3.15). The Georgii and Prost assemblages contained high percentages of loose teeth indicating poorer preservation but also the likelihood that most pigs were also slaughtered at these settlements as well (Figure 3.24). The discovery of partial skeletons of very young piglets at Gorodishche and Georgii also suggests that pigs were being reared at the settlements.

Similarly over 45% of the pig elements from the Troitsky sites in Novgorod came from the skull, mandibles and loose teeth, indicating that butchery often took place within the properties. However, percentages of upper limb bones gradually increased in the later medieval deposits, possibly indicating joints of fresh and preserved meat were more commonly brought to the urban properties (Figure 4.33). Although references to pigs are rare in the birch-bark documents, BBD 842, an early 12th-century document from the Troitsky excavations, includes in a list of items of tribute

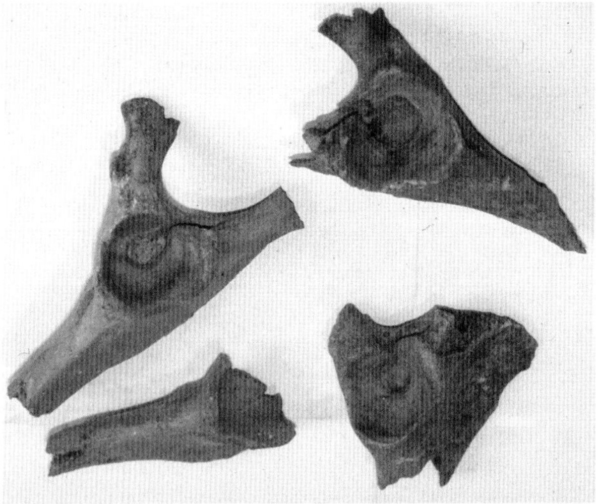

Figure 11.2: Examples of butchered pig pelves. Photograph by Krish Seetah

two backbones of pig, a clear indication that joints of pig were sometimes acquired. The low numbers of foot bones from all the Novgorod sites may also indicate that trotters were often previously removed before the remaining parts of the carcass were brought to the properties, although further sieving is again required to establish whether the low numbers of these small bones was not simply the consequence of poor retrieval.

Butchery marks were recorded more frequently on pig bones from the Troitsky sites than at Gorodishche, probably partly due to better preservation, but also perhaps to the more frequent use of knives for filleting. As expected, butchery marks were found more frequently on the major meat-bearing bones, particularly the scapula and pelvis (Figures 5.5 and 11.2). Although a few of the butchery classifications recorded in Novgorod were not encountered in Gorodishche, there were a lot of similarities in the location and nature of the butchery marks. There was a lot of evidence for the segmentation of the carcass, often through the severing of the shafts of limb bones and through the blades of scapulae (Tables 5.36). In two cases from the Troitsky sites there was clear evidence that the scapulae had been hung perhaps during smoking and salting.

Meat was commonly preserved by smoking and salting and salt production and trade was important in north-west Russia (Smith and Christian 1984). Staraya Russa was the centre of large-scale salt production in the region during the medieval period and therefore Novgorod had access to large relatively local supplies of salt (Logunov 1997). References to salt have been found on many birch-bark documents in Novgorod (Rybina 2001). Although many of the documents refer to salted fish, there are a number that indicate that salt itself was an item of tribute. One late 14th-century document reveals that salt was purchased from a German source (BBD 282). It is likely that salt was required for the preservation of the meat of cattle, sheep/goat and pigs, particularly to ensure supplies through the winter. Indeed it is feasible that fresh meat may have been less commonly consumed than preserved meat.

The use of pig skins for leather-working was fairly uncommon in medieval Russia (Kurbatov 2012) and there was little skinning evidence on the skull or foot bones. None of the leather fragments submitted for species identification derived from pig skins (Solovyov 2012; Sully 2012).

Pig mortality patterns

Tooth ageing analysis from Gorodishche and earlier layers from the Troitsky sites have shown that up to 30% of the mandibles belonged to pigs culled in their first year (Figures 3.16, 4.33–4.34). In both the Troitsky and Desyatinny-1 assemblages (Figure 4.56), pigs aged between 6–12 months of age were commonly slaughtered, probably indicating a focus on autumn slaughter. Neonatal pigs have been very rarely recorded in Novgorod, which perhaps indicates they were not commonly kept within the town, although further sieving is needed to confirm this. The proportion of first year mortalities decreased significantly (to c. 5%) in the 13th–15th-century assemblage from Troitsky XI (Figure 4.35), suggesting that pigs became less intensively exploited. Most of the other pigs recovered from Novgorod and its hinterland were culled in their second and third years with jaws of third year animals becoming more common in the later medieval assemblages. Few pigs of over three–four years of age were represented either in the mandibles or in the epiphyseal fusion data (Figures 3.17, 4.35–4.36), as is usually the case in archaeological assemblages. Pig populations can endure high rates of immature slaughter.

The sizes of pigs

The vast majority of the measured pig bones and teeth from Novgorod and its hinterland sites fell comfortably within the size range of domestic pigs rather than the larger wild boar (Figures 6.32–6.35). The pigs from the 9th–10th-century assemblages from Gorodishche, Georgii and Prost were on average slightly larger than those from Novgorod. And more bones of smaller pigs were found in later medieval deposits from Troitsky XI. The decrease may reflect the presence of a larger proportion of sows in the later assemblages (although this needs to be confirmed by analysis of sexing data). Another possibility is that pigs generally became slightly smaller in the later medieval period, perhaps reflecting the decrease in prime woodland habitats in the region.

Sheep and goats

Abundance

At Gorodishche, 57% of the specifically identified elements of sheep and goat belonged to sheep. At Georgii, four sheep but no goat bones were recorded and at Prost a single element of sheep was positively identified. Both sheep and goats were recorded in all three Troitsky sites. Sixty-three per cent of the mandibles with deciduous fourth premolars and 62% of 92 measured bones with positive species identifications belonged to sheep (Figures 4.38–4.40, 6.36–6.39), suggesting that sheep were present in greater quantities than goats but that goats were not uncommon within the ovicaprid assemblage. At Minino, 90% of the positively distinguished elements belonged to sheep. The high percentage of sheep at Minino could reflect their greater suitability to the local ecological conditions, where pollen evidence showed substantial development of more open conditions including dry pasture suitable for sheep during the medieval period (Spiridonova and Aleshinskaya 2012).

Goats formed a much higher proportion of the sheep/goat assemblage in Novgorod than in most European towns and proto-urban sites. For example, in the 11th-century deposits at Eketorp, Sweden, sheep provided over 96% of the diagnostic sheep and goat elements (Boessneck *et al.* 1979). In a 17th-century assemblage from Tornio, Finland, only 1% of the diagnostic elements were identified as goat (Puputti 2008). Interpretations of the relative abundance of sheep and goat are complicated by the presence of worked horn cores on many urban sites. These tend to be biased towards the large horn cores of male goats, whose horns were preferentially selected. For example at Viljandi, Estonia, 90% of the horn cores and skulls belonged to goat, whereas only 34% of the measured metacarpals were classified as goat (Rannamäe 2010). Similarly, in medieval levels in Riga, Latvia, 18 of the 25 goat elements were horn cores compared with only six of sheep. In contrast, sheep provided 21 of the 28 other elements identified to species (Maltby *et al.* 2019). Novgorod has also produced substantial numbers of goat horn cores, probably including many that were imported as raw material for working (Figure 6.39) but goats were well represented by other bones as well and it is likely that some of them were kept in the town itself.

However, sheep/goat elements overall were poorly represented in the assemblages from Novgorod and sites in its immediate hinterland, providing less than 10% of the total of cattle, sheep/goat and pig in nearly all of the assemblages investigated both in the recent project and in earlier excavations (Table 11.1). The wet meadowland around Novgorod would have been much better suited for cattle grazing and the substantial forests would not have favoured keeping sheep in large numbers either.

Although sheep and goat were always outnumbered by pig, the percentages of these species were closer in some of the spits dated to the 14th century from Troitsky XI (Figure 4.11). This may be a consequence of the significant decrease in woodland and the concomitant increase in arable in the area at the end of the 13th century (Spiridonova and Aleshinskaya 2012). However, the same trend was not observed in the Desyatinny sample, where sheep/goat were consistently less well represented than in the Troitsky assemblage and the percentage of pigs increased in the later medieval levels (Figure 4.53).

Sheep/goat are also poorly represented on most medieval sites in north-west Russia, often forming <10% of cattle, pig and sheep/goat NISP counts (Table 11.1). In contrast, at Minino, sheep/goat formed 44% of the cattle, pig and sheep/goat counts, reflecting the development of agricultural land and the increase in dry pasture more suitable for sheep.

Sheep/goat element representation and carcass processing

At Gorodishche loose teeth, mandibles and tibiae were the best represented sheep/goat elements (Table 3.18). These are sturdy elements that are often the most frequently recorded in archaeological assemblages. Loose teeth and small bones such as the carpals, tarsals and phalanges were better represented here than in the Troitsky assemblages because of more efficient retrieval methods. The presence of large numbers of head and foot bones in the Troitsky assemblages suggests that whole carcasses were often processed within the properties. However, comparisons of the counts of upper limb bones and metapodials in the Troitsky XI sequence (Figure 4.43) showed that upper limb bones became relatively more abundant in the later levels, indicating that more joints of meat could have been brought in. However, in Spits 4–2, metapodials increased significantly, perhaps indicating the import of these bones with skins. References to sheep and goat skins as tribute feature relatively frequently in the birch-bark documents (Rybina 2001). Goat skins were used in the manufacture of some types of footwear in Novgorod (Solovyov 2012) and both sheep and goat were identified in leather samples examined by Sully (2012).

Over 30% of the goat and sheep horn cores bear evidence of butchery associated with their separation from the skulls. However, there do not appear to have been large concentrations of horn cores within any of the properties that could indicate largescale horn-working similar to those on the Duboshin site in the Slavensky Quarter (Smirnova 2001, 83). Many of the other butchery marks were similar to those found on pigs including evidence that some limb bone shafts and scapula blades had been severed during the segmentation process. Fine incisions, particularly on the astragalus and mandible were more commonly encountered than in the case of pig.

Sheep/goat mortality patterns

Analysis of the sheep/goat mandibles was handicapped by the fact that older mandibles were not identified specifically to sheep or goat and the two species may have been exploited differently.

At Gorodishche, over half the sheep and goat mandibles were from lambs and kids under two years old but 31% of the mandibles were from adults over three years of age, a slightly higher proportion of adult animals than encountered in the earliest period of Novgorod. However, the sample was small (19 jaws). Both immature lambs and kids were represented on the three Troitsky sites and at least 59% of the mandibles in the 10th–12th-century assemblages were from animals that were culled before two years of age (Figures 4.44–4.45). They were also prevalent in the Desyatinny assemblage (Figure 4.57). This indicates that exploitation was focused on meat production. Percentages of mandibles of immature animals decreased to 52% in the smaller 12th–13th-century sample from Troitsky XI and to 45% in the most recent layers, indicating that meat exploitation may have become gradually less intensive. Dairy production could also have been a factor as a number of mandibles of lambs and kids under three months old were also recovered from both Troitsky and Desyatinny.

Evidence for wool production in Novgorod comes in a variety of forms. In addition to the references to sheep and goat skins in the birch-bark documents (Rybina 2001), wool-cleaning racks have been found on an 11th-century property on a Troitsky site and looms, carding combs and other wooden objects associated with wool production are common finds (Kublo 2007b). Analysis of textiles from the Nerevsky excavations showed those made of wool outnumbered those of linen and silk (Thompson 1967, 96–98). Most wool probably came from local sheep.

The importance of wool production is also evident in the sheep/goat mandible assemblage from the Troitsky sites. Relatively high proportions (20%–40%) of sheep and goat over three years old were found throughout the sequence (Figures 4.44–4.45). Many of these sheep were breeding stock and some may have been dairy producers, but they would also have produced several annual fleeces. There was also a modest increase in the percentage of mature adults in the upper spits of Troitsky XI, which may indicate that wool production was becoming more important from the middle of the 13th century onwards. Further samples of this date are, however, required to confirm this trend and in any case the main focus of sheep exploitation continued to be on meat production.

The sizes of sheep and goat

Evidence of the stature of sheep and goat was limited by the small sample sizes. Withers height estimates based on Teichert's (1975) and Schramm's (1967) conversion factors for the lengths of sheep and goat metapodials respectively (Table 6.26) showed that sheep from Novgorod were quite tall with an average withers height of 59.9 cm. The goats tended to be a little taller averaging 66.4 cm.

Horses

Abundance

Horse remains provided 8–9% of the domestic mammal elements from Gorodishche, Prost and Georgii (Figure 3.22), although their totals included several associated bone groups, the presence of which implies that some of their carcasses were not fully processed for food.

Horse bones formed a small but fairly consistent percentage of the domestic mammal elements in the Troitsky assemblages ranging from 2% to 5% of the identified domestic mammal fragments (Figures 4.3–4.7). They were better represented in the Desyatinny-1 assemblage where they provided 9% of the assemblage of domestic mammals (Table 4.75).

Comparisons of the relative abundance of horse and cattle on other Russian urban sites show some variations (Maltby 2013). Unusually high ratios of horse were recorded at Grodno and Staraya Ryazan (Tsalkin 1956), although it is not clear whether any of these counts included partial skeletons, which would significantly bias the results. The Russian results lie in stark contrast to those from many urban sites in the northern Baltic, where horse bones have been found in most urban deposits in very small numbers (Maltby 2013; Maltby *et al.* 2019; Makowiecki *et al.* 2019).

The Novgorod Chronicles sometimes refer to large numbers of horses (Michell and Forbes 2014). In AD 1217 Novgorod forces captured 700 horses from an unnamed enemy. In AD 1234 they took 300 horses from the Lithuanians and in AD 1398 they confiscated 3,000 from Dvina. Locally, the Chronicles reported that large numbers of horses died of disease in various years (*e.g.* AD 1115; 1158; 1203; 1291; 1444).

Horse element representation and carcass processing

The presence of several associated bone groups resulted in horse ribs and vertebrae being better represented at Gorodishche (Figure 3.19). Of particular interest was the deposition of complete horse skulls (most associated with mandibles) at regular intervals at the bottom of the defensive ditch before it was infilled in the late 9th century (Figure 3.25). All these horses were male and ranged between five and 15 years of age (see Sablin Chapter 3). Bit-wear damage to the second premolars indicated that many of them had been used for riding. These may have been horses acquired by the Prince of Novgorod's armed forces. There is no sign of wounds on any of the skulls and these may have been horses in their prime that were selected for sacrifice. Although less fragmented and less frequently butchered than cattle (Figure 5.3), 5% of the horse bones from Gorodishche bore processing marks, including those made during dismemberment and filleting, indicating some consumption of horsemeat.

The Troitsky horse assemblages included higher percentages of lower limb bones but lower percentages of cranial elements than other domestic species (Figures 4.50–4.51). The high percentages of tarsals and carpals suggest that they were introduced to the properties whilst attached to the radii and metatarsals that were required for bone-working (Maltby 2016a).

Although the levels of fragmentation and the relative frequency of butchery marks on horse bones was much lower than for cattle, over 16% of the horse bones from the Troitsky sites bore processing marks (Figure 5.3). Some of these were skinning and bone-working marks but others were the consequence of dismemberment and filleting and included marks very similar to those observed on cattle (Tables 5.36–5.58). Some horse carcasses were butchered for meat in Novgorod despite Christian decrees forbidding its consumption (Bartosiewicz 2003, 117). The Novgorod Chronicles record occasions when horseflesh was consumed in times of crisis, for example in AD 1128 and AD 1230 (Michell and Forbes 1914, 11, 80). Horse hides were also exploited for leather. Several phalanges bore cut marks indicative of initial skinning.

Horse mortality patterns

Nearly all the ageable horse bones from Gorodishche and the Novgorod sites came from adult animals (Table 4.61), indicating that the horses had been required for other duties before they died. Horses were mainly valued for riding and as beasts of burden including pulling ploughs, carts and sleighs. Horse vertebrae with ankylosing spondylitis were found in Gorodishche (Figure 11.3) and Desyatinny-1. This condition can occur in animals that have been frequently ridden or have borne heavy loads. Their military importance should also not be underestimated. The horse heads deposited in the Gorodishche ditch may have been from cavalry. Birch-bark documents allude to horses much more frequently than to any other species (Rybina 2001), reflecting their high status and economic value. Various documents reveal occasions when horses

Figure 11.3: Pathologically fused horse lumbar vertebrae from Gorodishche. Photograph by Ellen Hambleton

were bought and sold, ordered for military and agricultural activities, exchanged for debts and stolen. Plant macrofossil and insect analyses have shown that horses were stabled in various Novgorod properties (Monk and Johnston 2012; Reilly 2012).

The sizes of horses

The complete horse limb bones from excavations at Gorodishche between 1979 and 1998 had average estimated withers heights of 139 cm, the size of large ponies (Figure 6.53) but these were c. 6 cm larger than the average size of those recorded on later medieval sites in Russia (Tsalkin 1956). The complete metapodials from the 2000–2004 excavations in Gorodishche were from even larger horse with average heights of 144 cm. The horse skulls from the ditch were estimated to be of similar size (see Sablin in Chapter 6). The large size of the horses from Gorodishche may reflect the high status of the settlement and the presence of horses associated with the military forces.

The sizes of horses from Novgorod were more typical of those found on other Russian medieval sites with estimates of withers heights from 93 complete limb bones from the Troitsky sites averaging 133.4 cm with a range of between 117 cm and 153 cm (Figure 6.53).

Dogs and cats

Abundance

Although no complete skeletons were recovered, several associated groups of dog bones were recovered from Gorodishche and Troitsky, indicating that dog carcasses were rarely processed. Even including associated groups, dogs contributed only 1% of the domestic mammal bones from Gorodishche and cats were even less common at 0.3%. Cats were not identified at Prost and Georgii and dog was only recorded at Georgii (Figure 3.22). Dogs were also poorly represented on the Troitsky sites contributing only 1% of the domestic mammal counts. Cats were slightly better represented than dogs, forming around 2% of these counts including some in associated groups (Figure 4.3). Dogs provided 3% of the domestic mammals at Desyatinny-1. The bones again included some from partial skeletons. Cats provided less than 1% of the total (Figure 4.53). It is to be expected that dogs and cats would not be commonly found amongst food waste. However, the presence of cats and particularly dogs is evidenced by the high frequency of gnawed bones in the assemblages from both settlements, demonstrating that they had easy access to food waste. During the famine of AD 1230, dogs were reported as scavenging on the corpses of humans as well (Michell and Forbes 1914, 77).

Dog and cat element representation and carcass processing

The presence of partial skeletons accounted for a greater proportion of ribs and vertebrae in the dog assemblage from Gorodishche (Figure 3.20). Upper limb bones predominated in both the dog and cat assemblages from the Troitsky sites where the small bones of the feet in particular were poorly represented in the hand-collected assemblages (Figure 4.52).

Butchery marks were observed on two dog bones from Gorodishche and on 5% of the dog bones from the Troitsky sites. These included filleting as well as skinning marks. Similarly, several cat bones from the Troitsky sites have evidence for skinning and two bore filleting marks. The Novgorod Chronicles reported that dogs and cats (along with horses) were eaten during a famine in AD 1230 (Michell and Forbes 1914, 67; Riha 1970). By highlighting this, one assumes that the consumption of these species was not normal practice. Dogs would have had a wide range of uses including shepherding, guarding and hunting. A late 14th-century birch-bark document (BBD 135) refers to a hound (Rybina 2001, 129). Cats would have been useful in the control of vermin within Novgorod as well as being companion animals.

Dog and cat mortality data

Substantial numbers of unfused bones from both the Troitsky and Desyatinny sites indicate that many of the cats in Novgorod failed to attain adulthood. Some may have been culled to control the stray cat population and some of these and other cats have been culled immature for skinning. Similarly nearly a third of the dog limb bones from the Troitsky sites had unfused epiphyses (Table 4.67) indicating that there was quite a high rate of immature mortality. Again this may indicate measures of population control. However, nearly all the bones from Desyatinny were from adult dogs.

The sizes of dogs and cats

The sizes of dogs found on the Troitsky sites ranged from animals as small as a beagle (shoulder height 33 cm) to as large as a German shepherd (69 cm). Most dogs, however, were of medium size with shoulder heights that ranged between 40 cm and 50 cm (Figures 6.55 and Figure 11.4). Metrical analysis from the Desyatinny site also indicated

Figure 11.4: *Dog skulls from Troitsky illustrating the wide range in the sizes of dogs. Photograph by Sheila Hamilton-Dyer*

that most of the dogs were of medium size. There were a few smaller and larger dogs, which could indicate the presence of specialised breeds (Zinoviev 2013).

Wild mammals

Abundance

There is little evidence for extensive exploitation of wild mammals in the hinterland of Novgorod during the 9th and 10th centuries. Despite extensive sieving, bones of nine species provided only 2% of the mammal sample at Gorodishche. Hare were the most commonly identified, followed by beaver and squirrel. Elk, wolf, bear, stoat, marten and vole were also identified (Figure 7.1). Wild mammal bones were a little more abundant in the smaller assemblage from Georgii. Most of these belonged to elk, although fur-bearing species (beaver, squirrel, bear and lynx) were also recorded (Figure 7.1). The high percentage (22%) of wild species in the Prost assemblage was due to the discovery of an associated group of foot bones of pine martens (Figure 7.5).

Wild mammals formed 1% or less of the total identified mammal species recovered by hand from the Troitsky excavations, and only 3% of the sieved sample (Figure 7.7). Comparably low percentages were obtained from assemblages from the Desyatinny-1 (0.4%) (Table 7.25), Nerevsky (1.0%) and Slavensky sites (0.2%) (Maltby 2013).

In stark contrast excavations at Minino have produced an assemblage in which wild species provided nearly 65% of the identified mammal remains with fur-bearing species being prominent (Figures 8.2–8.3). Therefore, the significance of Novgorod's role in the international fur trade is best reflected in the composition of the animal bone assemblage at Minino, located in one of the major hunting areas, rather than in the town itself. The trade was in furs and pelts, not carcasses. Animal bones found in Novgorod principally represent local consumption practices and are unlikely to produce evidence for the import of pelts, particularly as many of those pelts were destined for subsequent export.

Elk

Elk meat appears to have been consumed only occasionally at Gorodishche and Georgii (Figure 7.1). Elk was probably the most important of the wild mammals in providing meat in Novgorod, although they only provided less than 1% of the identified mammal elements from any of the sites. Butchery marks were observed on a third of the elk elements. Good and poor meat bones are both present and it is possible that sometimes complete elk carcasses were brought to the town for processing. However foot bones are well represented (Figure 7.4) and some of these may have been brought to Novgorod whilst still attached to the hides. A 12th-century birch-bark document (BBD 153) requested the purchase of an elk skin from a currier (Rybina 2012, 451).

There was very limited evidence for the use of elk antlers in any of the assemblages examined in this project, as antler objects were previously removed for artefact studies. A worked offcut of elk antler was recorded at Gorodishche and another was found in the Troitsky assemblage. Split elk skulls found in Gorodishche and Troitsky had evidence that the antlers had been sawn off (Figure 7.3). However, artefact studies have shown that elk antler was commercially important being commonly used in the manufacture of combs in particular (Smirnova 2001). Worked antler offcuts have been found in abundance within some properties indicating specialist production. As in the case of goat horns, it is likely that large numbers of antlers were brought into the town. Only about half of over 300 single-sided antler combs examined by Smirnova (2005, 9–15) were made from elk antlers, whereas 44% came from red deer, a species that was not local to Novgorod, indicating that the import of antler combs was common.

Elk provided over 4% of the medieval mammal assemblage in the Minino assemblage, with their abundance increasing in the later levels (Figures 8.3–8.4), showing they formed a higher proportion of the meat supply than in Novgorod.

Hare

Hares were the most common wild mammal represented at Gorodishche (Figure 7.1), although they contributed only 1% of the 1979–1998 assemblage. They were also the most common wild species represented in the 9th and 10th-century assemblage from the 2000–2004 excavations, although they became less common in the 11th–12th-century layers (Table 7.7). Hares were also found in small numbers at Georgii.

Hare bones were also the most common of the wild mammals represented on the Troitsky sites, particularly in the later levels (Figure 7.7). However, they contributed only 0.3% of the identified mammal assemblage. Many of their bones, however, may have been overlooked during normal retrieval. Cut marks made during dismemberment and filleting were observed on 9% of their bones.

Hare provided only 1% of the mammal bones in the medieval assemblage from Minino (Figure 8.3). Hares seem to have been only a minor supplement to the meat diet on all the sites investigated.

Beaver

There is abundant evidence from documentary sources that beavers were captured in vast numbers in Russia for their pelts in medieval Russia (Martin 1986). Their bones have provided substantial proportions of zooarchaeological assemblages from some villages and towns in the forest zone of Eastern Europe, particularly those of 10th–12th-century date (Makarov 2012b).

Most of the evidence for beaver exploitation in Novgorod itself comes from the birch-bark documents. Several are concerned with the acquisition of beaver skins via tribute or purchase. For example, BBD 420, dated to the mid-13th century, mentions the sale of 40 beaver pelts (Rybina 2001, 129).

The commercial importance of beavers is not reflected in the zooarchaeological assemblages from Novgorod. Bones of beaver provided only 0.3% of the

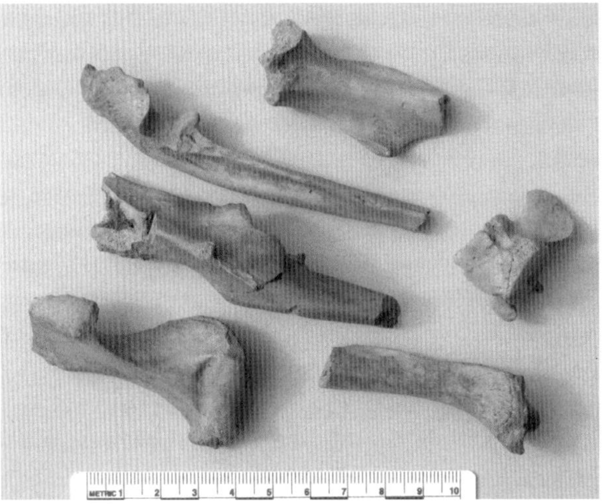

Figure 11.5: Selection of beaver bones butchered for meat from Novgorod. Photograph by Sheila Hamilton-Dyer

Troitsky mammal assemblages (Table 7.9) and were even rarer in the Desyatinny-1 assemblage (Table 7.26). Bones from all parts of the body were represented indicating that whole carcasses were processed in the town. Foot bones were under-represented (Figure 7.10) and were possibly removed with the skins, although retrieval biases may also have been a factor. Skinning marks were recorded on a few skull and mandible fragments but most of the butchery marks were associated with dismemberment and filleting (Figure 11.5). The zooarchaeological evidence therefore indicates that beaver meat was occasionally eaten in the town. There is also evidence that beavers were butchered for meat at Gorodishche.

In contrast, beavers were the best represented species in the Minino assemblage providing 35% of the mammal assemblage (Figure 8.3). The assemblage again included many good meat-bearing bones. Animals hunted for their pelts therefore also provided significant amounts of meat for the local community. A number of other settlements in the Beloozero and Vologda regions have also produced assemblages containing substantial percentages of beaver bones (Makarov 2009a).

However the percentage of beaver declined significantly to 22% of the identified mammals in the 12th–13th-century deposits at Minino (Figure 8.5; Table 8.2). Possible over-exploitation was indicated by changes in their mortality patterns (Figure 8.6). The decline of beavers at Minino coincided with the decline in beavers in the Novgorod assemblages, where their bones are largely absent from the Troitsky XI spits dated to the 13th century and later (Figure 7.8). A similar decline has been noted on the Nerevsky site (Tsalkin 1956; Maltby 2013). References to beavers in birch-bark documents also largely disappear after the early 13th century (Rybina 2001). Over-exploitation and woodland clearance for farming are both likely to have been factors in their decline. This was the beginning of a long period of decline in beavers in European Russia from where they had largely disappeared by the 19th century because of over-exploitation.

Other wild mammals

At Minino, in addition to beaver, large numbers of bones of squirrel and marten were recovered along with a small number of bones of other fur-bearing species such as otter, polecat, fox and bear (Figure 8.5). A number of other settlements in the

Beloozero and Vologda regions have also produced assemblages containing substantial percentages of bones of fur-bearing species (Makarov 2009). Hunting for fur-bearing species was an important aspect of the inhabitants of these settlements, although they were also concerned with farming and other economic activities.

The pivotal role of Novgorod in the international fur trade during the medieval period is well known through documentary sources (Martin 1986; Makarov 2006; 2009a; 2009b; 2012b). Many thousands of pelts were collected every year from the forest zones of northern Russia with squirrels being particularly important. There were frequent references to squirrel pelts in birch-bark documents (Rybina 2001; 2007). A recently discovered example dating to the late 12th century (BBD 1021) speaks of a transaction by a manager of a boyar involving approximately 3,100 squirrel furs (Schaeken 2017).

Evidence for hunting fur-bearing animals has also been found amongst the artefact assemblages. In particular blunt arrowheads were made from a variety of materials including wood, iron and osseous materials. These were designed to stun their prey without damaging the skins. Over a hundred of these have been found in Novgorod and they have also been discovered at Gorodishche (Rybina 2007; Makarov 2012b). Several were also found at Minino including three in male graves, showing that the importance of hunting was commemorated in funerary rites (Makarov 2009a; 2012b).

However, the vast importance of fur-bearing species in the regional economy has not been reflected in the animal bone assemblages from Novgorod or in slightly earlier settlements in its immediate hinterland. Only three squirrel bones were recovered during normal excavations on the Troitsky sites. Sieving produced another bone and it is possible that further sieving would produce more bones of squirrels and other small fur-bearing species such as pine marten, otter, polecat, stoat, fox, lynx, bear, wolf and sable, which have only been found, if at all, in very small numbers in excavations of the town (Maltby 2013). Similarly only 11 squirrel bones and one each of stoat and marten were recovered from the 1993–1998 excavations at Gorodishche despite an extensive sieving programme. Squirrels were absent from the 2000–2004 assemblage, but two bones of marten and 11 from fox were recovered (Table 7.7). As noted above, Prost produced the best evidence for the processing of pelts, at least on a small scale, in the accumulation of foot bones from a minimum of four pine martens (Figure 7.5). The small assemblage from Georgii produced bones of beaver, lynx, squirrel and a bear metacarpal with clear skinning marks (Figure 7.1). It would appear that inhabitants of settlements in the Novgorod hinterland were involved in acquiring pelts in the 9th and 10th centuries, albeit on a much smaller scale than in the Minino region.

Nine third phalanges were the only bear bones identified in the Troitsky assemblage, two of which have been cut. Some of these could have been amulets (one of the phalanges was perforated), but it is probable that others were attached to bearskins acquired by the inhabitants of the town. Several birch-bark documents indicate that bearskins were traded items (Rybina 2001).

A few bones of wild boar were also recorded on the Troitsky sites based on their exceptionally large size. It is also possible that some of the pig bones from the Desyatinny site belonged to wild boar. However, it seems unlikely that wild boar was a common resource in Novgorod. Bones assigned to wild boar also formed less than 1% of the mammal bones from Minino (Figure 8.3).

Although combs made from red deer and reindeer antlers have been identified in Novgorod (Smirnova 2005), no bones of these species were catalogued in these deposits apart from three elements of reindeer from the Troitsky sites (Figure 7.8). Reindeer was also represented in small numbers at Minino (Figure 8.3).

Prost was the only site in this survey to produce a bone of a wild large bovid. This possibly belonged to an aurochs (*Bos primigenius*), a species that may not have become extinct in Russia until the 13th century (van Vuure 2005). Perhaps more likely is that it belonged to a bison (*Bison bonasus*), which can still be found in very small numbers in Eastern Europe.

The Minino assemblage included several species of wild mammals not found in the sites in and around Novgorod. These included roe deer, badger, polecat, wolverine and otter (Figures 8.3 and 8.5). None of these were represented by large numbers of bones but they include further species probably captured principally for their skins and fur. These included a skeleton of a male polecat placed in a burial of a woman (Figure 8.1).

The discovery of the macaque skull from a late 12th-century deposit in Gorodishche indicates the import of an exotic North African species to this high status site (Figure 7.6). The macaque may have been a gift to the Prince or part of a menagerie (Brisbane *et al.* 2007b).

Comparisons with several assemblages from other Russian and eastern European towns have shown that variations in the presence and relative abundance of different wild mammal species is dependent upon a variety of factors including climate, vegetation and the social status of the settlement. For example, bones of seals occur only on the most northerly sites such as Staraya Ladoga, Turku, Torino and Eketorp (Tsalkin 1956, Tourenen 2008, Puputti 2008, Boessneck *et al.* 1979), whereas they have not been found in Novgorod. Beaver bones have been found in most Russian towns (Tsalkin 1956; Lantseva and Lapshin 2001) although they are notably absent from the Romanov Dvor site in Moscow, where most of the deposits are of post-medieval date (Glazunova *et al.* 2009), perhaps again indicating the decline in beaver population through over-exploitation.

Birds (Mark Maltby and Sheila Hamilton-Dyer)

Detailed analysis of the bird bones can be found in Chapter 9. The exploitation of birds for food on the hinterland sites in the 9th and 10th centuries was much less important than the meat obtained from domestic mammals and fishing. The inhabitants of Gorodishche focused their exploitation on the capture of waterfowl, particularly ducks and geese, which would have been in plentiful supply in the neighbouring lake and river particularly in the summer (Figures 9.1 and 9.5). The same was probably the case for the other hinterland settlements as well, although their assemblages were tiny. Although it is feasible that some domesticated ducks and geese were kept, domestic fowl was the only species for which there was convincing evidence that birds were kept in captivity. Large galliforms, particularly the capercaillie, were also hunted for food providing the inhabitants of Gorodishche with occasional meals. The wings of eagles and possibly other raptors may have been imported for the exploitation of feathers.

Although the assemblage from Novgorod was much larger, the retrieval of bird bones was severely hampered by the excavation methods. This undoubtedly created biases with smaller species likely to be significantly under-represented. Retrieval bias was also evident in the relative abundance of different elements collected. Nevertheless the analysis of the assemblage produced some important insights into how the residents of Novgorod exploited birds. As at Gorodishche, domestic fowl was the only species of poultry that occurred in large numbers (Figure 9.6). The sizes of the ducks and geese were no larger than wild mallard and greylag geese respectively. Therefore there was no conclusive evidence that any of the ducks or geese were kept in captivity, although domestic ducks and geese are often no larger than their wild counterparts (Poland 2018). Genetic analysis has also suggested that domestic geese were established in some parts of Russia during the early medieval period (Honka *et al.* 2018) and one suspects that domestic geese were kept in Novgorod as well.

Ageing evidence indicated that some chickens were kept in the town and the presence of medullary bone in many of the leg bones strongly suggested that many hens were kept for egg production (Figures 9.17–9.18). There is increasing evidence to suggest that domestic fowl became more important in the later medieval period, at least in relation to other birds, but this needs to be assessed on a wider range of assemblages from different areas of the town. Metrical analysis confirmed that adult hens outnumbered males, many of the latter being killed as young birds (Figure 9.19). Most domestic fowl were of small stature. Although there was no clear evidence for diachronic changes in their sizes, there were some indications that some larger birds were present in small numbers (Figures 9.19–9.21).

Ducks continued to feature prominently in the Troitsky assemblages with particularly high levels encountered in the lower spits of Troitsky IX–XI (Figure 9.6). The residents of that area of Novgorod clearly took advantage of the large numbers of ducks available in the hinterland. As at Gorodishche, several species of duck were brought into the town. Most of the bones belonged to large ducks, the vast majority of which were probably mallards, a very common resident in the vicinity today. Most of the smallest duck bones were a good match for teal. These were also well represented particularly in sieved samples. Several other species of duck were also exploited in smaller numbers. These are likely to have included garganey, wigeon, goldeneye and tufted duck (Figure 11.6). These are all migratory species and are still commonly hunted in the Novgorod region, particularly in Spring and Autumn. Other wetland birds captured for food included at least two species of geese, swan, crane, stork, heron, grebe, cormorant and several species of wader, including woodcock and snipe. Large game birds,

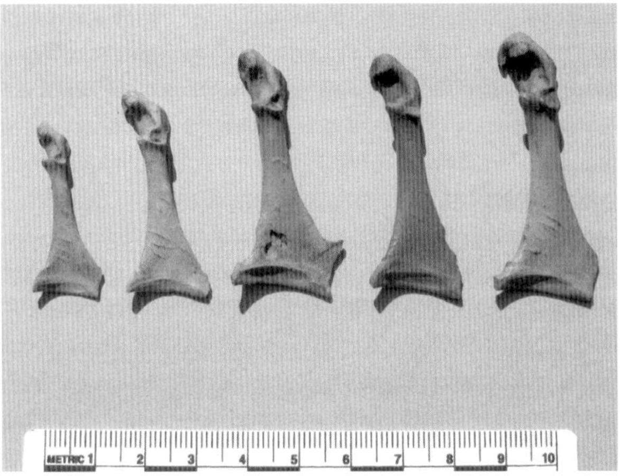

Figure 11.6: *Duck coracoids from Troitsky sites. The variation in size shows that several species were exploited. Photograph by Sheila Hamilton-Dyer*

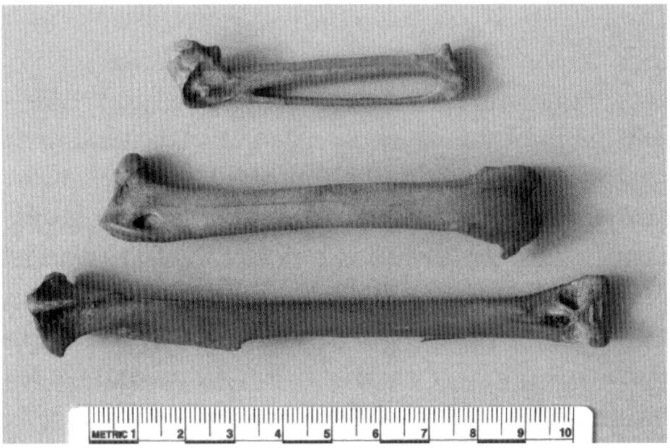

Figure 11.7: *Female goshawk bones from Troitsky sites. Photograph by Sheila Hamilton-Dyer*

particularly capercaillie, were also hunted. Bones of smaller game birds, such as hazel grouse and partridge, were also present in small numbers.

Several species of raptors were identified. These included eagle, goshawk, sparrowhawk, harrier, kestrel, hobby and merlin. Of these, goshawk and sparrowhawk were the most frequent and included some partial skeletons. Most of these remains were probably from birds kept for falconry. This is supported by the bias towards larger females in the goshawk and sparrowhawk assemblages (Figure 11.7). Jesses associated with keeping of raptors and other birds in captivity have been discovered in Novgorod (Figure 11.8) and Tver (Hamilton-Dyer 2002; Zinoviev 2017). Eagles were not commonly kept by falconers and the presence of butchery marks on some of the wing bones indicated that their feathers may have been targeted.

Some taxa such as gulls, buzzards, ravens, smaller corvids, owls, pigeons and small passerines could have been resident in or close to the town, taking advantage of the food available amongst urban refuse. However, some of them could have been caught during falconry and butchery marks indicated that the carcasses of some of the corvids at least had been exploited for meat.

Overall, species representation in most of the Novgorod assemblages was slightly more varied than those from Gorodishche indicating perhaps greater diversity in the diets of the boyar families and greater interest in falconry. The latter may also be reflected by the increase in the proportion of raptors in the 11th–12th-century assemblage from Gorodishche.

The bird assemblage from Gorodishche and Novgorod are unusual in several respects, particularly in relation to the high frequency of ducks. However, they

Figure 11.8: Buzzard tarsometatarsus with leather jess from the Nerevsky excavations indicating this bird was kept in captivity. Photograph by Sheila Hamilton-Dyer

are not unique. For example, a small settlement at Struga-2 in eastern Belarus produced an avian assemblage of 209 bird bones of 10th–13th-century date, 87% of which were identified as ducks including 60% which were of mallard-size. Chickens provided only 9% of the assemblage (Zelenkov 2008; Gorobets and Kovalchuk 2017). In assemblages from recent excavations at Staraya Ladoga, which date mainly between the 8th and 10th centuries, bones of mallard-sized ducks slightly outnumbered domestic fowl, and smaller species of duck were also well represented. However, the assemblage differs from those at Gorodishche and Novgorod by having much larger percentages of capercaillie and grouse (Shaymuratova *et al.* 2019).

Most avian assemblages from medieval Europe have been dominated by bones of domestic fowl and geese. The species represented in Novgorod were similar to those found in the post-medieval deposits of the Romanov Dvor excavations in Moscow (Glazunova *et al.* 2009). However, there, bones of geese (mainly domestic) outnumbered (26%) both domestic fowl (18%) and mallard-sized ducks (8%). Smaller duck species and other waterfowl were poorly represented but large landfowl formed a more substantial portion of the assemblage than in Novgorod with black grouse (20%) and capercaillie (11%), being particularly well represented.

Domestic fowl bones have formed over half the total bird bones identified from medieval sites in Estonia and Latvia (Maltby *et al.* 2019). They are also dominant in many Polish urban assemblages including Gdansk in Poland (Makowiecki and Gotfredson 2002; Makowiecki *et al.* 2019).

Fish (Mark Maltby and Sheila Hamilton-Dyer)

A detailed analysis of the fish bones found during the project in the town of Novgorod itself, the hinterland sites and Minino has been provided in Chapter 10. This section will summarise that information but will also draw on other forms of evidence.

Extensive sieving at Gorodishche and Prost demonstrated that fish were already an important commodity in the Novgorod hinterland in the early medieval period. Fish bones provided over 32% of all the bones recorded from Gorodishche. The species represented were largely the same as those recovered from later deposits in Novgorod itself. Cyprinids were the most common, followed by zander and pike (Figure 10.1). Perch was poorly represented. The same species were found at Georgii and Prost, with most of the identified elements belonging to cyprinids. Zander, cyprinids and pike were also present in the small sample from Vasilievskoye (Table 10.26). The local fish stocks were clearly heavily exploited by the 9th century and continued to be an important resource thereafter. A few bones of whitefish, catfish and sturgeon were found at Gorodishche. The presence of catfish and whitefish in particular may indicate that the acquisition through trade of prized fish may have begun prior to the foundation of Novgorod.

There can be little doubt that fish continued to be an important element in Novgorodians diets after the foundation of the town. Unfortunately, traditional excavations methods in Novgorod itself have not been conducive to the recovery of fish bones. They provided only 2% of the total elements retrieved by hand from the three Troitsky sites. The contrast with the make-up of the sieved assemblages is stark. Fish elements contributed 75% of the total fragments (Maltby and Hamilton-Dyer 2001). It is unsurprising that the hand-collected sample was dominated by larger bones of large fish such as pike and zander (Figure 10.2). Similar biases can be seen in the assemblages of hand-collected material recovered from the Nerevsky and Desyatinny sites (Table 10.13).

The fish species identified in the more reliable sieved samples from the Troitsky sites are again dominated by cyprinids. Most probably belonged to bream but here, and at Gorodishche, bones of roach, dace, chub, ide and silver bream were also recorded. Additional cyprinid species were identified in the Nerevsky assemblage (Sychevskaya 1965). Zander and pike of various sizes were also quite commonly represented. Bones of perch and occasionally ruffe were also found (Figure 10.2).

Because of these abundant local resources and Novgorod's distance from the sea it is not surprising that marine fish are absent from these Novgorod assemblages; imported herring apparently did not appear in Novgorod until the early 15th century (Smith 1977, 63). It is highly likely that a higher proportion of fish protein figured in the diet than is shown by the recovered remains; in the 1970s the annual commercial catch from Lake Ilmen was about 2,000–3,000 tons (Berka 1990). Much of the catch would have been eaten fresh but salt-dried and smoked fish would have been laid in for the winter. Smoke houses can still be seen around the shores of Lake Ilmen today. Fish roe, other than the famous sturgeon caviar, would also have been preserved for later use and leaves no trace in the bone record.

It is interesting to compare the evidence of fishing drawn from the zooarchaeological evidence with the information that can be gleaned from the birch-bark documents. Fish are the most frequent of animal foodstuffs mentioned in the documents but the species are not always recorded. The prestigious sturgeon is listed as a tribute payment in BBD 259, a late 14th-century document from the Nerevsky site (Rybina 2001,

129). By the time of Novgorod's foundation the sturgeon had already become uncommon in the Baltic and Lake Ladoga region (Kolman *et al.* 2011). Very few sturgeon remains have been found in Novgorod even amongst the hand-collected material (Figures 10.2 and 11.9).

Figure 11.9: *Two of the very few sturgeon bones found during the project. Photograph by Sheila Hamilton-Dyer*

Named species that occur most frequently in the birch-bark documents are the salmonids (whitefish, salmon and salmon-trout) (Rybina 2001; Brisbane and Maltby 2002). Whitefish are mentioned in at least three documents. The earliest (BBD 831) is from the Troitsky excavations, and is dated to the 12th century. It lists ten whitefish amongst other items of tribute. A 14th-century document (BBD 144) seeks the acquisition of 50 whitefish. Another 14th-century document (BBD 280) lists five whitefish and, according to some translations, five salmon-trout (Rybina 2001, 129). Salmon and whitefish are both listed in a 13th-century document (BBD 147). BBD 169 (late 14th century) requests ten salmon from two people. Rare and special fish such as sturgeon, salmon and whitefish are likely to have been reserved for tribute payments and hence they appear more frequently in the documents than in the archaeological assemblages. Indeed bones of imported fish have been found very rarely in the Novgorod assemblages. In addition to the few bones of sturgeon, the eel bones were probably from preserved fish, possibly imported from the Baltic. Some of the large wels catfish could also have been imported. However, fish supplies came overwhelmingly from local sources. Later writings suggest that most people in north-west Russia consumed the same fish as found in the excavations; bream, zander, pike and perch (Smith 1977), and these are still the most frequently available in the markets in Novgorod today.

However, despite the importance of local fish, they appear infrequently in birch-bark documents. Only one (BBD 169) mentions pike. Bream are also listed in this early 15th-century document. One hundred bream also feature in BBD 964, a late 14th-century document found on a site inside the Kremlin (Dekker 2018, 118). References to preserved fish are found in several documents. For example, BBD 258 (late 14th century) lists a total of 16 dried and three lightly-salted salmon. Smoked and salted fish would have been important commodities to store for winter consumption.

Evidence for fishing in Novgorod can be found in the abundant finds of artefacts such as fishing rods, nets, floats, basket traps, weights and fish-hooks (Rybina 2007, 124–130). Many fish would have been caught in nets from rafts and boats, and these too have been found in Novgorod (Dubrovin 2007a).

The analysis of fish bones from selected medieval contexts at Minino revealed that the remains were dominated by perch and, to a lesser extent, cyprinids whilst pike were moderately represented (Figure 10.4). Concentrations of scales of cyprinids, pike and perch in some pits show that large numbers of locally-caught fish were sometimes processed at the same time, probably for preservation and storage. Only one bone of zander was identified and this almost certainly was imported. The few sturgeon represented are also likely to have been imported. Pike, perch and cyprinids were deposited in some of the graves at Minino providing a further insight into the importance of fishing in the community.

Further afield, fish remains from archaeological excavations in the area around Pskov list 19 species (Smith 1977), of which sturgeon and salmon were considered to be imports. The main species were pike, perch and bream. Other cyprinids, zander, wels catfish and whitefish are also listed as at Novgorod. There are minor differences – burbot has been found in Pskov while Novgorod has ruffe. The latter is a small species and, as most sites have not employed sieving, the bones of this and other small species may not have been recovered. The burbot is a minor component of the Lake Ilmen ichthyofauna today (Savenskova *et al.* 2010). A multi-disciplinary survey by Salmina (2016) has shed further light on the importance of fishing in the Pskov region in the 16th century and has shown how documentary sources and archaeological fieldwork can be used in conjunction with zooarchaeological evidence to gain a deep understanding of fishing practices.

In medieval Tallin, Estonia, excavations have found a mixture of freshwater and marine fish, including herring, flounder and cod, again showing how fishing varied according to local supplies and also through access to the Baltic fish-trading network (Lõugas *et al.* 2016).

Similarly, in recent excavations in medieval Riga, Latvia, bones of haddock (*Melanogrammus aeglefinus)*, cod (*Gadus morhua*), perch, zander, salmon, pike and sturgeon were identified, indicating access to marine and migratory species as well as freshwater stocks. In contrast the fish bones from the inland site at Karski Castle, Latvia, belonged mainly to cyprinids, perch and pike caught locally and it was not until the later medieval period that these supplies were supplemented by imports of cod and possibly zander (Maltby *et al.* 2019).

From a slightly later period, the fish processed in a 16th-century merchant's house in Helsinki, Finland, were mainly perch, herring, pike, fourhorn sculpin and cyprinids, again showing a mix of freshwater and marine resources (Mannermaa 2016).

These and other analyses of fish remains from medieval sites in north-eastern Europe have shown that most of the fish on all the settlements came from local stocks, and that inter-settlement variations largely reflect which species were abundant nearby. As elsewhere, the inhabitants of Novgorod, Gorodishche and Minino relied principally on their local sources with only very occasional supplements of fish imported from further afield.

EVALUATION OF THE PROJECT AND ITS RESULTS

Sampling and the assemblages

It must be borne in mind that the sampling strategies formulated were governed largely by pragmatic considerations of time, resources and available expertise and assistance, which, as outlined in Chapter 2, varied from year to year. The main disappointment is that it has not been possible as yet to devise comprehensive sampling strategies for the Novgorod sites. Although we have been able to demonstrate that a selective sieving programme would greatly enhance the recovery of bones, particularly those of fish, we have not been able to influence the general excavation methods, which due to a variety of internal and external pressures have continued to be carried out in their traditional manner. To be fair, a comprehensive sampling programme would involve a very substantial investment in facilities, personnel and time, which have not as yet become available. The backlog of material that could accumulate prior to processing would create a major logistical problem.

We did, however, thanks to Peter Gaidukov's co-operation, acquire an almost complete sequence of animal bones recovered by hand from the Troitsky XI site. This allowed us the opportunity to compare assemblages from between the late 10th and early 15th centuries on one site. This has enabled general comparisons of species abundance, element representation, butchery evidence, ageing and metrical data to be made throughout the sequence. We remain sceptical about the policy of digging in spits. Although we accept that generally deeper deposits are older, the subtleties of the stratigraphic interpretation were not within our control. We were able to separate bones from two different properties within this sequence and made some basic comparisons between the assemblages found within them through time. However, it was not possible to carry out more detailed intra-site studies within these properties. It has not been possible, for example, to compare assemblages from within various buildings with those from outdoor contexts. Such comparisons could potentially provide further insights into how space was used within the properties and where refuse was disposed.

The control of sampling on other Troitsky sites was even more limited. However, we did obtain hand-collected samples from the lower spits of trenches IX and X, which have been compared with similar samples from Troitsky XI.

The methods of recording were modified slightly during different seasons, although essentially they remained the same as those described in Chapter 2 for the 1993 season. The summary recording allowed large samples to be scanned. By the end of the project, nearly 58,000 mammal bone fragments (NISP) had been counted from the Troitsky sites, including over 37,500 identified to species. Detailed records of nearly 4,500 bird bones were also made. This is a large sample by any standards. Other statistics are as impressive. Gnawing damage, for example, was noted on over 10,000 elements. Nearly 1,750 mandibles of cattle, sheep/goat and pig have provided tooth eruption data and have been recorded using Grant's (1982) methodology. Epiphyseal fusion data were recorded on over 7,450 mammal and 1,950 bird bones. Metrical data were recorded on over 3,400 mammal bones. More than a single measurement was available on many of these. Over 5,200 separate measurements were taken on nearly 2,200 cattle bones for example. Butchery marks were noted on over 8,400 mammal bones from Novgorod and detailed butchery records were made on over 1,600 mammal bones from some of the lower spits of Troitsky XI. Therefore, although the sampling methods and contextual control prevented the detailed analysis of intra-site variability and limited the amount of information that could be gained from the analysis of species abundance and element representation, we obtained a very substantial dataset that enabled us to investigate mortality profiles, butchery methods and the stature of the animals exploited in this area of the medieval town. Together with the bones from the Desyatinny-1 site examined by Andrei Zinoviev, this is the first substantial collection of animal bones analysed from Novgorod since Tsalkin's (1956) work over 60 years ago and some aspects such as gnawing and butchery have been recorded for the first time. This therefore is the first zooarchaeological, as opposed to purely zoological study, of animal bones from the medieval town.

The Gorodishche samples have been fully recorded and have produced a sample that is less well preserved than the Novgorod assemblage but much more carefully retrieved with greater contextual integrity. It has been possible to carry out intra-site variability studies and the surviving samples provide a much more balanced picture of the relative importance of bird, fish and mammals than at Novgorod. The excavations from 1979 and the 1990s provided a substantial assemblage of nearly 17,000 bones dated to the 9th and 10th centuries, of which over 8,100 mammal, bird and fish bones were identified. It is a pity that many of the later medieval deposits at Gorodishche were too admixed to provide useful assemblages contemporary with some of the larger deposits from Novgorod. However, some of the 11,800 mammal and bird bones (including nearly 5,000 identified specimens) from the 2000–2004 excavations analysed by Mikhail Sablin do date from the 11th and 12th centuries and are contemporary with the early phases at Novgorod.

The sample from Minino provided an opportunity to compare the assemblages from Novgorod and its hinterland with a site located deep in the interior of the forest zone at the edge of the Novgorod territory. Although recording and quantification methods for the mammal bones by Olga Krylovich and Arkady Savinetsky were not identical, they were close enough for valid comparisons and contrasts to be made with the assemblages from Novgorod and the hinterland sites. It is unfortunate that not enough time was available for more fish bones to be recorded by Sheila Hamilton-Dyer but her methods of identification and recording were identical to those carried out at Gorodishche. Ideally, it would have been useful to have acquired more assemblages from other non-urban and/or low status sites contemporary with and much closer to Novgorod. However, no material from excavations of such sites were available at the time.

The results

We believe that the zooarchaeological studies carried out for this project have significantly enhanced our understanding of the residents of Veliky Novgorod and its territory. Given a more consistent retrieval policy, the potential to further these preliminary findings, in Novgorod in particular, is enormous.

However, we have already established that there were significant variations in the exploitation of animals within the region and between different periods. For example, it seems that the inhabitants of the later medieval properties of Troitsky XI, whilst continuing to eat a lot of beef and pork, consumed a little more mutton, goat meat and chicken but fewer ducks than their forebears. They continued, and possibly increased, their use of dairy products and eggs. Wild mammals such as elk and hare continued to provide occasional supplements to the diet but the consumption of beaver hams discontinued after beaver populations declined through over-exploitation for their pelts. However, Novgorodians continued not to be averse to eating horsemeat and occasionally even ate dogs in times of stress. The inhabitants of Novgorod kept horses, dogs, cats and chickens in the town and it is probable that some cattle, pigs and goats were resident in the town as well. Pigs were kept within the earlier settlement at Gorodishche. Many of the domestic animals consumed in the Troitsky properties appear to have been brought there for slaughter, or at least a lot of the carcass processing took place within them. However, there is some evidence that joints of meat, particularly of sheep and pigs were also acquired particularly in the later period. Exactly how important fishes were in their diet will require further research but it is likely that the resources of Lake Ilmen and the local rivers were exploited at least as intensively as they were in the 9th and 10th centuries by the inhabitants of Georgii, Prost and Gorodishche. Cyprinids, pike, zander and perch were frequently eaten. Occasionally, some of the boyars acquired prized fish such as sturgeon, salmon, large catfish, whitefish and eels through trade and tribute.

Meanwhile deep in the forest zone near the borders of the Novgorod Lands, the inhabitants of Minino relied much more heavily on the meat of some of the wild mammals they captured for their skins. They also commonly ate beef but consumed more mutton and less pork than the residents of Novgorod. They also fished in the local lakes exploiting mainly perch and cyprinids but rarely ate zander, because the species did not inhabit these lakes. They placed fish, wild mammals and hunting equipment in some of their graves.

The key role that Novgorod played in the international trade in furs is reflected in the make-up of the zooarchaeological assemblage at Minino, which is dominated by fur-bearing species, particularly squirrels and beavers. The pelts of animals captured in the forests of Novgorodian lands were transported to Novgorod and beyond in huge numbers. However, the importance of the fur trade is not reflected in the faunal assemblages from Novgorod itself, as most processing took place elsewhere and evidence for the importance of furs and skins of wild species relies mainly on documentary evidence. However, the decline in the trade in beaver pelts in the 12th–13th centuries is reflected widely. The numbers of beaver bones declined markedly in the Minino assemblage and they virtually disappeared in the Novgorod assemblages. Over-exploitation, which is reflected in changes in beaver mortality patterns at Minino, together with the clearance of woodland for agriculture and pasture, were significant factors in their decline.

The inhabitants of Novgorod consumed many of the same species of mammals, birds and fish as those eaten elsewhere in north-eastern Europe but there were notable variations in the relative importance of different species. For example, they ate more beef than many of their contemporaries in other towns. Goats may have been consumed more frequently but lamb and mutton was much less important than in many towns. Amongst the bird assemblage, although chickens were kept in Novgorod, wild ducks and large landfowl appear to have been exploited more commonly than in many towns in the region. Novgorod also lay beyond the Baltic trading network in marine fish and relied instead mainly on local freshwater sources.

Our studies into the exploitation of animals in medieval Novgorod region have shown that zooarchaeological analysis can be enhanced when it is combined with other sources of evidence, including in this case, the birch-bark documents and other documentary sources, artefacts and environmental archaeological data (pollen; plant macrofossils; wood; leather; insects). The superb preservation conditions in Novgorod have provided a wealth of data not often available elsewhere. None of these sources of evidence provide complete answers but by using a holistic approach, we can get a much more detailed and nuanced appreciation of how humans exploited animals.

Future studies

This analysis has demonstrated that detailed zooarchaeological studies can provide insights into the lives of the residents of medieval Novgorod. There is great potential for further research using traditional zooarchaeological methods. Within Novgorod itself, much more detailed inter- and intra-site comparisons are required to study variations in the diet of different areas within the town. Most of the recent faunal analyses in Novgorod have been focused on assemblages from sites on the western (Kremlin) side of the town. On the east (Market) side, there are two areas where foreign merchants resided. These Swedish (Gotlander) and German (St Peterhof) enclaves displayed their ethnic identity in their material culture. They imported and used ceramic tableware, which has been found much less frequently elsewhere in the town, where the inhabitants continued to use wooden vessels and local ceramic wares. Thus both native and foreign groups maintained their cultural identity (Gaimster 2001; 2006; 2012). It would be interesting to investigate whether this cultural diversity was also reflected in the choices of food consumed by these different communities. Further excavations targeted on these areas could be designed to obtain good samples of both plant and animal remains to explore dietary variations. Beyond the town, there is a need to investigate more assemblages from rural settlements in Novgorod's hinterland, to explore possible variations in diet and animal exploitation.

More detailed examination of the distribution of bones both within individual properties and in areas outside them would undoubtedly provide further insights into how space was utilised within them, particularly in relation to carcass processing and rubbish disposal. Chronological trends in the relationships between humans and animals also need further investigation. There are opportunities to study butchery patterns in more detail and very little analysis of pathology has been carried out. For any of these studies to be successful, there needs to be a significant increase in sieving of the deposits from the excavated sites and a sampling policy that allows assemblages from different parts of the same properties to be recorded separately and in sufficient detail. Further multidisciplinary studies should also be encouraged to allow more comprehensive interpretations of how humans and animals interacted within the town and beyond.

Since the project began, many advances have been made in scientific archaeology and the superb preservation of organic materials in Novgorod is already providing materials for such studies. A fruitful area for research into the diets of Novgorodians would be the analysis of carbon and nitrogen stable isotopes of human remains. Stable isotope studies have enhanced our understanding of the diets of the inhabitants of medieval Tallinn, for example (Lightfoot *et al.* 2016). There are huge samples of animal bones available to carry out analyses of stable isotopes and ancient DNA. Isotope studies on domestic mammals from medieval Poland have, for example, shed new light on animal husbandry strategies (Scull *et al.* 2019). Such studies need to be carried out in conjunction with more traditional faunal analyses to achieve better impact.

It has been a privilege to work on this fantastic archaeological resource and collaborate with so many colleagues from Russia, Britain and other parts of Europe. We believe we have significantly increased our understanding of the roles animals played in the lives of the people of medieval Novgorod and its lands. There is much more to learn and we look forward to seeing how this initial assessment can be developed to provide deeper insights into human relationships with animals in medieval Russia.

REFERENCES

Albarella, A. (2005) 'Meat production and consumption in town and country', in Giles, K. and Dyer, C. (eds), *Town and Country in the Middle Ages: Contrasts, Contacts and Interconnections, 1100–1500*, London: Maney, 131–147.

Albarella, U., Johnstone, C. and Vickers, K. (2008) 'The development of animal husbandry from the Late Iron Age to the end of the Roman period: a case study from South-East Britain', *Journal of Archaeological Science* 35, 1828–1848.

Albarella, A. and Payne, S. (2005) 'Neolithic pigs from Durrington Walls, Wiltshire: a biometrical database', *Journal of Archaeological Science* 32, 589–599.

Aleksandrovskaya, E.I., Aleksandrovsky, A.L., Gaidukov, P.G. and Krenke, N.A. (2001) 'Woodland, meadow, field and town layout: the evidence from analyses of the earliest cultural deposits and buried soil in Novgorod', in Brisbane, M. and Gaimster, D. (eds), *Novgorod: The Archaeology of a Russian Medieval City and its Hinterland*, London: British Museum Occasional Papers 141, 15–21.

Aleksashina, K.V. (1950) 'Faunal remains from a number of city-sites in the Upper Volga region', *Materialy i issledovaniya po arkheologii SSSR* (Materials on the Archaeology of the USSR) 13, 148–151. (in Russian)

Allen, K.R. (1966) 'Some methods for estimating exploited populations', *Journal of the Fisheries Research Board of Canada* 23(10), 1553–1574.

Allen, K.R. and Kirkwood, G.P. (1988) 'Marine mammals', in Gulland, J.A. (ed.), *Fish Population Dynamics (2nd Edition)*, Chichester: Wiley and Sons, 251–269.

Andreyeva, E.G. (1977) 'The Fauna of the "Krutik" settlement (based on the faunal remains from the excavations of the Byelozyorskaya Expedition), in *Byelozyorskaya Ves' (po materialam poseleniya Krutik IX–X vv.)* [Byelozyorskaya Village (based on materials from the Krutik Settlement for the 9th and 10th centuries), Petrozavodsk, 182–186. (in Russian)

Angelstam, P. (2004) 'Habitat thresholds and effects of forest landscape change on the distribution and abundance of black grouse and capercaillie', *Ecological Bulletins* 51, 173–187.

Anon. (1958) *Documents Concerning the Social and Economic History of North-western Russia in the Late-14th and Early-15th Centuries Volume 2*, Moscow: USSR Academy of Sciences. (in Russian)

Antipina, E.E. (1997) 'Methods for the reconstruction of aspects of livestock-breeding in the southern part of Eastern Europe in the Bronze Age', *Rossiiskaya arkheologiya* (Russian Archaeology) 3, 20–32. (in Russian)

Antipina, E.E. (2003) 'Methodological problems in the study of faunal remains from archaeological sites', in *Noveishie arkheozoologicheskie issledovaniya v Rossii* (Latest Archaeozoological Research in Russia), Moscow: Yazyki slavyanskoi kultury (Languages of Slavonic Culture), 7–33. (in Russian)

Armitage, P.L. and Clutton-Brock, J. (1976) 'A system for classification and description of the horn cores of cattle from archaeological sites', *Journal of Archaeological Science* 3, 329–348.

Ascádi, J. and Nemeskéri, J. (1970) *History of Human Life Span and Mortality*, Budapest: Akadémiai Kiedó.

Atlas of the Novgorod Region (1982) *Atlas Novgorodskoye Oblasti*, Moscow: Geodesic and Cartographic Ministry of the Soviet Union.

Baker, P. and Worley, F. (2019) *Animal Bones and Archaeology: Recovery to Archive*, Swindon: Historic England.

Balčiauskienė, L., Juškaitis, R. and Atkočaitis, O. (2005) 'The diet of the tawny owl (*Strix aluco*) in south-western Lithuania during the breeding period', *Acta Zoologica Lituanica*, 15(1), 13–20.

REFERENCES

Bartosiewicz, L. (2003) 'A millennium of migrations: proto-historic mobile pastoralism in Hungary', *Bulletin of the Florida Museum of Natural History* 44, 101–130.

Bartosiewicz, L., Van Neer, W. and Lentacker, A. (1997) *Draught Cattle: Their Osteological Identification and History*, Tervuren: Annales du Musée Royal de l'Afrique Centrale, Sciences Zoologiques 281.

Baskin, L.M. (2009) 'Status of regional moose populations in European and Asiatic Russia', *Alces* 45, 1–4.

Bauthian, I., Iljinsky, I., Fokin, S., Julliard, R., Gossmann, F. and Ferrand, Y. (2006) 'Survival rates of Russian woodcocks', *Proceedings of the 6th Woodcock and Snipe Workshop, November 2003 in Nantes*, Wageningen: Wetlands International, 62–65.

Benecke, N. (1986) *Die Entwicklung der Haustierhaltung in Sudlichen Ostseeraum Weimarer*, Thuringen: Monographien zur Ur- und Frühgeschichte 18.

Berka, R. (1990) *Inland Capture Fisheries of the USSR*, Rome: Food and Agriculture Organisation Fishing Technical Paper 311.

Bibikova, V.I. (1950) 'Fauna from the Pekunovo city-site', *Materialy i issledovaniya po arkheologii SSSR* (Materials and Research on Archaeology in the USSR) 13, 133–147. (in Russian)

Birnbaum, H. (1981) *Novgorod the Great: Essays in the History and Culture of a Medieval City-State*, Columbus: Slavica.

Birnbaum, H. (1989) *Novgorod and Dubrovnik: Two Slavic City-Republics and their Civilization*, Zagreb: Jugoslavenska Akademija Znanosti i Umjetnosti.

Birnbaum, H. (1996) *Novgorod in Focus*, Columbus: Slavica.

Blokin, Y.Y., Mezhnev, A.P. and Fokin, S.Y. (2006) 'Woodcock hunting bag statistics in Russia since 1996', *Proceedings of the 6th Woodcock and Snipe Workshop, November 2003 in Nantes*, Wageningen: Wetlands International, 17–23.

Blokin, Y.Y., Solokha, A., Gorokhovsky, K., Mezhnev, A.P. and Fokin, S.Y. (2015) 'Hunting bags of woodcock, snipes and other waders in Russia', *Wetlands International Woodcock and Snipe Specialist Group Newsletter* 41, 13–18.

Bockstoce, J.R. and Botkin, D.B. (1983) 'The historical status and reduction of the Western Arctic bowhead whale population by the pelagic whaling industry', *Report of the. International Whaling Commission* Special issue 5, 107–142.

Boessneck, J. (1956) 'Ein Beitrag zur Errechnung der Widerristhöhe nach Metapodienmassen bei Rindern', *Journal of Animal Breeding and Genetics* 68(1), 75–90.

Boessneck, J., Muller, H.-H. and Teichert, M. (1964) 'Osteologische Untersuchmerkmale zwischen Schaf (*Ovis aries* Linné) und Ziege (*Capra hircus* Linné)', *Kuhn-Archiv* 78, 1–29.

Boessneck, J., von den Driesch, A. and Stenberger, L. (1979) *Eketorp: Befestigung und Siedlung auf Őland, Schweden: die Fauna*, Stockholm: Almqvist and Wiksell International.

Bogatkina, G. (1995) 'Archaeological research into the materials from the city-site of Indakar', *Proceedings of the Investigation of the Idnakar Fortified Settlement (IX–XIII cent)*, Izhevsk: Udmurtiya Institute for History, Art and Literature, Russian Academy of Sciences, 141–158. (in Russian)

Bolotova, N.L., Konovalov, A.F., Borisov, M. and Dumnich, N.V. (2010) 'Natural and anthropogenic factors of the establishing of invasive fish populations in the aquatic ecosystems of the Vologda Region', *Russian Journal of Biological Invasions* 1(4), 251–263.

Borchtchevski, V.G. (2017) 'Does beginning of Spring represent a critical period in annual cycle of capercaillie (*Tetrao urogallus*, Tetronidae, Galliformes)?', *Biology Bulletin* 44(7), 751–760.

Bovy, K. (2012) 'Why so many wings? A re-examination of avian skeletal part representation in the south-central Northwest Coast, USA', *Journal of Archaeological Science* 39, 2049–2059.

Brauner, A.A. (1916) 'Horses in burials in burial-mounds in the Tiraspol District of the Kherson Province', *Zapiski obshchestva selskogo khozyaistva Yuzhnoi Rossii* (Notes of the Agricultural Society of Southern Russia) 86, 1–49.

Brisbane, M. (ed.) (1992) *The Archaeology of Novgorod, Russia: Recent Results from the Town and its Hinterland*, Lincoln: Society for Medieval Archaeology Monograph 13.

Brisbane, M. (2013a) 'Early trading centres of the Baltic: a view from the English coast', in Makarov, N.A., Mastykova, A. and Khokhlov, A. (eds), *Archaeology of the Baltic Region*, Saint Petersburg: Nestor-Historia, 161–166.

REFERENCES

Brisbane, M. (2013b) 'Baltic beads and beaver: motivations for medieval settlement expansion in Northwestern Russia', in Pope, P.E. and Lewis-Simpson, S. (eds), *Exploring Atlantic Transitions: Archaeologies of Transience and Permanence in New Found Lands*, Woodbridge: Boydell Press, 15–24.

Brisbane, M. and Gaimster, D. (eds) (2001) *Novgorod: The Archaeology of a Russian Medieval City and its Hinterland*, London: British Museum Occasional Papers 141.

Brisbane, M. and Hather, J. (eds) (2007) *Wood Use in Medieval Novgorod*, Oxford: Oxbow Books.

Brisbane, M. and Maltby, M. (2002) 'Love letters to bare bones: a comparison of two types of evidence for the use of animals in medieval Novgorod', *Archaeological Review from Cambridge* 18, 99–119.

Brisbane, M., Hambleton, E., and Maltby, M. (2007a) 'An African monkey at the court of the Novgorod princes' in Nosov, E.N. and Musin, A. (eds), *The Origins of the Russian State: Proceedings of the International Scientific Conference October 4–7, 2005, Veliky Novgorod, Russia*, St Petersburg: Russian Academy of Sciences, 74–81.

Brisbane, M., Hambleton, E., Maltby, M. and Nosov, E. (2007b) 'A monkey's tale: the skull of a macaque found at Ryurik Gorodishche during excavations in 2003', *Medieval Archaeology* 51, 185–191.

Brisbane, M., Makarov, N.A. and Nosov, E.N. (eds) (2012) *The Archaeology of Medieval Novgorod in Context*, Oxford: Oxbow Books.

Brisbane, M., Hambleton, E., Hamilton-Dyer, S. and Maltby, M. (2015) 'Reflections on the ecological setting and environmental impact of Medieval settlement expansion in northern Russia', in Gaidukov, P. (ed.), *Towns and Villages of Medieval Russia*, Moscow: Russian Academy of Sciences, 36–44.

Bull, G. and Payne, S. (1982) 'Tooth eruption and epiphyseal fusion in pigs and wild boar', in Wilson, B., Grigson, C. and Payne, S. (eds), *Ageing and Sexing of Animal Bones from Archaeological Sites*, Oxford: British Archaeological Reports (British Series) 109, 55–71.

Chapman, D.G. (1974) 'Estimation of population size and sustainable yield of sei whales in the Antarctic', *Report of the. International Whaling Commission* 24, 82–90.

Clark, W.S. 1999. *A Field Guide to the Raptors of Europe, the Middle East and North Africa*, Oxford: Oxford University Press.

Conenna, I., Valkama, J. and Chamberlain, D. (2017) 'Interactive effects of climate and forest canopy cover on goshawk productivity', *Journal of Ornithology* 158(3), 799–809.

Corrales, C., Pavlovska, M. and Höglund, J. (2014) 'Phylogeography and subspecies status of black grouse', *Journal of Ornithology* 155(1), 13–25.

Cramp, S. and Simmons, K.E.L. (eds) (1977) *The Birds of the Western Palearctic Volume I*, Oxford: Oxford University Press.

Cramp, S. and Simmons, K.E.L. (eds) (1980) *The Birds of the Western Palearctic Volume II*, Oxford: Oxford University Press.

Cramp, S. and Simmons, K.E.L. (eds) (1983) *The Birds of the Western Palearctic Volume III*, Oxford: Oxford University Press.

Danilchenko, V.P. (1984) 'Osteological material from the city-site of Seletsk', *KSIA* (Short Reports from the Moscow Institute of Archaeology) 178, 29–34. (in Russian)

Davis, S.J.M., Svensson, E.M., Albarella, U., Detry, C., Götherström, A., Pires, A.E. and Ginja, C. (2012) 'Molecular and osteometric sexing of cattle metacarpals: a case study from 15th century AD Beja, Portugal', *Journal of Archaeological Science* 39.

Dekker, S. (2018) *Old Russian Birchbark Letters*, Leiden: Brill Rodopi.

De Lury, D.B. (1947) 'On the estimation of biological populations', *Biometrics* 3, 145–167.

Dinesman, L.G., Kiseleva, N.K., Savinetsky, A.B. and Khassanov, B.F. (1999) *Secular Dynamics of Coastal Zone Ecosystems of the Northeastern Chukchi Peninsula*, Tübingen: Mo Vince.

Dinesman L.G. and Savinetsky, A.B. (2003) 'Quantitative analysis of bones in the cultural levels of ancient human settlements', in *Noveishie arkheozoologicheskie issledovaniya v Rossii* (Latest Archaeozoological Research in Russia), Moscow: Yazyki slavyanskoi kultury (Languages of Slavonic Culture), 34–55. (in Russian)

Dobrynina, I.N. and Kharitonov, S.P. (2006) 'The Russian waterbird migration atlas: temporal variation in migration 1445 1451 routes', in Boere, G.C., Galbraith, C.A. and Stroud, D.A. (eds), *Waterbirds Around the World*, Edinburgh: The Stationery Office, 582–589.

Dubrovin, G.E. (2007a) 'Water transport', in Brisbane, M. and Hather, J. (eds), *Wood Use in Medieval Novgorod*, Oxford: Oxbow Books, 229–262.

Dubrovin, G.E. (2007b) 'Land transport', in Brisbane, M. and Hather, J. (eds), *Wood Use in Medieval Novgorod*, Oxford: Oxbow Books, 209–228.

Dumnich, N.V., Konovalov, A.F. and Bolotova, N.L. (2003) 'Overfishing effects of Zander (*Stizostedion lucioperca* L.) in Beloe Lake North-West Russia', *Percis III, The Third International Percid Fish Symposium*, Madison: University of Wisconsin.

Ekdahl, S. (1998) 'Horses and crossbows: two important warfare advantages of the Teutonic Order in Prussia', in Nicholson, H. (ed.), *The Military Orders, II: Welfare and Warfare*, Farnham: Ashgate, 119–151.

Erkind, A. (2011) 'Barrels of fur: natural resources and the state in the long history of Russia', *Journal of Eurasian Studies* 164–171.

Fennell, J. (1983) *The Crisis of Medieval Russia, 1200–1304*, London: Longman.

Filipowiak, W. (1979) 'Problematik der archäologischen Forschungen auf dem Beispiel des frümittelalterlichen Wolins', in Kubasiewicz, M. (ed), *Archaeozooology*, Szczecin: Szczecin Agricultural Academy, 85–96.

Gabler, K.-O. (1985) *Osteologische Unterscheidungsmerkmale zwischen Mähnenspringer (Ammotragus lervia), Hausschaf (Ovis aries) und Hausziege (Capra hircus)*, München: Ph.D. thesis.

Gaimster, D. (2001) 'Pelts, pitch and pottery: the archaeology of Hanseatic trade in Medieval Novgorod', in Brisbane, M. and Gaimster, D. (eds), *Novgorod: The Archaeology of a Russian Medieval City and its Hinterland*, London: British Museum Occasional Papers 141, 67–78.

Gaimster, D. (2006) 'Pottery imported from the West: reception and resistance', in Orton, C. (ed.), *The Pottery from Medieval Novgorod and its Region*, London: University College London, 135–143.

Gaimster, D. (2012) 'The Hanseatic cultural signature: exploring globalization on the micro-scale in late Medieval Northern Europe', *European Journal of Archaeology* 17(1), 60–81.

Gaginskaya, A., Starikov, D. and Kouzov, S. (2014) 'Status of the breeding population of great cormorants in the Russian part of the Gulf of Finland in 2012', in Bregnballe, T., Lynch, J., Parz-Gollner, R., Marion, L., Volponi, S., Paquet, J.-Y., Carss, D.N. and van Eerden, M.R. (eds), *Breeding Numbers of Great Cormorants* Phalacrocorax carbo *in the Western Palearctic, 2012–2013 IUCN-Wetlands International Cormorant Research Group Report*, Aarhus University: Scientific Report from the Danish Centre for Environment and Energy 99, 171–174.

Gersmann, K.-H. and Grimm, O. (eds) (2018) *Raptor and Human: Falconry and Bird Symbolism Throughout the Millennia on a Global Scale*, Neumünster: Wachholtz Murmann.

Glazunova, A., Kaliakin, V., Krenke, N., Lantseva, M. and Tsepkin, E. (2009) 'The archaeozoological collection from the excavations at Romanov Dvor', in Krenke, N., *Archaeology of the Romanov Dvor: Early Beginnings and History of the Center of Moscow*, Moscow: Institute of Archaeology, 161–175. (in Russian)

Gorobets, L. and Kovalchuk, O. (2017) 'Birds in the medieval culture and economy of the East Slavs in the 10th–13th centuries AD', *Environmental Archaeology* 22(2), 147–165.

Grant, A. (1982) 'The use of tooth wear as a guide to the age of domestic ungulates', in Wilson, B., Grigson, C. and Payne, S. (eds), *Ageing and Sexing of Animal Bones from Archaeological Sites*, Oxford: British Archaeological Reports (British Series) 109, 91–108.

Grayson, D.K. (1984) *Quantitative Zooarchaeology*, Orlando: Academic Press.

Grigson, C. (1982a) 'Sex and age determination of some bones and teeth of domestic cattle: a review of the literature', in Wilson, B., Grigson, C. and Payne, S. (eds), *Ageing and Sexing of Animal Bones from Archaeological Sites*, Oxford: British Archaeological Reports (British Series) 109, 7–24.

Grigson, C. (1982b) 'Sexing Neolithic domestic cattle skulls and horn cores', in Wilson, B., Grigson, C. and Payne, S. (eds), *Ageing and Sexing Animal Bones from Archaeological Sites*, Oxford: British Archaeological Reports (British Series) 109, 25–36.

Gromova, V.I. (1949) 'A history of horses (the *Equus* species) in the Old World, Part 1', *Trudy Paleontologicheskogo instituta* (Proceedings of the Palaeontology Institute) 17, 1–374. (in Russian)

REFERENCES

Gromova, V.I. (1953) *Osteological Distinctions Between Genera Capra (Goats) and Ovis (Sheep). Guide to Identification of Bones from Excavations*, Moscow: Soviet Union Academy of Sciences Press. (in Russian)

Guillemain, M., Van Wilgenburg, S.L., Legagneux, P. and Hobson K.A. (2014) 'Assessing geographic origins of teal (*Anas crecca*) through stable-hydrogen (d^{2}H) isotope analyses of feathers and ring-recoveries', *Journal of Ornithology* 155(1), 165–172.

Guzmán, J.L., Ferrand, Y. and Arroyo, B. (2011) 'Origin and migration of woodcock *Scolopax rusticola* wintering in Spain', *European Journal of Wildlife Research* 57(3), 647–655.

Halperin, C.J. (1999) 'Novgorod and the Novgorodian Land', *Cahiers du monde russe* 40(3).

Halstead, P., Collins, P. and Isaakidou, V. (2002) 'Sorting the sheep from the goat: morphological distinctions between mandibles and mandibular teeth of adult *Ovis* and *Capra*', *Journal of Archaeological Science* 25, 545–554.

Hambleton, E. (1999) *Animal Husbandry Regimes in Iron Age Britain*, Oxford: British Archaeological Reports (British Series) 282.

Hamilton-Dyer, S. (2002) 'The bird resources of medieval Novgorod, Russia', in Bochenski, Z.M., Bochenski, Z. and Stewart, J. (eds), 'Proceedings of the 4th Meeting of the ICAZ Bird Working Group, Krakow, Poland, 11–15 September, 2001', *Acta zoologica cracoviensia* 45 (special issue), 99–107.

Hamilton-Dyer, S. (2009) 'Fish remains', in Makarov, N.A. (ed.), *The Archaeology of Northern Rus 900–1300 AD: Volume 3*, Moscow: Russian Academy of Sciences, Institute of Archaeology, 25–29.

Hamilton-Dyer, S., Brisbane, M. and Maltby, M. (2017) 'Fish, feather, fur and forest: exploitation of wild animals in medieval Novgorod and its territory', *Quaternary International* 460, 97–107.

Hamilton-Dyer, S. and Maltby, M. (2000) 'The animal bones', in Barrett, J., Freeman, P. and Woodward, A., *Cadbury Castle, Somerset: The Later Prehistoric and Early Historic Archaeology*, London: English Heritage Archaeological Report 20, 278–291.

Harcourt, R.A. (1974) 'The dog in prehistoric and early historic Britain', *Journal of Archaeological Science* 127, 56–63.

Harlin, E.-K., Mannermaa, K. and Ukkonen, P. (2019) 'Animal bones from medieval and early modern Saami settlements in Finnish Lapland', in Mannermaa, K., Manninen, M.A., Pesonen, P. and Seppänen, L. (eds), *Helsinki Harvest: Proceedings of the 11th Nordic Conference on the Application of Scientific Methods in Archaeology*, Helsinki: The Archaeological Society of Finland Monograph 7, 149–177.

Helama, S., Huhtamaa, H., Verkasalo, E. and Läänelaid, A. (2017) 'Something old, something new, something borrowed: new insights to human-environment interaction in medieval Novgorod inferred from tree rings', *Journal of Archaeological Science: Reports* 13, 341–350.

Higham, C.F.W. (1967) 'Stock rearing as a cultural factor in prehistoric Europe', *Proceedings of the Prehistoric Society* 33, 84–106.

Hokhlova, T.Y. and Artemjev, A.V. (2002) 'Reassessment of the southern Limit for whooper swans breeding in Northwest Russia', *Waterbirds* 25 (Special Publication 1: Proceedings of the Fourth International Swan Symposium 2001), 67–73.

Honka, J., Heino, M.T., Kvist, L., Askeyev, I.V., Shaymuratova, D.N., Askeyev, O.V., Askeyev, A.O., Heikkinen, M.E., Searle, J.B. and Aspi, J. (2018) 'Over a thousand years of evolutionary history of domestic geese from Russian archaeological sites analysed using ancient DNA', *Genes* 9, 367 doi:10.3390/genes9070367.

Ilyashenko, E. and Markin, Y. (2012) 'Changing of the Eurasian Crane staging area distribution in the European part of Russia from 1982 to 2007', in Harris, J. (ed.), *Proceedings of the Cranes, Agriculture, and Climate Change Workshop at Muraviovka Park. Russia, 28 May–3 June 2010*, Baraboo, Wisconsin: International Crane Foundation, 89–100.

Inostrantsev, A.A. (1882) *Doistoricheskii Chelovek Kamennogo Veka Poberezhya Ladozhskogo Ozera* (Prehistoric Stone-Age Man on the Shores of Lake Ladoga), Saint Petersburg. (in Russian)

Iwaszczuk, U. (2014) 'Animal husbandry on the Polish territory in the Early Middle Ages', *Quaternary International* 346, 69–101.

Izyumova, S.A. (1959) 'On the history of leather-working and shoe-making in Novgorod the Great', in *Materials of the Novgorod Archaeological Expedition, Volume II*, Moscow: Nauka, Materials and Research on the Archaeology of the USSR LXV, 199–222.

Izyumova, S.A. (1967) 'Leather-working and shoemaking', in Thompson, M.W., *Novgorod the Great: Excavations at the Medieval City 1951–1962 Directed by A.V. Artsikhovsky and B.A. Kolchin*, London: Eyelyn, Adams and MacKay, 82–84.

Jacobsen, B.V. and Sonerud, G.A. (2003) 'Synchronous switch in diet and hunting habitat as a response to disappearance of snow cover in Tengmalm's Owl *Aegolius funereus*', *Ornis Fennica* 70, 78–88.

Johnsgard, P.A. (2016) *Swans: Their Biology and Natural History*, University of Nebraska, Zea E-Books 38. http://digitalcommons.unl.edu/zeabook/38.

Johnstone, C. (2004) *A Biometric Study of Equids in the Roman World*, York: Department of Archaeology, University of York Ph.D. Thesis. http://www.york.ac.ukdepts/arch/pgstudents/Johnstone_thesis/CJohnstonePhDvol1.pdf.

Jones, G. and Sadler, P. (2012) 'Age at death in cattle: methods, older cattle and known-age reference material', *Environmental Archaeology* 17, 11–28.

Jonsson, L. (1986) 'Finska gäddor och Bergenfisk – ett försök att belysa Uppsala fiskimport under medeltid och yngre Vasatid', in Cnattingius, N. and Nevéus, T. (eds), *Från Östra Aros till Uppsala. En samling uppsatser kring det medeltida Uppsala* (Uppsala stads historia IV), Uppsala: Historiekomm, 122–139.

Khoroshev, A.S. (2007) 'Household objects', in Brisbane, M. and Hather, J. (eds), *Wood Use in Medieval Novgorod*, Oxford: Oxbow Books, 263–277.

Khoroshev, A.S. and Sorokin, A.N. (1992) 'Buildings and properties from the Lyudin End of Novgorod', in Brisbane, M. (ed.), *The Archaeology of Novgorod, Russia*, Lincoln: Society for Medieval Archaeology Monograph 13, 107–159.

Khoroshev, A.S., Sorokin, A.N. and Petrov, M.I. (2001) 'Property layout in medieval Novgorod in the tenth to fifteenth centuries', in Brisbane, M. and Gaimster, D. (eds), *Novgorod: The Archaeology of a Russian Medieval City and its Hinterland*, London: British Museum Occasional Papers 141, 23–30.

Kiesewalter, L. (1888) *Skelettmessungen an Pferden als Bietrag zur Theoretischen Beurteilungslehre des Pferdes*, University of Leipzig Dissertation.

King, A. (1999) 'Diet in the Roman world: a regional inter-site comparison of the animal bones', *Journal of Roman Studies* 12, 168–202.

Kirikov, S.V. (1960) *Izmenenina Zhivotnogo Mira v Prirodnykh Zonakh SSSR* (Changes in the Animal World in the Natural Zones of the USSR Volume 2: The Forest Zone and Forest-Tundra), Moscow: USSR Academy of Sciences. (in Russian)

Kolman, R., Kapusta, A. and Morzuch, J. (2011) 'History of the sturgeon in the Baltic Sea and Lake Ladoga', in Williot, P., Rochard, E., Desse-Berset, N., Kirschbaum, F. and Gessner, J. (eds), *Biology and Conservation of the European Sturgeon Acipenser sturio L. 1758: The Reunion of the European and Atlantic Sturgeons*, New York: Springer, 221–226.

Konovalov, A., Kaldma, K., Bokotey, A., Brossault, P., Chapalain, F., Dmitrenok, M., Dzyubenko, N., Sellis, U., Strazds, M., Strenna, L., Treinys, R., Zielinski, P. and Väli, Ű. (2105) 'Spatio-temporal variation in nestling sex ratio among the black stork *Ciconia nigra* populations across Europe', *Journal of Ornithology* 156(2), 381–387.

Kouzov, S. and Loseva, A. (2016) 'Modern distribution, new breeding and moulting places of the Mute Swan (*Cygnus olor* Gmelin) in the Leningrad region in 2005–2015', *Biological Communications* 1, 116–136.

Kratochvil, Z. (1969) 'Species criteria on the tibia of *Ovis ammon f. aries* L. and *Capra aegagrus f. hircus* L.', *Acta Veterinaria Brno* 38, 483–490.

Krylovich, O.A. (2004) *Changes in the Hunted Animals of North-western Russia in Recent Millennia*, Moscow: Moscow State University Dissertation. (in Russian)

Kubasiewicz, M. (1975) 'Zur interpretation von zooarchäologischen materialen aus grossen frümittelalterlichen siedlungen', in Clason, A. (ed.), *Archaeozoological Studies*, Amsterdam: Elsevier, 240–247.

Kublo, E.K. (2007a) 'Footwear production', in Brisbane, M. and Hather, J. (eds), *Wood Use in Medieval Novgorod*, Oxford: Oxbow Books, 158–164.

Kublo, E.K. (2007b) 'Spinning and weaving', in Brisbane, M. and Hather, J. (eds), *Wood Use in Medieval Novgorod*, Oxford: Oxbow Books, 136–157.

Kumari, E.V. (ed.) (1972) *Geese in the USSR*, Tartu: Academy of Sciences of the Estonian SSR. (in Russian with English summary)

Kurbatov, A.V. (2012) 'Leather-working in north-west Russia', in Brisbane, M., Makarov, N.A. and Nosov, E.N. (eds), *The Archaeology of Medieval Novgorod in Context*, Oxford: Oxbow Books, 381–415.

Lantseva, M.E. and Lapshin, V.A. (2001) 'Results of identification of osseous mammalian remains from excavations in 1994–1997 at Tver Kremlin', in *Tver Kremlin: Complex Archaeological Source Study (according to data from excavation site Tver Kremlin-11, 1993–1997)*, Saint Petersburg: Evropeysky Dom, 171–180. (in Russian)

Lehikoinen, A., Saurola, P., Byholm, P., Lindén, A. and Valkama, J. (2010) 'Life history events of the Eurasian sparrowhawk *Accipiter nisus* in a changing climate', *Journal of Avian Biology* 41, 627–636.

Lepiksaar, J. (1983) *Osteologia: I. Pisces*, Göteborg: privately printed.

Libois, R.M. and Hallet-Libois, C. (1988) *Éléments pour l'identification des restes crâniens des poissons dulçaquicoles de Belgique et du nord de la France: 2 – Cypriniformes*. Juan-les-Pins: APCDA Fiches d'Ostéologie Animale pour l'Archéologie, Série A: Poissons, No. 4.

Libois, R.M., Hallet-Libois, C. and Rosoux, R. (1987) *Éléments pour l'identification des restes crâniens des poissons dulçaquicoles de Belgique et du nord de la France: 1 – Anguilliformes, Gastérostéiformes, Cyprinodontiformes et Perciformes*, Juan-les-Pins: APCDA, Fiches d'Ostéologie Animale pour l'Archéologie, Série A: Poissons, No. 3.

Lie, R. (1988) 'Animal bones', in Schia, E. (ed.), *De Arkeologiske Utgravninger I Gamlebyen, Oslo. Bind 5. 'Midets Tomt – Søndre Felt'*, Oslo: Alvheim and Eide, 153–196.

Lightfoot, E., Naum, M., Kadkas, V. and Russow, E. (2016) 'The influence of social status and erhnicity on diet in mediaeval Tallinn as seen through stable isotope analysis', *Estonian Journal of Archaeology* 20, 81–107.

Logunov, E. (1997) A history of salt production in Russia', *Science Tribune*, January 1997. http://www.tribunes.com/tribune/sel/logu.htm.

Lõugas, L., Kadakas, V. and Kadakas, U. (2016) 'Fishery in prehistoric and medieval Tallinn, Estonia', *Environmental Archaeology*, 21(4), 361–368.

Ludwig, T. and Klaus, S. (2017) 'Habitat selection in the post-breeding period by hazel grouse *Tetrastes bonasia* in the Bohemian Forest', *Journal of Ornithology* 158(1), 101–112.

MacKinnon, M. (2001) 'High on the hog: linking zooarchaeological, literary, and artistic data for pig breeds in Roman Italy', *American Journal of Archaeology* 105(4), 649–673.

Makarov, N.A. (ed.) (2001) *Kubenskoye ozero: vzglyad skvoz tysyacheletiya. Shest lyet raskopok Mininskogo arkheologicheskogo kompleksa* (Lake Kubenskoye: a view through the centuries: Six years of excavations at the Minino archaeological complex), Vologda: 'Drevnosti Severa' (Antiquities of the North). (in Russian)

Makarov, N.A. (2006) 'Traders in the forest: the northern periphery of 'Rus' in the medieval trade network', in Tracy, J., Reyerson, K., Noonan, T. and Stavrou, T. (eds), *Pre-modern Russia and its World: Essays in Honor of Thomas S. Noonan*, Wiesbaden: Harrassowitz, 115–134.

Makarov, N.A. (ed.) (2007) *The Archaeology of the Rural Areas of Northern Rus AD 900–1300: Medieval Settlements and Burial Grounds in the Lake Kubenskoye Region Volume 1*, Moscow: Nauka. (in Russian with English summaries)

Makarov, N.A. (ed.) (2008) *The Archaeology of the Rural Areas of Northern Rus AD 900–1300: Medieval Settlements and Burial Grounds in the Lake Kubenskoye Region Volume 2: Material Culture and Chronology*, Moscow: Nauka. (in Russian with English summaries)

Makarov, N.A. (ed.) (2009a) *The Archaeology of the Rural Areas of Northern Rus AD 900–1300: Medieval Settlements and Burial Grounds in the Lake Kubenskoye Region Volume 3: Environment, Society and Cultural Pattern*, Moscow: Nauka. (in Russian with English summaries)

Makarov, N.A. (2009b) 'Rural settlement and trade networks in northern Russia, AD 900–1250', in Mango, M. (ed.), *Byzantine Trade, 4th–12th Centuries: The Archaeology of Regional and International Exchange. Papers of the 38th Symposium of Byzantine Studies, University of Oxford, 2004*, Farnham: Publications of the Society for the Promotion of Byzantine Studies 14, 443–461.

Makarov, N.A. (2012a) 'The Minino Project: the investigation of a group of medieval sites in the Byeloozero region of northern Russia', in Brisbane, M., Makarov, N.A. and Nosov, E. (eds), *The Archaeology of Medieval Novgorod in Context: Studies in Centre/Periphery Relations*. Oxford: Oxbow Books, 40–57.

Makarov, N.A. (2012b) 'The fur trade in the economy of the northern borderlands of medieval Russia', in Brisbane, M., Makarov, N.A. and Nosov, E. (eds), *The Archaeology of Medieval Novgorod in Context: Studies in Centre/Periphery Relations*, Oxford: Oxbow Books, 381–390.

Makarov, N.A., Zakharov, S.D. and Zaitseva, I.E. (2000) 'Rural settlements on Lake Kubenskoye in the 12th and 13th centuries – from their birth to their abandonment, *Rus v XIII veke: kontinuitet ili razryv traditsii?* (Medieval Russia in the 13th Century: Continuity or a Break in Traditions?), Moscow, 65–72. (in Russian)

Makowiecki, D. (2003) 'Culture and fishes in the past', in Guzmán, A.F., Polaco, Ó.J. and Aguilar, F.J. (eds), *Presence of the Archaeoichthyology in México: Proceedings of the 12th meeting of the Fish Remains Working Group of the International Council of Archaeozoology, Guadalajara, México, September 2003*, Guadalajara: INAH México, 97–100.

Makowiecki, D. and Gotfredson, A.B. (2002) 'Bird remains of Medieval and Post-Medieval coastal sites at the Southern Baltic Sea, Poland', in Bochenski, Z.M., Bochenski, Z. and Stewart, J. (eds), 'Proceedings of the 4th Meeting of the ICAZ Bird Working Group, Krakow, Poland, 11–15 September, 2001', *Acta zoologica cracoviensia* 45 (special issue), 65–84.

Makowiecki, D. and Makowiecka, M. (2014) 'The character of animal exploitation and the environment at the Polish/Prussian frontier in the medieval period: a case study', *Archaeologica Baltica* 20, 91–114.

Makowiecki, D., Zabiliska-Kunek, M., Seetah, K., Jarzebowski, M. and Pluskowski, A. (2019) 'Farming, hunting and fishing in medieval Prussia: the zooarchaeological data', in Pluskowski, A. (ed.), *Environment, Colonization and the Baltic Crusader States (Terra Sacra I)*, Turnhout: Brepols, 333–376.

Maltby, M. (1989) 'Urban-rural variation in the butchering of cattle in Romano-British Hampshire', in Serjeantson, D. and Waldron, T. (eds), *Diets and Crafts in Towns*, Oxford: British Archaeological Reports (British Series) 199, 75–106.

Maltby, M. (2002) 'Animal bones in archaeology: how archaeozoologists can make a greater contribution to British Iron Age and Romano-British archaeology', in Dobney, K. and O'Connor, T. (eds), *Bones and the Man: Studies in Honour of Don Brothwell*, Oxford: Oxbow Books, 88–94.

Maltby, M. (2007) 'Chop and change: specialist cattle carcass processing in Roman Britain', in Croxford, B., Ray, N., Roth, R. and White, N. (eds), *TRAC 2006: Proceedings of the 16th Annual Theoretical Roman Archaeology Conference, Cambridge 2006*, Oxford: Oxbow Books, 59–76.

Maltby, M. (2010) *Feeding a Roman Town: Environmental Evidence from Excavations in Winchester, 1972–1985*, Winchester: Winchester Museums Service.

Maltby, M. (2012) 'From *Alces* to zander: a summary of the zooarchaeological evidence from Novgorod, Gorodishche and Minino', in Brisbane, M., Makarov, N.A. and Nosov, E.N. (eds), *The Archaeology of Medieval Novgorod in Context: Studies in Centre/Periphery Relations*, Oxford: Oxbow Books, 351–381.

Maltby, M. (2013) 'The exploitation of animals in towns in the Medieval Baltic trading network: a case study from Novgorod', in Makarov, N.A., Mastykova, A. and Khokhlov, A. (eds), *Archaeology of the Baltic Region*, Saint Petersburg: Nestor-Historia, 229–244.

Maltby, M. (2017a) 'Horseflesh and beaver pelts: aspects of faunal studies in Medieval Novgorod and its region', in Choyke A. and Jatitz, G. (eds), *Animaltown: Beasts in Medieval Urban Space*, Oxford: British Archaeological Reports (International Series) 2818, 27–34.

Maltby, M. (2017b) 'From bovid to beaver: animal exploitation in medieval North-West Russia', in Albarella, U., Russ, H., Vickers, K. and Viner, S. (eds), *The Oxford Handbook of Zooarchaeology*, Oxford: Oxford University Press, 230–244.

Maltby, M. (2019) 'Exploiting animals in the medieval north-western frontier', in Pluskowski, A. (ed.), *Ecologies of Crusading, Colonization and Religious Conversion in the Medieval Baltic (Terra Sacra II)*, Turnhout: Brepols, 211–222.

Maltby, M. and Hamilton-Dyer, S. (1995) 'Animal bone from excavations in Novgorod', *Novgorod and Novgorod Region History and Archaeology*, 9, 129–157. (in Russian)

Maltby, M. and Hamilton-Dyer, S. (2001a) 'Animal bone studies in Novgorod and its hinterland', in Brisbane, M. and Gaimster, D. (eds), *Novgorod: The Archaeology of a Russian Medieval City and its Hinterland*, London: British Museum Occasional Papers 141, 119–126.

Maltby, M. and Hamilton-Dyer, S. (2001b) 'Auch Fleisch und Fisch standen auf dem Speiseplan: Studien an Tierknochen aus Novgorod und seinem Umland', in Müller-Wille, M., Yanin, V., Nosov, E.N. and Rybina, E.A. (eds) *Novgorod: das mittelalterliche Zentrum und sein Umland in Norden Rußlands*, Wacholtz Verlag: Studien zur Siedlungsgeschichte und Archäologie der Osteegbiete, 369–380.

Maltby, M., Pluskowski, A. and Seetah, K. (2009) 'Animal bones from the industrial quarter at Malbork, Poland: towards an ecology of a castle built in Prussia by the Teutonic Order', *Crusades* 8, 191–212.

Maltby, M., Pluskowski, A., Rannamäe, E. and Seetah, K. (2019) 'Farming, hunting and fishing in medieval Livonia: the zooarchaeological data' in Pluskowski, A. (ed.), *Environment, Colonization and the Baltic Crusader States (Terra Sacra I)*, Turnhout: Brepols, 137–173.

Mannermaa, K. (2016) 'Fish bones from the Old Town of Helsinki (Finland) sixteenth–seventeenth century', *Environmental Archaeology* 21(4), 369–380.

Martin, J. (1986) T*reasure of the Land of Darkness: The Fur Trade and its Significance for Medieval Russia*, Cambridge: Cambridge University Press.

Mathys, L., Zimmermann, N.E., Zbinden, N. and Suter, W. (2006) 'Identifying habitat suitability for hazel grouse *Bonasa bonasia* at the landscape scale', *Wildlife Biology* 12(4), 357–366.

Melin, C., Szafranska, P.A., Bogumila, J. and Barton, K. (2006) 'Biogeographical variation in the population density of wild boar (*Sus scrofa*) in western Eurasia', *Journal of Biogeography* 33, 803–811.

Men'shikova, S.V. (2006) 'Hunting habitats of harriers in agricultural landscapes of the Leningrad region', in *Status of Raptor Species in eastern Fennoscandia: Proceedings of the Workshop, Kostomuksha, Karelia, Russia, November 8–10, 2005*, Petrozavodsk: KarRC RAS, 105–113.

Michell, R. and Forbes, N. (1914) *The Chronicles of Novgorod 1016–1471*, London: Royal Historical Society.

Mischenko, A.L. and Sukhanova, O.V. (1998) 'Waders of the Novgorod region: peculiarities of their distribution and important breeding areas', *International Wader Studies* 10, 285–290.

Monk, M. and Johnston, P. (2012) 'Perspectives on non-wood plants in the sampled assemblage from the Troitsky Excavations of Medieval Novgorod', in Brisbane, M., Makarov, N.A. and Nosov, E.N. (eds), *The Archaeology of Medieval Novgorod in Context: Studies in Centre/Periphery Relations*, Oxford: Oxbow Books, 283–320.

Morales, A. and Rosenlund, K. (1979) *Fish Bone Measurements*, Copenhagen: Copenhagen Zoological Museum.

Musin, A. (2018) 'The rise of Novgorod revisited', *Slavia Antiqua* 59, 165–194.

Nadjafzadeh, M., Voigt, C.C. and Krone, O. (2016) 'Spatial, seasonal and individual variation in the diet of white-tailed eagles *Haliaeetus albicilla* assessed using stable isotope ratios', *Ibis* 158, 1–15.

Naumov, N.P. (1955) *Ekologiya zhivotnykh* (Animal Ecology), Moscow: Nauka. (in Russian)

Nehring, A. (1884) 'Fossile pferde aus deutschen Diluvial-Ablagerungen und ihre Beziehungen zu denlebenden pferden. Ein Beitrag zur Geschichte des Hallspferdes', *Landwirthshaf Jahr.* 13, 81–160.

Neronov, V.M., Khlyap, L.A., Bobrov, V.V. and Warshavsky, A.A. (2008) 'Alien species of mammals and their impact on natural ecosystems in the biosphere reserves of Russia', *Integrative Zoology* 3, 83–94.

Nosov, E.N. (1992) 'Ryurik Gorodishche and the settlements to the North of Lake Ilmen', in Brisbane, M. (ed.), *The Archaeology of Novgorod, Russia*, Lincoln: Society for Medieval Archaeology Monograph 3, 5–66.

Nosov, E.N. (2001) 'A typology for towns on the River Volkhov', in Brisbane, M. and Gaimster, D. (eds), *Novgorod: The Archaeology of a Russian Medieval City and its Hinterland*, London: British Museum Occasional Papers 141, 5–10.

Nosov, E.N. (2007) 'Thirty years of systematic excavations at Ryurik Gorodishche: the results and prospects', in Nosov, E.N. and Musin, A. (eds), *The Origins of the Russian State: Proceedings of the International Scientific Conference October 4–7, 2005, Veliky Novgorod, Russia*, St Petersburg: Russian Academy of Sciences, 23–58.

O'Connor, T. (2010) 'Livestock and deadstock in Early Medieval Europe from the North Sea to the Baltic', *Environmental Archaeology* 15, 1–15.

Orten Lie, R. (2018) 'Falconry, falcon-catching and the role of birds of prey in trade and as alliance gifts in Norway (800–1800 AD) with an emphasis on Norwegian and later foreign participants in falcon-catching', in Gersmann, K.-H. and Grimm, O. (eds), *Raptor and Human: Falconry and Bird Symbolism Throughout the Millennia on a Global Scale*, Neumünster: Wachholtz Murmann, 727–786.

Orton, C. (ed.) (2006) *The Pottery from Medieval Novgorod and its Region*, London: University College London.

Payne, S. (1985) 'Morphological distinctions between mandibular teeth of young sheep, *Ovis*, and Goat, *Capra*', *Journal of Archaeological Science* 12, 139–147.

Pchelintsev, V.G. (2006) 'Distribution and abundance of some raptor species in the Leningrad region', in *Status of Raptor Species in eastern Fennoscandia: Proceedings of the Workshop, Kostomuksha, Karelia, Russia, November 8–10, 2005*, Petrozavodsk: KarRC RAS, 120–124.

Perkhavko, V.B. (2012) *The Medieval Russian Merchant Class*, Moscow: Kuchkovo Field Publishers. (in Russian)

Petrov, M.I. (2011) 'Estates of medieval Novgorod: the question of criteria for comparison', *Novgorod and Novgorod Land, History and Archaeology, vol. 25*, 137–147. (in Russian)

Petrov, M.I. (2014) 'The estate complex of the early 13th century on the Nikolsky-2007 site, Novgorod', *Pskov and Pskov Land, History and Archaeology*, vol. 29(59), 235–244. (in Russian)

Phillips, R. and Rix, M. (1988) *A Guide to the Freshwater Fish of Britain, Ireland and Europe*, London: Treasure Press.

Pihlström, H. (2013) 'The cormorants of Lake Ladoga in the early 20th century: Arctic invaders or continental colonists?', *Memoranda Society Fauna Flora Fennica* 89, 17–24.

Piličiauskienė, G. and Masiulienė, I. (2013) 'Animal breeding and butchering: a glimpse from Old Klaipėda', *Archaeologia Baltica* 16, 168–185.

Pluskowski, A., Makowiecki, D., Maltby, M., Rannamäe, E., Lõugas, L., Maldre, L., Daugnora, L., Black, S., Müldner, G. and Seetah, K. (2019) 'The Baltic crusades and ecological transformation: the zooarchaeology of conquest and cultural change in the Eastern Baltic in the second millennium AD', *Quaternary International* 510, 28–43.

Pluskowski, A., Seetah, K. and Maltby, M. (2010) 'Potential osteoarchaeological evidence for riding and the military use of horses at Malbork Castle, Poland', *International Journal of Osteoarchaeology* 20, 335–343.

Pluskowski, A., Seetah, K., Maltby, M., Banajea, R., Black, S. and Kalnins, G. (2018) 'Late-Medieval horse remains at Cēsis Castle, Latvia, and the Teutonic Order's equestrian resources in Livonia', *Medieval Archaeology* 62(2), 351–359.

Poland, G. (2018) *A methodological approach to the identification of duck and goose remains from archaeological sites with an application to Roman Britain*, University of Sheffield Ph.D. Thesis.

Price, N. (2000) 'Novgorod, Kiev and their satellites: the city-state model and the Viking Age polities of European Russia', in Hansen, M.-H. (ed.), *A Comparative Study of Thirty City-state Cultures: An Investigation Conducted by the Copenhagen Polis Centre*, Copenhagen: The Royal Danish Academy of Science and Letters, 263–275.

Prummel, W. (1997) 'Evidence of hawking (falconry) from bird and mammal bones', *International Journal of Osteoarchaeology* 7(4), 333–338.

REFERENCES

Prummel, W. (2018) 'The archaeological-archaeozoological identification of falconry: methodological remarks and some Dutch examples', in Gersmann, K.-H. and Grimm, O. (eds), *Raptor and Human: Falconry and Bird Symbolism throughout the Millennia on a Global Scale*, Neumünster: Wachholtz Murmann, 467–478.

Prummel, W. and Frisch, H.-J. (1986) 'A guide for the distinction of species, sex and body side in bones of sheep and goat', *Journal of Archaeological Science* 13, 567–577.

Puputti, A. (2008) 'A zooarchaeogy of modernizing human-animal relationships in Tornio, northern Finland, 1620–1800', *Post-medieval Archaeology* 42, 304–316.

Pūtys, Z. and Zarankaitė, J. (2010) 'Diet of the great cormorant (*Phalacrocorax carbo* Sinensis) at the Juodkrantė colony, Lithuania', *Acta Zoologica Lituanica* 20(3), 179–189.

Rannamäe, E. (2010) *A Zooarchaeological Study of Animal Consumption in Medieval, Viljandi*, Lund: University of Tartu MA Thesis.

Rannamäe, E. and Lõugas, L. (2019) 'Animal exploitation in Karski and Viljandi (Estonia) in the late Iron Age and medieval period', in Pluskowski, A. (ed.), *Ecologies of Crusading, Colonization and Religious Conversion in the Medieval Baltic (Terra Sacra II)*, Turnhout: Brepols, 61–76.

Reilly, E. (2012) 'Fair and foul: analysis of sub-fossil insect remains from Troitsky XI–XIII, Novgorod (1996–2002)', in Brisbane, M., Makarov, N.A. and Nosov, E.N. (eds), *The Archaeology of Medieval Novgorod in Context*, Oxford: Oxbow Books, 265–282.

Reynolds, A. and Suggs, B. (2001) 'Building construction in medieval Novgorod: the results of excavations in Troitsky Sites XI and XII, 1088', in Brisbane, M. and Gaimster, D. (eds), *Novgorod: The Archaeology of a Russian Medieval City and its Hinterland*, London: British Museum Occasional Papers 141, 31–46.

Riha, T. (1970) *Readings in Russian Civilization: Russia before Peter the Great 900–1700*, Chicago: University of Chicago Press.

Riker, U.E. (1979) *Methods for the Assessment and Interpretation of Biological Indices of Fish Populations*, Moscow: Pishchevaya promyshlennost (Food Industry).

Rixson, D. (1989) 'Butchery evidence on animal bones', *Circaea* 6(1), 49–62.

Rowley-Conwy, P. and Owen, A. (2011) 'Grooved Ware feasting in Yorkshire: late Neolithic animal consumption at Rudston Wold', *Oxford Journal of Archaeology* 30(4), 325–367.

Rutz, C., Bijslma, R.G. Marquiss, M. and Kenward, R.E. (2006) 'Population limitation in the northern goshawk in Europe: a review with case studies', *Studies in Avian Biology* 31, 158–197

Rybina, E.A. (1978) *Archaeological Essays on the History of Novgorod Trade in the 10th–14th Centuries*, Moscow: MSU Publishers. (in Russian)

Rybina, E.A. (1992) 'Recent finds from excavations in Novgorod', in Brisbane, M. (ed.), *The Archaeology of Novgorod, Russia: Recent Results from the Town and its Hinterland*, Lincoln: Society for Medieval Archaeology Monograph 13, 160–192.

Rybina, E.A. (2001a) *Trade in Medieval Novgorod. Historical and Archaeological Essays*, Novgorod: Novgorod University. (in Russian)

Rybina, E.A. (2001b) 'The birch-bark letters: the domestic economy of medieval Novgorod', in Brisbane, M. and Gaimster, D. (eds), *Novgorod: The Archaeology of a Russian Medieval City and its Hinterland*, London: British Museum Occasional Papers 141, 127–131.

Rybina, E.A. (2007) 'Fishing and hunting', in Brisbane, M. and Hather, J. (eds), *Wood Use in Medieval Novgorod*, Oxford: Oxbow Books, 124–135.

Rybina, E.A. (2012b) 'Evidence concerning craft production in the birch-bark documents', in Brisbane, M., Makarov, N.A. and Nosov, E.N. (eds), *The Archaeology of Medieval Novgorod in Context*, Oxford: Oxbow Books, 448–454.

Safonov, V.G. (1966) 'Morphological features and the structure of the beaver population', *Byulleten' Moskovskogo Obshchestva Ispytatelei Prirody Otdel Biologicheskii* (Bulletin of the Moscow Society, Biology Section) 71(4), 5–19. (in Russian)

Salmi, J.A., Auvinen, H., Raitaniemi, J., Kurkilhati, M., Lilja, J. and Maikola, R. (2015) 'Perch (*Perca fluviatilis*) and pikeperch (*Sander lucioperca*) in the diet of the great cormorant (*Phalacrocorax carbo*) and effects on catches in the Archipelago Sea, southwest coast of Finland', *Fisheries Research* 164, 26–34.

REFERENCES

Salmina, E. (2016) 'Fishing settlements in the Pskov region in the 16th century (according to archaeological data and written sources)', *Environmental Archaeology* 21(4), 402–410.

Salvagno, L. and Albarella, U. (2017) 'A morphometric system to distinguish sheep and goat postcranial bones', *PLOS One* 12(6), e0178543. https://doi.org/10.1371/journal.pone.0178543

Savenskova, T., Lemeshko, N. and Asnova, T. (2010) 'Ilmen Lake Brief', in Aladin, N. and Plotnikov, I. (eds), *ILBM Challenges for three Russian lakes: Ilmen, Chudsko-Pskovskoe/Peipsi-Pihva and Ladoga, Final Review Meeting and International Symposium, November 2010*, Shiga University: ILBM-Governance Project.

Savientsky, A. and Krylovich, O. (2009) 'Archaeozoological remains and changes in populations in utilized mammals', in Makarov, N.A. (ed.), *The Archaeology of Northern Rus 900–1300 AD: Volume 3*, Moscow: Russian Academy of Sciences, Institute of Archaeology, 16–24. (in Russian)

Savinetsky, A. and Krylovich, O. (2011) 'On the history of the spread of the black rat (*Rattus rattus* L.) in northwestern Russia', *Biology Bulletin* 38(2), 203–207.

Schramm, Z. (1967) 'Long bones and height in withers of goat', *Roczniki Wyzsej Szkoły Rolniczej w Poznanin* 36, 89–105. (in Polish with English and Russian summary)

Schaeken, J. (2012) 'The birchbark documents in time and space – revisited', in Zilmer, K. and Jesch, J. (eds), *Epigraphic Literacy and Christian Identity. Modes of Written Discourse in the newly Christian European North* (Utrecht Studies in Medieval Literacy, 4), Turnhout: Brepols, 201–224.

Schaeken, J. (2017) 'Comments on birchbark documents found in the twenty-first century', *Russian Linguistics* 41, 123–149.

Schmôlcke, U. and Zachos, F.E. (2005) 'Holocene distribution and extinction of the moose (*Alces alces*, Cervidae) in Central Europe', *Mammalian Biology* 70(6), 329–344.

Scull, C., Müldner, G. and Makowiecki, D. (2019) 'Carbon and nitrogen stable isotope evidence for animal husbandry and environmental change in medieval Kulmerland', in Pluskowski, A. (ed.), *Ecologies of Crusading, Colonization and Religious Conversion in the Medieval Baltic (Terra Sacra II)*, Turnhout: Brepols, 83–94.

Serjeantson, D. (2002) 'Goose husbandry in medieval England, and the problem of ageing goose bones', Proceedings of the 4th Meeting of the ICAZ Bird Working Group, *Acta Zoologica Cracoviensia* 45, 39–54.

Shaymuratova, D., Askeyev, I. and Askeyev, O. (2019) 'The studies of archaeological bird remains from medieval Staraya Ladoga: new results and interpretations', in Mannermaa, K., Manninen, M.A., Pesonen, P. and Seppänen, L. (eds), *Helsinki Harvest: Proceedings of the 11th Nordic Conference on the Application of Scientific Methods in Archaeology*, Helsinki: Monographs of the Archaeological Society of Finland 7, 93–114.

Shchadilov, Y.M., Rees, E.C., Belousova, A.V. and Bowler, J.M. (2002) 'Annual variation in the proportion of whooper swans and Bewick's swans breeding in Northern European Russia', *Waterbirds* 25 (Special Publication 1: Proceedings of the Fourth International Swan Symposium 2001), 86–94.

Silver, I.A. (1969) 'The ageing of domestic animals', in Brothwell, D. and Higgs, E.S. (eds), *Science in Archaeology* (2nd Edition), London: Thames and Hudson, 283–302.

Smirnova, L. (2001) 'The working of antler, bone and ivory in Novgorod: a study of a craft industry', in Brisbane, M. and Gaimster, D. (eds), *Novgorod: The Archaeology of a Russian Medieval City and its Hinterland*, London: British Museum Occasional Papers 141, 79–84.

Smirnova, L. (2005) *Comb-Making in Medieval Novgorod (950–1450): An Industry in Transition.* Oxford: British Archaeological Reports (International Series) 1369.

Smith, R.E.F. (1977) *Peasant Farming in Muscovy*, Cambridge: Cambridge University Press.

Smith, R.E.F. and Christian, D. (1984) *Bread and Salt: A Social and Economic History of Food and Drink in Russia*, Cambridge: Cambridge University Press.

Sobociński, M. and Makowiecki, D. (1992) 'Zwierzęce szczątki kostne z wykopalisk w Gniewie nad Wislą: Stanowisko 2 – Stare Miasto (Animal bone remains from excavations at Gniew on the Vistula River: Stand 2 – Old Town)', *Roczniki Akedemii w Poznaniu* 237, 161–195. (in Polish)

Sokolov, N.N. (1926) 'A geomorphological description of the area in which the River Volkhov and Lake Ilmen are situated: relief, deposits and historical development', *Materials on the Investigation of the River Volkhov and its Basin* 7, Leningrad. (in Russian)

REFERENCES

Solovyov, D.I. (2012) 'Leather objects from Troitsky XI, Novgorod', in Brisbane, M., Makarov, N.A. and Nosov, E.N. (eds), *The Archaeology of Medieval Novgorod in Context*, Oxford: Oxbow Books, 416–419.

Spiridonova, E. and Aleshinskaya, A. (2012) 'Results of palynological investigations of the archaeological sites in the Lake Ilmen and Lake Kubenskoye study areas', in Brisbane, M., Makarov, N.A. and Nosov, E.N. (eds), *The Archaeology of Medieval Novgorod in Context*, Oxford: Oxbow Books, 10–39.

Sully, D. (2012) 'Preliminary identification of leather fragments from Novgorod excavations (1991–2001)', in Brisbane, M., Makarov, N.A. and Nosov, E.N. (eds), *The Archaeology of Medieval Novgorod in Context*, Oxford: Oxbow Books, 421–425.

Sychevskaya, E.K. (1965) 'Fish of medieval Novgorod', *Sovetskaya Arkheologiya* 1, 236–256. (in Russian)

Tarabardina, O.A. (2007) 'Building timber in medieval Novgorod based on materials from the Troitsky XI and Troitsky XII excavations', in Brisbane, M. and Hather, J. (eds), *Wood Use in Medieval Novgorod*, Oxford: Oxbow Books, 106–116.

Teichert, M. (1975) 'Osteometrische Untersuchungen zur Berechnung der Widerristhohe bei Schafen', in Clason, A.T. (ed), *Zooarchaeological Studies*, Amsterdam: Elsevier, 51–69.

Thomas, R., Holmes, M. and Morris, J. (2013) '"So bigge as bigge may be": tracking size and shape change in domestic livestock in London (AD 1220–1900)', *Journal of Archaeological Science* 40(8), 3309–3325.

Thompson, M.W. (1965) *Novgorod the Great: Excavations at the Medieval City 1951–1962 Directed by A.V. Artsikhovsky and B.A. Kolchin*, London: Eyelyn, Adams and MacKay.

Timchenko, N.G. (1972) *Kistorii okhoty i zhivotnovodstva v Kievskoi Rusi [Srednee Podnyeprovie]*, [Towards a History of Hunting and Animal Breeding in Kievan Rus (Middle Reaches of the Dnieper)]. Kiev: Naukova dumka. (in Russian)

Tornberg, R., Mönkkönensami, M. and Kivelä, M. (2009) 'Landscape and season effects on the diet of the goshawk', *Ibis* 151(2), 396–400.

Tourunen, A. (2008) *Animals in an Urban Context: A Zooarchaeological Study of the Medieval and Post-medieval Town of Turku*, Turun Yliopisto: University of Turku Ph.D. Thesis.

Troianovsky, S.V. (1998) 'About some results of excavations in the Novgorod Kremlin, 1992–1996', *Novgorod and Novgorod Land, History and Archaeology* 12, 58–70. (in Russian)

Tsalkin, V.I. (1956) 'Material concerning the history of animal husbandry and hunting in medieval Russia', *Materialy i issledovaniya po arkheologii SSSR* (Materials and Investigations of Archaeology, USSR) 51, Moscow: Academy of Sciences. (in Russian)

Tsalkin, V.I. (1960) 'Changes in metapodials and their significance for the study of cattle and horses in antiquity', *Byulleten' Moskovskogo Obshchestva Ispytatelei Prirody Otdel Biologicheskii* (Bulletin of the Moscow Society, Biology Section) 65, 109–126. (in Russian)

Tsalkin, V.I. (1971) 'Some results of animal bone studies in from excavations in Moscow', in Voronin, N.N. and Rabinovich, M.G., *Antiques of Moscow Kremlin. Data and Studies on Archaeology of USSR. Materials and Studies of Archaeology of Moscow*, Moscow: Nauka, 164–185. (in Russian)

Tsyganova, S.A. (1999) 'The western limit for the distribution of the sable in the Holocene Period', *VI Syezd teriological obshchestva* (6th Congress of the Teriological Society), Abstracts for Papers, Moscow.

Ubelacker, D.H. (1978) *Human Skeletal Remains*, Chicago: Aldine.

Van Neer, W., Noyen, K. and de Cupere, B. (2002) 'On the use of endosteal layers and medullary bone from domestic fowl in archaeozoological studies', *Journal of Archaeological Science* 29, 123–134.

van Vuure, C.T. (2005) *Retracing the Aurochs: History, Morphology and Ecology of an Extinct Wild Ox*, Sofia: Pensoft.

Vereshchagin, N.K. and Rusakov, O.S. (1979) *Kopytnye Severo-zapada SSSR (istoriya, obraz zhizni i khozyaistvennoye ispolzovanie)* (Ungulates of the North-West of the USSR (History, Way of Life and Utilization), Leningrad: Nauka. (in Russian)

Vitt, V.O. (1952) 'Loshadi Pazyrykskikh Kurganov (Horses of the Pazyryk Kurgans)', *Sovetskaja Archeologija* 16, 163–205. (in Russian)

Volkopyalov, B.P. (1968) *Pig-keeping*, Leningrad: Kolos. (in Russian)
von den Driesch, A. (1976) *A Guide to the Measurement of Animal Bones from Archaeological Sites*, Peabody Museum Monograph 1: Harvard University.
von den Driesch, A. and Boesssneck, J. (1974) 'Kritische Anmerkungen zur Widerristhöhenberechnung aus Längenmassen vor- und frühgeschichtlicher Tierknochen', *Säugetierkundliche Mitteilungen* 22, 325–348.
Wheeler, A. (1978) *Key to the Fishes of Northern Europe*, London: Frederick Warne.
Wheeler, A. and Jones, A.K.G. (1989) *Fishes*, Cambridge: Cambridge University Press, Cambridge Manuals in Archaeology.
Wigh, B. (2001) *Animal Husbandry in the Viking Age Town of Birka and its Hinterland: Excavations in the Black Earth 1990–95*, Stockholm: Birka Studies 7.
Yanin, V.L. (1992) 'The archaeological study of Novgorod: an historical perspective', in Brisbane, M. (ed.), *The Archaeology of Novgorod, Russia: Recent Results from the Town and its Hinterland*, Lincoln: Society for Medieval Archaeology Monograph 13, 67–106.
Yeremeyev, I.I. (2012) 'The natural environment and settlement patterns of the Lake Ilmen region in the last third of the first millennium AD', in Brisbane, M., Makarov, N. and Nosov, E. (eds), *The Archaeology of Medieval Novgorod in Context*, Oxford: Oxbow Books, 139–151.
Zakharov, S.D. (2012) 'Buildings and structures of the Minino archaeological complex' in Brisbane, M., Makarov, N. and Nosov, E. (eds), *The Archaeology of Medieval Novgorod in Context*, Oxford: Oxbow Books, 58–75.
Zasurtsev, P.I. (1967) *Novgorod as Discovered by Archaeologists*, Moscow: Nauka. (in Russian)
Zeiler, J.T. (2010) 'Hunting the hunters: owls and birds of prey as part of the falconers' game bag', in Prummel, W., Zeiler, J.T. and Brinkhuizen, D.C. (eds), *Birds in Archaeology: Proceedings of the 6th Meeting of the ICAZ Bird Working Group in Groningen*, Barkhuis: Groningen, 165–168.
Zelenkov, N.V. (2008) 'Ostatki ptic iz dvuh arheologicheskih pamjatnikov Respubliki Belarus [Bird remains from two archaeological sites in the Republic of Belarus]', *Ornitologija* 35, 126–128. (in Russian)
Zinoviev, A.V. (2008) 'A study of animal bones from archaeological excavations in Saint-Assumption Trinity Monastery (Staritsa, Tver District, Russia)', *Report of Tver Archaeological and Restoration Research Centre* No. 10. (in Russian)
Zinoviev, A.V. (2012) 'Study of the medieval dogs from Novgorod, Russia (X–XIV century)', *International Journal of Osteoarchaeology* 22(2), 145–157.
Zinoviev, A.V. (2013) 'Of friends and food: dogs in medieval Novgorod the Great', *Archaeologia Baltica* 17, 152–157.
Zinoviev, A.V. (2016) 'Early falconry in Russia: recent finds in Novgorod the Great and Tver', in Szymak, U. and Sianko, P. (eds), *Falconry – Its Influence on Biodiversity and Cultural Heritage*, 63–68.
Zinoviev, A.V. (2017) 'Leather straps with avian tarsometatarsi from the medieval Russia: Jesses or amulets?', *Environmental Archaeology* 22(2), 171–174.
Žulkus, V. and Daugnora, L. (2009) 'What did the Order's brothers eat in the Klaipėda Castle? (The historical and zooarchaeological data)', *Archaeologia Baltica* 12, 74–87.